Samuel Pierpont Langley

Researches on Solar Heat and Its Absorption by the Earth's

Atmosphere

A report on the Mount Whitney expedition

Samuel Pierpont Langley

Researches on Solar Heat and Its Absorption by the Earth's Atmosphere
A report on the Mount Whitney expedition

ISBN/EAN: 9783337329334

Printed in Europe, USA, Canada, Australia, Japan

Cover: Foto ©berggeist007 / pixelio.de

More available books at **www.hansebooks.com**

UNITED STATES OF AMERICA.

WAR DEPARTMENT.

PROFESSIONAL PAPERS OF THE SIGNAL SERVICE.
No. XV.

RESEARCHES ON SOLAR HEAT

AND

ITS ABSORPTION BY THE EARTH'S ATMOSPHERE.

A REPORT OF THE MOUNT WHITNEY EXPEDITION.

PREPARED UNDER THE DIRECTION OF

BRIG. AND BVT. MAJ. GEN. W. B. HAZEN,
CHIEF SIGNAL OFFICER OF THE ARMY.

BY

S. P. LANGLEY,
DIRECTOR OF THE ALLEGHENY OBSERVATORY, WITH THE APPROVAL OF ITS TRUSTEES.

PUBLISHED BY AUTHORITY OF THE SECRETARY OF WAR.

WASHINGTON:
GOVERNMENT PRINTING OFFICE.
1884.

12535——No. XV

NOTE

The publication of this Professional Paper is to be considered merely as a means of bringing it before the attention of the scientific world, and not in any way as an indorsement of the views or theories therein set forth.

PREFACE.

An apology for placing this work before the scientific correspondents of this office so long after the date of the observations would be due if the great labor involved in the final preparation of the manuscript for the printer were not apparent to all.

Professor Langley is not only too well known to require introduction, but this work contains within itself sufficient evidence of his worth, of his skill in original investigations, and of his perseverance in overcoming obstacles. It should be said that the aid given him, which he so gracefully acknowledges in the text, was necessarily limited. A large part of the expense of the outfit was generously borne by a friend of the Allegheny Observatory.

The suitability of the site chosen for these investigations led Professor Langley to recommend that it be declared a Government reservation, and the President having favorably considered this recommendation, it is now available for researches in this and similar fields of inquiry.

The subject herein treated is one of great importance and value to the meteorological work of the Signal Office, and it is esteemed a privilege to publish it in connection with the Professional Papers of this Service.

W. B. H.

TABLE OF CONTENTS.

REPORT OF THE MOUNT WHITNEY EXPEDITION.

BY

S. P. LANGLEY,

DIRECTOR OF THE ALLEGHENY OBSERVATORY.

LETTER OF TRANSMITTAL.

—

ALLEGHENY OBSERVATORY,
Allegheny, Pa., December 21, 1883.

GENERAL: In transmitting the following report on the expedition to Mount Whitney, it appears proper that some account of the inception of its plan should be presented to the public.

Investigations carried on here for some years had, in 1880, led to conclusions of interest to astronomy and meteorology, which it was found desirable to verify by experiments on a very elevated mountain.

The considerable expenditure needed for the special instrumental outfit of an expedition for that purpose had been provided by the liberality of a citizen of Pittsburg, and other preparations commenced at that time. The bearings of its objects on meteorological knowledge becoming known to you, the expedition then received material assistance from the Signal Service, and proceeded under your official direction, in July, 1881, to Mount Whitney, in Southern California. Its results are so intimately connected with the previous investigations referred to at the Allegheny Observatory, and with others undertaken there since on its own account in elucidation of them, that they are hardly separable.

The donor of the principal means for the expedition, desiring only that its results shall appear in the form likely to be of widest use, without reference to any private interest, and the trustees of this Observatory concurring, I have the honor to now address to you the report of the expedition, and with it an account of whatever in this Observatory's own researches is needed in elucidation of it.

Leaving it to Capt. O. E. Michaelis, of the Ordnance Department, to make such a report as he may think necessary upon the faithful performance of their military duties by the escort and Signal Service observers, I desire to acknowledge the obligations of the expedition to him, not only in his official capacity, but for his valued voluntary services as an observer, which I have elsewhere spoken of.

I had every reason to be satisfied with Sergeants Dobbins and Namry of the Signal Service, and I should add that Corporal Lamouette, of the Eighth Infantry, rendered very intelligent and acceptable help beyond his immediate line of duty.

I have elsewhere acknowledged the important aid received through Mr. Frank Thomson, Vice-President of the Pennsylvania Railroad Company, and also the assistance rendered by Professor Pickering, of the Harvard College Observatory.

Permit me to take this opportunity of expressing my personal thanks for the aid which the object I have had so much at heart, has received from you in every way.

I have the honor to be, very respectfully, yours,

S. P. LANGLEY,
Director of the Allegheny Observatory.

General W. B. HAZEN, U. S. A.,
Chief Signal Officer, Washington, D. C.

INTRODUCTION.

—

If the observation of the amount of heat the sun sends the earth is among the most important and difficult in astronomical physics, it may also be termed the fundamental problem of meteorology, nearly all whose phenomena would become predictable, if we knew both the original quantity and kind of this heat; how it affects the constituents of the atmosphere on its passage earthward; how much of it reaches the soil; how, through the aid of the atmosphere, it maintains the surface temperature of this planet; and how, in diminished quantity and altered kind, it is finally returned to outer space.

Meteorologists have till lately occupied themselves more with the secondary effects of this solar radiation than with the considerations just referred to, though this primary study will at least enable us to survey subordinate and familiar phenomena from a more general point of view, and will correct some errors. The knowledge that the solar heat finds its way in more easily than out, and the inference that our atmosphere acts like the glass of a hot-bed in raising the temperature of the soil—even this knowledge, imperfect and misleading as it may be when thus stated, has been most useful in giving us a key to subsidiary phenomena. It seems doubtful, however, whether even the meaning of this has always been clearly apprehended, when we find Sir John Herschel (a distinguished meteorologist as well as an eminent astronomer) saying "* * * the climate of the moon must be very extraordinary; the alternation being that of unmitigated and burning sunshine fiercer than an equatorial noon, continuing for a whole fortnight, and the keenest severity of frost far exceeding that of our polar winters, for an equal time. * * * The surface of the full moon exposed to us must necessarily be very much heated, *possibly* to a degree much exceeding that of boiling water."

It is here evidently implied that a planet at the earth's (or moon's) distance from the sun would merely suffer great vicissitudes of temperature, if deprived of its atmosphere, while yet that the mean temperature of the cycle of day and night would not be greatly altered; and though the labors of Tyndall and others have given us some idea of the way in which our own atmosphere may act, not merely as a conservator against the vicissitudes of radiation in day or night, but in raising this mean temperature itself, we have had till very recently scarcely any just conception of the processes by which it does so, or of the surprising extent to which we are indebted to its action.

According to the results of experiments made in the years 1880 and 1881 at the Allegheny Observatory, all the thermal phenomena we have been alluding to, and on which the existence of organic life depends, depend in turn on a little-regarded property of our atmosphere (*selective absorption*), without which, though it retained all its present constituents and transmitted all the heat it does now, the temperature of the soil, even in the tropics at mid day under a vertical sun, would fall to some hundreds of degrees below zero.

Observations made later confirmed the opinion that, so far from the temperature of the soil being chiefly due to the direct solar rays, these rays alone are far too feeble to melt the mercury in our thermometer bulbs,[*] and that this direct solar radiation is actually insignificant, compared with the temperature which the atmosphere through this selective absorption educes from it.

According to the present view of physicists, the solar energy is conveyed to us in vibrations varying from a wave-length of less than .0003 mm. to one indefinitely greater (the longest measured in

[*] I think a similar statement has been made by Mr. Ericsson.

11

the present investigation being about .005 mm.), some of which vibrations (those only whose wave-lengths are from 0.0004 mm. to .0007 mm., or violet to red) affect the eye with the sensation of light; all of which, so far as is known, produce chemical action; all of which, without exception, convey heat.

"Light," "chemical action," "heat," then, are not qualities inherent in the ray, but names given to the different manifestations of one and the same radiant energy, which is interpreted to us in terms depending upon the wave-lengths of the ray, and on the medium through which it passes or on which it falls.

Let us, to gain clearer conceptions, suppose one of these rays isolated * from the rest, and, as an example, let it be one whose wave-length is about 0.0004 mm., which, when it falls on the retina, gives the sensation of "violet light," which, falling on certain salts of silver, darkens them ("chemical" action), and which falling on a sufficiently sensitive thermometer covered with lamp-black would be absorbed by the latter and cause "heat."

Considering now the particular ray instanced, in reference to its heating power alone, we observe, in view of what has just been said, that everything we know about it we know through some particular medium on which it acts, and that what we learn about it is generally true only of this ray and not of another. Our thermometer, for instance, according as its bulb is covered with white lead or lampblack, gives a wholly different account of the amount of heat in it; and if we could measure the heat in this ray above the earth's absorbing atmosphere, and again at the earth's surface, we should find a notable difference, showing that only the smaller part of it is transmitted. (In this particular case we find that if the sun's heat were all of this quality the soil would receive only about 40 per cent. of it.)

If, now, we consider some other ray, for instance one at the other extremity of the spectrum, whose wave-length is over .0020 mm. ("dark heat"), we find its visual, chemical, and heating effects altogether different. Though quite as energetic as the first, it is invisible (i. e., to us, though it may affect some other than the human retina); it has no chemical action on the previous substances (though it has on certain others); and, as regards its heat, it will very possibly be insensible to the surface which absorbed the first, while the same instrument, if its bulb be coated with some other substance, may reveal its presence. Finally, we observe, by methods to be described, that more than nine-tenths of the heat in this last case is transmitted by our atmosphere; so that if the sun's heat were all of this quality, the soil would receive (normally) over 90 per cent. of it.

Like facts could be learned of an unlimited number of heat rays. Each differs from the others, not only in amount but in kind. We know the heat of the ray only through its action on media; and everything we know of those media through which the rays collectively pass (e. g., the atmosphere), or on which they collectively fall (e. g., the surface of the soil or the thermometer bulb), shows that these distinguish between different kinds of heat with an actually infinite minuteness of discrimination, letting one kind pass and holding back another, as though by an intelligent choice. It is to this action that the name of "*selective absorption*" has been given.

The foregoing general considerations lead at once to others of practical import: for instance, to the conclusion that the thermometer must be a very imperfect measurer of radiant heat; and we are led also to ask how far all our present conclusions as to such heat (derived as they chiefly are through the thermometer) may require revisal. Probably most of those who use it, while aware that there are varieties of heat which it cannot discriminate, suppose that it still gives the total amount correctly. It will appear, however, more fully later that it not only gives an inadequate amount for the heat which actually falls on it, but that in estimating the amount of heat emitted, as in the case of the sun, its use leads to gross errors, in a matter of fundamental importance.

From what has preceded, we are by no means to conclude that the thermometer can be dispensed with, but that its indications need here to be interpreted through observations made by

---

* As the wave-length changes continuously, not abruptly, we cannot by any physical means actually isolate an absolutely homogeneous ray, such as our differential formulæ consider; and what is here said is to be understood as true, with a more and more close approximation, as the width of our actual heat-pencil is made less. As all our actual observation must be on heat-pencils of sensible width, this restriction is important and should not be forgotten.

some instrument which can discriminate between the different kinds of heat. And this is, above all, necessary when we are trying to estimate the amount of solar heat before absorption (the *solar constant*).

Could we ascend above the atmosphere, this heat might be directly measured. Evidently, since this is impossible, and since we can only observe the portion which filters down to us after absorption, we must add to this observed remnant a quantity equal to that which the atmosphere has taken out, in order to reproduce the original amount.

To find what it has taken out, we must study the action in detail, and, from the knowledge thus gained, frame a rule or formula which shall enable us to infer the loss, since we cannot directly determine it.

It is because the exact determination of the solar constant thus *presupposes a minute knowledge of the way in which the sun's heat is affected by the earth's atmosphere;* and because every change in our atmosphere comes from this same heat, that the solution of the problem interests meteorology as well as astronomical physics.

We have just seen that nothing less than a complete knowledge of the laws under which the atmosphere is governed by solar heat, would enable us to frame the exact rule for finding the latter, but though such knowledge really exceeds human powers, most observers have contented themselves with a simple and primitive hypothesis, in using which they really ignore the infinite complexity of the problem here presented us, and assuming that it is as simple as we could wish it to be, proceed to compute the solution by such a formula as it would be most convenient to us if nature would follow. Thus, owing to the temptation to accept as still sufficient any time-honored scientific dogma, which has respectable sponsors, the simple formula established over a century ago by Bouguer and consecrated by the use of a Herschel and a Pouillet, to whom it embodied all the knowledge of their time, is commonly used to-day by observers, who have only to look about them to see that it has long ceased to express the facts known to our own.

To justify this language, let us consider what the problem appears to be at a first glance, and what the first suggestion is for solving it. If a beam of sunlight enters through a crevice in a dark room, the light is partly interrupted by the dust particles in the air, the apartment is visibly illuminated by the light reflected from them, and the direct beam having lost something by this process, is not so bright after it has crossed the room as before it entered it. If a quarter of the light was thus scattered, the beam after it crossed the room would be but three fourths as bright as when it entered it, and if we were to trace the now diminished beam through a second apartment altogether like the other, it seems at first reasonable to suppose that the same proportion, or three-fourths of the remainder, would be transmitted, and so on, and that the light would be the same kind of light as before, and only diminished in amount. The assumption originally made by Bouguer[*] and followed by Herschel and Pouillet was that it was in this manner that the solar heat was interrupted by our atmosphere, and that by using such a simple progression the original heat could be calculated.[†]

Now, it is no doubt true that a very sensible portion of light and heat are scattered by an analogous process in our atmosphere; but we have in our present knowledge to consider that heat is not a simple emanation, but a compound of an infinite number of radiations, and that these are affected in an infinite diversity of ways by the different atmospheric agents, the grosser dust particles affecting them nearly all alike, or with a general absorption; the minuter ones beginning to act selectively, or, on the whole, more at one end of the spectrum than another; smaller particles,

[*] Bouguer, Traité de la lumière. Paris, 1760.

[†] Let us divide in imagination any homogeneous absorbing medium into successive strata of absolutely identical thickness and chemical constitution.

Let A be a source of radiant heat whose intensity is reduced by passage through the first stratum to (let us suppose) $\frac{3}{4}$ (or 0.75) of the first. Then, since the second stratum is identical in constitution and amount with the first and must (it is assumed) have an identical effect, it will absorb $\frac{1}{4}$ of what enters it, and A $\times \frac{3}{4} \times \frac{3}{4}$, $\frac{3}{4}$A or A $\cdot 0.75^2$, will emerge from the second, $\frac{3}{4}$ of this $\frac{3}{4}^3$A or A 0.75^3, will emerge from the third, and so on, the percentage transmitted by the unit of thickness, the "*coefficient of transmission*" being evidently the common ratio of a geometrical progression, so that if the original heat be A and the coefficient of transmission p, the amount of heat after passing through x strata will be Ap^x.

To apply this principle to the estimate of the heat outside the atmosphere (i. e., before absorption), let S be a small

whether of dust or mist, and smaller still, forming a probably continuous sequence of more and more selective action down almost to the actual molecule, whose action is felt in the purely selective absorption of some single ray.

The effect of the action of the grosser particles then is to produce a general and comparatively indifferent absorption of all rays, so that the spectrum after such an absorption would simply seem less bright or less hot. The effect of the smaller ones is, as has just been said, to act more at one end of the spectrum than another, with a progressive absorption, so that the quality of the radiation is sensibly affected as well as its quantity. The effect of the molecular absorption is to fill the spectrum with evidences of the selective action in the form of the dark telluric lines, taking out some kinds of light and heat and not others, so that after absorption what remains is not only less in amount but *quite altered in kind*. Between these three examples of absorption, we repeat, an unlimited number of others must exist; but we shall need here for simplicity to treat the whole as coming under one or the other of these three types, a procedure already more accurate than the primitive one followed by Herschel and Pouillet, but which we recognize to be still but a convention, which is imposed on us provisionally by the actual complexity of nature.

It will be seen now more clearly that the whole process, still in almost universal use, is founded on a pure assumption, for no one has actually been without our atmosphere to see what the absorption is, and it is simply taken for granted that the same proportion of heat will be absorbed by one like stratum as by another. On actually trying the experiment, however, with media in the laboratory, Melloni long since observed that like proportions were *not* absorbed by like strata; and the reason was found in the fact that radiant heat is not a simple emanation, but the sum of an infinity of diverse ones, each with its own separate rate of absorption. It follows that the coefficient of transmission is truly constant only in the case of the absolutely homogeneous ray, which the thermometer cannot in the least discriminate, and hence, that the original heat of the sun, and the amount absorbed, cannot be ascertained correctly by this instrument and this rule. Physicists have been slow, however, as we say, in making this application of Melloni's principle to the present case, but have continued to deduce the solar constant from thermometric observations, in which the heat is either treated as absolutely homogeneous, or in which its non homogeneity is scarcely recognized as a factor of importance.[*] This neglect to make what seems so pertinent an application of Melloni's observation, even after it had been explained and extended (by Biot), will seem more explicable, when it is remembered that no direct means of measuring

portion of the earth's surface and EK the upper surface of the atmosphere, which is here supposed, for simplicity, t

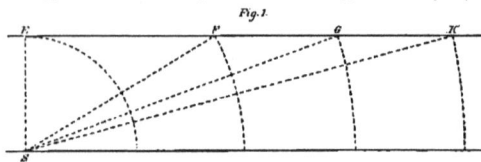

Fig. 1.

Path of Rays in the Atmosphere.

be of uniform density and constitution. (The effects of the actually unequal density of successive strata can, it is assumed, be calculated and allowed for.) Let S be the observer's station, then LS would be the direction of a ray were the sun in the zenith where the absorption is least. FS, GS, KS, the lengths of the paths of the rays as the sun sinks lower—lengths easily computable ; and, to fix our ideas, let FS

2 FS, GS 3 KS, KS 4 FS, etc. The original heat A would, if the sun were in the zenith, become Ap after passing through one stratum (FS); and, according to what has been assumed, it would become (if the sun's zenith distance were FSF) Ap² after absorption by the two strata between F and S, Ap³ after absorption by the three strata between G and S, etc. A, the original heat, and p, the coefficient of transmission, are unknown; but if we make an observation of the heat actually reaching S along FS (let us call this heat b) and again later in the day along KS (calling this second observed quantity c), we have in the particular case supposed

$$Ap² = b$$
$$Ap³ = c$$

whence A and p both become known, and it is evidently easy to extend the solution to the general case of any number of strata, s. Ap², t, then, in the exponential formula of Pouillet, and of later investigators, whose fundamental (and erroneous) assumption is, that the coefficient of transmission (p) is a constant. If it be not a constant (and I shall prove that it is not), the whole superstructure falls to the ground.

[*] Exceptions to this remark, however, are to be made in favor of the work of Principal Forbes and M. Crova.

the absorption in even approximately homogeneous rays till very recently existed, and that departure from the old formula which ignores the difficulties, involves their recognition, and the devisal of new processes to meet them.

The writer has demonstrated that in neglecting to observe approximately homogeneous rays we not only commit an error, but an error which always has the same sign, and that the absorption thus found is always too small. He accordingly devoted much time to the construction of an instrument (the bolometer, which will be described in its place) for the special study of such heat rays, and, with this, observations were carried on in the years 1880 and 1881 at Allegheny, with the conclusions which have just been stated. With this instrument the heat in some approximately homogeneous ray (that is in some separate pencil of rays of nearly the same wave-length) is measured in the pure and normal spectrum at successive hours of the day, and the calculation of the absorption on Bouguer's principle (justly applicable to strictly homogeneous waves) gives the heat outside the atmosphere in this approximately homogeneous portion with a degree of approximation, depending on the actual minuteness of the part examined. The process is then repeated on another limited set of rays, and another, until the separate percentage and the separate original heat is found for each heat pencil directly or by interpolation, and then finally the whole heat, by the summing of its parts, the result being that the solar constant is much greater than it was believed to be, and the absorption of the atmosphere much greater.

With whatever pains we measure, however, we remain at the mercy of the fluctuations of our lower air, and are compelled to make assumptions which we would gladly avoid. Thus, we are compelled to assume that the absorptive powers of the air are the same throughout the day, though this is at least doubtful, even in the case of the most absolutely pure sky. We are obliged to assume that like masses of air produce like absorptions, which is doubtful, even when the ray absorbed is sensibly homogeneous, and we must assume that the air above us is disposed in concentric strata, while our observations tell us little of its true disposition. On these and many other points, we know just enough to distrust our own enforced assumptions, without being able to positively verify or disprove them. Besides such difficulties as these arising from our ignorance, we are met with almost insuperable physical ones coming from the incessant clouds, mist, and changes of our lower atmosphere, which the observer knows only too well, and which make it literally true that not one day of unexceptionable conditions is to be found in an average year, while yet daily observations must be commenced with every clear morning, since we never know which is the day which may prove fair to its close.

These remarks must be borne in mind in reading the account of the preliminary observations at Allegheny, on the absorption of the heat in the spectrum, given in the following chapter—observations which it is necessary to supply here, as they were the immediate cause of the expedition and are intimately connected with its work.

The meteorological reader is asked to bear in mind throughout, that (in the opinion at least of the present writer) the master-key to some of the most important problems of his science is to be found in the hitherto unrecognized study of the *selective* absorption of our atmosphere for heat,

RESEARCHES ON SOLAR HEAT AND ITS ABSORPTION BY THE EARTH'S ATMOSPHERE

CHAPTER I

PRELIMINARY OBSERVATIONS ON SELECTIVE ABSORPTION AT ALLEGHENY DURING 1880 AND 1881.

The heat in the spectrum formed by a prism is not only diminished in an uncertain degree by absorption in its substance, but is dispersed in a manner differing with every prism and exactly expressible by no known formula. The spectrum formed by a reflecting grating, on the contrary, is nearly free from absorption, and may be strictly normal, so that measurements with the grating possess the inestimable advantage of enabling us to fix the wave-length of every ray measured; but, while the average heat in the grating spectrum is, at best, less than one tenth that in the prismatic, the latter is itself, when taken in portions so narrow as to be approximately homogeneous, almost insensible.

As the best thermo-pile was found incapable of measuring heat in such narrow portions of the grating spectrum, I was led to the invention of an instrument for this purpose, the bolometer (βολή, ωτρον), whose construction will be found described in the Proceedings of the American Academy of Arts and Sciences, Vol. XVI (1881). With this apparatus the experiments on the diffraction spectrum were resumed; the first entirely unquestionable evidence of measurable heat, in a width so small as to be properly described as linear, having been obtained on October 7, 1880. Nearly the whole year 1880 passed in modifications of the instrument, or in the making of these measures which gave promise from the first of bringing results of value.

When we have first with this measured the heat directly in the normal spectrum formed by a grating, we can return with advantage to the prism, whose indications now become intelligible.

In these first measures, which were carried to a wave-length of .001 mm.,* I employed two of the admirable gratings of Mr. Rutherfurd, one containing 17,296 lines to the inch, or 681 to the millimeter, and the other one-half that number, both ruled upon speculum metal, and I used a slit at a distance of 5 m. without any collimator, keeping the grating normal to the optical axis. It will be seen, then, that the rays passed through no absorbing medium whatsoever, except the sun's atmosphere and our own.

The rays from the grating fell upon a concave speculum (whose principal focal distance was about one meter), and from this were concentrated upon the mouth of the bolometer, forming a narrow spectrum, which passed down the case of the instrument and fell upon the bolometer thread. As this thread moves along the spectrum parallel to the Fraunhofer lines, its coincidence with one of them is notified by a lowering of its temperature and a deflection of the galvanometer. The instrument is, of course, equally sensitive to the invisible radiation as to the visible. It is important to observe that no screen is interposed between the bolometer and the grating, for the temperature of the screen itself, as it is replaced or withdrawn, will certainly affect such measure-

---

* Through these measures the unit of wave-length will be the micron (μ) = .0001 mm., or 10,000 times the unit of Angström. Thus the wave-length of Fraunhofer's "A" is here written 0s.76.

ments as these. Through the whole course of the experiment the bolometer is uninterruptedly exposed to radiations from the grating, whether reflected by it, or emanating from its own substance. The interruption of the solar radiation is affected at the other end of the train, 5 meters beyond the grating itself. In the gratings employed, one of the second spectra is very feeble, or almost lacking. The rays of the second spectrum are necessarily superposed on those of double the wave-length in the first; and as all evidence of solar radiation in the most sensitive apparatus at the sea level dies out near $\lambda = 0^\mu.3$ in the ultra violet, it follows that we can measure down in the first spectrum as far as $\lambda = 0^\mu.6$, or in fact further, without any fear whatever of our results being affected by the underlying second spectrum, even if that were a strong one. We have, therefore, knowing the amount of heat in the second spectrum at $0^\mu.5$, and knowing that our ultimate point of measurement at $1^\mu.0$ in the first spectrum overlies $0^\mu.5$ in the second, the means of asserting with confidence that no considerable error can be introduced from this cause, after an allowance has been made here for the minute effect of this actually weak second spectrum. An allowance is also made to reduce the effect to that which would have been observed with a grating so coarsely ruled as to cause no considerable deviation from the slit of any portion of the spectrum measured. The bolometer (in a constant position relative to the concave mirror, such that the optical axis of the latter bisected the angle between its central thread and the center of the grating), was moved, together with the mirror, by a tangent screw in arc, so that the spectrum appeared to traverse its face.

The actual angular deviation of any ray under examination was obtained from a divided circle, on which the arm carrying both mirror and bolometer moved. A particular description is not given, as the whole apparatus was replaced by a more perfect one later. That actually used will be intelligible by the sketch (Fig 2), where S is the slit, G the grating, M the concave mirror, B the bolometer, and C the divided circle.

The light came from the silvered mirror of a heliostat, passing through the slit, at a distance of about 5 m. from the grating, which was bolted immovably above the center of the circle of a massive dividing engine, with the grating's plane always perpendicular to the line joining its center and the slit. The mirror and the bolometer, with their attachments, were fastened to this movable circle.

An allowance has been made for the absorption of speculum metal and silver, but the absorption of the iron strips of the bolometer has only been indirectly allowed for. This has been done by comparison with the action of a bolometer, with lampblacked surface. The wave lengths are derived from the measured angles by the use of the formula

$$n s \lambda = \sin i + \sin r.$$

where n is the order of the spectrum, s the space between the lines of the grating, $\lambda =$ the wave-length of the ray, i the angle of incidence (in the present instance 0), and r the angle of diffraction.

In the early observations it appeared from the examination of the diffraction spectrum up to $\lambda = 1^\mu.0$, that the energy in the invisible part as far as this was much less than in the visible. Nothing definite is, even at this time, known to physicists as to the extent of the normal solar spectrum; but the prismatic spectrum is still very commonly supposed to be limited by theoretical considerations to an extent little greater than this; and one of those most conversant with the subject has treated this wave-length ($i. e.$, $1^\mu.0$) as marking the limit of everything known to exist.[†]

† Draper, "On the Phosphorograph of a Solar Spectrum, and on the Lines in the Infra-red Region," Proceedings of the American Academy, Vol. XVI, p. 233, December, 1880. He asks: "Do we not encounter the objection that this wave-length, $10,750^{m}$ ¹ (the limit of Captain Abney's map), is altogether beyond the theoretical limit of the prismatic spectrum?" Previous measurements of heat had, it will be remembered, been made by comparing its total amounts, in the visible and invisible prismatic spectrum, which gives us no knowledge as to wave-lengths in any case, and wave-lengths in the dark-heat region had been estimated, by hazardous extrapolations, from contradictory formulæ—formulæ which profess a theoretical basis, but contradict each other. Thus Müller finds, by Redtenbacher's formula, a wave-length of nearly $5^\mu.0$ for the extreme solar heat rays; Draper (as we have just seen), a wave-length of but $1^\mu.0$ for the same rays, &c. All these formulæ (Briot's, Cauchy's, &c.) agree well with the observations in the visible spectrum, which they have in fact been originally deduced from. They contradict each other thus grossly when used for extrapolating the place of the extreme infra-red rays, whose real place we give later from actual measures.

It seemed at first, then, improbable that the heat below the red should materially exceed, or even equal, that above it; for this would demand (since the heat shown by the last ordinate at $\lambda = 1''.0$ is very small) an extension of the curve of heat, as obtained from the grating, to a distance enormously beyond the furthest limit then assigned to the normal spectrum by experiment. The writer's further investigations, however, led him to believe that this immense and unverified extension really existed, and to thus confirm by independent means the statements of Tyndall and others as to the great heat in this region. He was unable to determine its exact limit with the

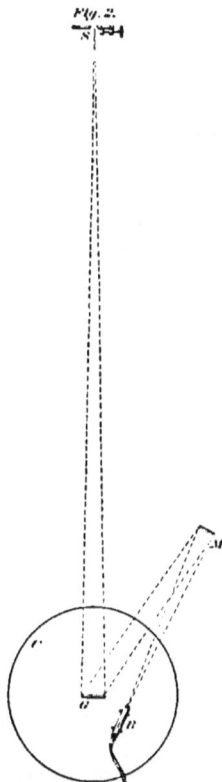

Fig. 2.

Arrangement of Apparatus for Measuring
Angular Direction of a Ray of Heat

grating as then used, on account of the overlapping spectra, but was some two years since led, from experiments not here detailed, to suspect the existence of solar heat at a distance of nearly four times the wave-length of the lowest visible line A ($\lambda = 0''.76$), or at $\lambda = 3''.0$.[*]

[*] See Comptes Rendus de l'Institut de France, July 1, 1881.

We receive all the solar radiations through an absorbing atmosphere, and it was the special object of these investigations to determine, not only the amount of heat in each ray, but the separate absorbent action of the atmosphere on each.

The great difficulty in this investigation, after the provision of a sufficiently delicate heat-measurer, lies in the varying amount of radiant energy which our atmosphere transmits, even for equal air-masses. The solar radiation is itself sensibly constant, but the variations in the radiant heat actually transmitted are notable, even from one minute to another under an apparently clear sky. The bolometer, in fact, constantly sees (if I may use the expression) clouds which the eye does not. That these incessant variations are in fact due to extraneous causes and not to the instrument itself has been abundantly demonstrated by measurements on a constant source of heat. Those taken, for instance, on a petroleum lamp, so placed as to give nearly the same galvanometer deflection as the sun did, were found to indicate a probable error, for a single observation, of less than one per cent. The variations from minute to minute (under a visually clear sky) amount, frequently, to ten times the probable instrumental error, and they can only be partly eliminated by repeating the observations a great number of times on many different days. It is probable, too, that there is a systematic change in the absorbent power even of a given air-mass as the sun approaches the horizon, but this point may be considered later. Actually, twenty-nine such days' observations have been made (as appears below) in the preliminary series, but it would be an error to suppose that this number was obtained without the sacrifice of a still larger number on which the apparatus was prepared, and the day spent without results, owing to the still more considerable atmospheric changes between morning and afternoon. Even of the twenty-nine days cited, and which may be considered exceptionally fair, it will be seen that in only ten cases did the sky continue sufficiently constant in the morning and afternoon to allow complete series to be taken.

It will be understood that we aim to make at least two sets of measures throughout the spectrum daily, one when the rays have been little absorbed (at noon), the other when they have been greatly absorbed (in the morning or afternoon). It will be understood, from what has preceded, that the exponential formula of Pouillet, founded on the assumption that like masses absorb like proportions (though misleading as applied to the complex radiations noted by the thermometer, and rigorously applicable only to strictly homogeneous rays), is yet more nearly applicable to those which form the subject of these experiments, for though these cannot be absolutely homogeneous, we may consider them as nearly so, as they are physically measurable by the most delicate means known. The mass of air, through which the rays pass, is taken proportional to secant z, for zenith distances less than 65°, and for those greater to

$$\frac{0.0174 \times \text{tabular refraction}}{\text{cosine apparent altitude}}$$

The unit mass of air is that for which secant $z = 1$, or that vertically above an observer at the sea level, and whose weight is represented by the mean barometric pressure of 760 mm., or 7.6 dm. The coefficient of transmission of heat for this unit atmosphere is here called a, so that heat, which was E before absorption, becomes Ea after absorption by one such unit stratum, and Ea^n after absorption by n strata.

It is convenient to employ in the preparatory computations, as the unit of mass of mercury in the barometer, one decimeter. If we choose to employ as our unit for the barometer, the whole height of the column at sea-level, we must then divide the value of the barometric pressure here given by 7.6. The mass of air through which the rays pass then being proportional to the actual barometric pressure may be expressed in units, each of which is represented by the pressure of one decimeter of mercury at the sea-level. Since we may take any unit we please, we may, if we wish to do so for any special purpose, treat this as the unit of air-mass, and call its coefficient of transmission by some special name. Thus if t were the coefficient of transmission for an air mass, represented by one decimeter of mercury, $t^{76} = a$, and either (for a homogeneous ray) gives the transmission for an entire atmosphere. The coefficient of transmission for one atmosphere (a) is then the proportion of the radiation transmitted by a sun in the zenith to an observer at the sea level (where the barometric pressure is 7.6 dm.), and this is here shown to be (under constant atmospheric conditions) constant for any given ray, but to vary greatly from one to another. Thus by reference to Table 6, we find of three solar rays, whose wave lengths are $0^\mu.375$, $0^\mu.600$, $1^\mu.000$ that of the ray whose wave-length is $0^\mu.375$ (in the ultra-violet), 61 per cent. of the original energy would be absorbed and 39 transmitted ; of wave-length $0^\mu.600$ in the orange, 36 per cent. would be absorbed and 64 transmitted ; of wave-length $1^\mu.000$ (in the infra red) 20 per cent. is absorbed and 80 transmitted, &c.

The following list shows the dates at which bolometer observations were made at Allegheny up to June, 1881, for the measurement of heat in the spectrum and the determination of atmospheric transmission, by the comparison of noon and afternoon measures. Those days on which noon measurements were taken, which were rendered useless for this purpose by subsequent changes in the condition of the sky or by other causes, are indicated by an asterisk. It will be seen that of twenty nine days of observation only ten could be fully utilized, and that all of the year 1880 may be considered to have passed in the experiment and practice which made the observations of 1881 effective.

Dates : 1880, November 12,* December 11,* December 18 :* 1881, January 12,* January 18,* January 28, February 2, February 3,* February 5,* February 17, February 19,* February 22,* February 26,* March 2,* March 10,* March 11,* March 25,* March 28,* April 7,* April 16,* April 22, April 23, April 28,* April 29, April 30, May 1,* May 26,* May 27,* May 28.

We will select as an example of an actual day's observations those of April 29, 1881. The record is made in a book prepared for the purpose, from which a copy of the original entry is here given.

1. Station : Allegheny.
2. Date : April 29, 1881.
3. Wet bulb, 11 .11 C.
4. Dry bulb, 12 .67 C.
5. Black bulb, 24 .44 C. } at 9 h. 15 m., a. m.
6. Barometer, 735 mm.
7. Temperature apparatus, 16 .0 C.
8. State of sky, *milky blue, with frequent clouds.*
9. Aperture of slit, 0.001 m.
10. Slit to grating, 1.85 m.
11. Grating to mirror, 1.091 m.
12. Mirror to bolometer, 1.15 m.
13. Grating used, No. 2, large.
14. Weak second spectrum thrown *west.*
15. Galvanometer used, "Elliott No. 3."
16. Bolometer used, "No. 1, old" (iron).
17. Rheostat used, "Mercury" and resistance box.
18. Setting on D, (south vernier), 125 .31'.
19. Deflection battery galvanometer, div. 250.
20. Constant of battery galvanometer, 0.00951.
21. Current of battery, 0.239 Ampere.
22. Reader and recorder at spectro-bolometer, F. W. V.
23. Reader at galvanometer, W. A. K.
24. Estimated weight, *all of equal value.*
25. REMARKS.—*The galvanometer drift was moderate and tolerably uniform. Its amount at the time of each deflection has been determined, and the necessary correction for it is included in each recorded reading.*

The conditions of observation which have changed during the day are the following. (When the original readings were on the Fahrenheit scale, they have been reduced to Centigrade.)

TABLE 1.

Time	9ʰ 15ᵐ a. m.	11ʰ 30ᵐ a. m.	1ʰ 35ᵐ p. m.	5ʰ 45ᵐ p. m.
Wet bulb	11 .11 C.	12 .22 C.	11 .67 C.	11 .94 C.
Dry bulb	12 .67 C.	14 .89 C.	15 .00 C.	14 .44 C.
Difference	1 .56 C.	2 .67 C.	3 .33 C.	2 .50 C.
Force of vapor	8 .92	8ᵐᵐ .99	8ᵐᵐ .29	8ᵐᵐ .92
Relative humidity, per cent	81. 5	71. 4	64 .7	72 .2
Sun thermometer (black bulb)	21 .41 C.	31 .11 C.	31 .11 C.	19 .17 C.
Excess of black bulb	11 .78 C.	16 .22 C.	16 .11 C.	4 .72 C.
Temperature apparatus	16 .0 C.	18 .5 C.		

* The black bulb thermometer, whose reading in the sun is recorded above, was mounted in a thin glass inclosure filled with air at atmospheric pressure. An excess of the thermometer of 11 .6 C. above the shade temperature corresponded to 1 calorie of the Pouillet pyrheliometer (not corrected).

The sky remained of apparently the same character during the day, namely, "milky blue with frequent clouds."

TABLE 2.

λ	0ᵐ.375	0ᵐ.40	0ᵐ.45	0ᵐ.50	0ᵐ.60	0ᵐ.70	0ᵐ.80	0ᵐ.90	1ᵐ.00
Setting south vernier	140 .21	135 .21	131 .18	129 .15	125 .02	120 .41	116 .08	111 .21	106 .11
8ʰ 02ᵐ to 8ʰ 33ᵐ a. m.	4	15	45	117	143	195	132	94	90
8ʰ 37ᵐ to 9ʰ 05ᵐ a. m.	15	29	85	135	169	198	138	92	82
Morning means	10 · 15	22	65	126	156	197	135	93	86
11ʰ 40ᵐ a. m. to 12ʰ 02ᵐ p. m.	15	31	135	140	252	243	152	103	88
12ʰ 02ᵐ to 12ʰ 31ᵐ p. m.	1 · 19	26	92	161	218	226	126	96	90
Noon means	15	29	114	151	235	235	139	100	89
5ʰ 07ᵐ to 5ʰ 20ᵐ p. m.	- 15 · 12	10 · 3	19	65 ·	91	119	71	70	59 · 73
5ʰ 22ᵐ to 5ʰ 30ᵐ p. m.	- 1	12	49	59	119	112	72	46	
Afternoon means	5	8	19	62	107	116	72	58	66

The arrows indicate the order in which the measurements were made.
The units are the arbitrary divisions of the scale of the galvanometer.

The state of the sky in these observations is the primary consideration. An absolute deep blue over the whole sky, except the horizon, such as may be seen in Colorado or in some parts of California, is almost unknown in Allegheny. The best days for our purpose are those where the blue of the sky is seen between passing clouds at times when these seem to sweep the sky nearly clean of all traces of mist or haze between them. Yet even here the blue is not, to a critical eye, the blue of the sky of the Colorado table-lands or the African desert. The word "milky," above employed, must be understood as used in comparison with the recollection of a nearly perfect sky.

The grating used, described as "No. 2, large," was the second of two large gratings ruled on Mr. Rutherford's engine by Chapman, giving 1¾ inches square of ruled surface. The galvanometer, "No. 3," was a very sensitive Thompson's reflecting galvanometer, by Elliott Bros., of the most recent construction.

When the astatic galvanometer is in a condition of great sensitiveness, it is in a condition of partial instability; and under these circumstances, minute changes in the temperature of the room and of the instruments will alter its directive force slowly from hour to hour, so that the image has a motion upon the scale due to this cause, quite independently of the diurnal variation, a motion which is observed whether the thermo-pile or bolometer be used, and to which extraneous causes in either of the last instruments contribute. This motion is here called "the drift." (It has been greatly diminished, by various improvements since the observations here described.)

The bolometer used, described as "No. 1," was composed of fifteen central strips of iron, exposing 60 square millimeters of surface. (Appendix No. I.) The setting on D, is that given by an eye-piece mounted in a cylinder like that in which the bolometer is incased, and interchangeable with it, so that the optical axis of the one and the thermal axis of the other may be made to coincide with precision. The battery galvanometer determines the strength of the current, which is kept constant during the day. The constant of the battery galvanometer is the number by which the deflection must be multiplied to give the current in Amperes. The readings are taken by exposing successively to the portions of the spectrum chosen, beginning in the violet end and going down to the infra red at 1ᵘ.0, and then returning to the violet so that two readings are taken on every point. It is the mean of these two sets of readings which is given above, and which constitutes a series.

The heat corresponding to $\lambda = 0^\mu.375$ is very feeble at the sea-level, even with a high sun, and almost disappears at the low sun observation. It is only, then, on days of unusual clearness that the heat in this ray, and others of shorter wave length, can be well observed, except at noon. The reader will observe that the mean time of the morning observations was at 8ʰ 16ᵐ a. m., when the sun's distance from the meridian was 3ʰ 11ᵐ, and that those of the afternoon were taken at the mean time 5ʰ 20ᵐ. p. m., when the sun's distance from the meridian was 5ʰ 23ᵐ. During the morning observations, then, the sun's rays had a smaller air mass to traverse than in the afternoon, and the result is shown in the larger galvanometer deflection. It will be observed that the absorption for any given ray depends, in theory, upon the mass of air traversed, and not on the length of the path; but that this mass, for zenith distances of less than 65°, being almost exactly proportional to the length of the path—that is, to sec z—we may use this expression in finding the mass. Evidently if the air be heavier, as shown by the barometer, the mass will be greater, and sensibly so in proportion to the increased weight, whence the mass of air traversed (in the case of the actual example, and others where the zenith distance is less than 65°) will be $M\beta = \sec z \times \beta$, where M is the length of the ray's path through the atmosphere, that from a zenith sun being unity, and β is the barometric reading (here expressed in decimeters). For zenith distances greater than 65°, we use the formula, derived from that of La Place, where

$$M = \frac{0.0171 \times \text{tabular refraction}}{\cos. \text{ app. altitude}}.$$

which gives the mass with more than sufficient correctness for our purpose, even when the sun is approaching the horizon. From the observed times, then, we find respectively for the morning and afternoon observations

Sun's hour angle $= 3^h\ 11^m,\ 0^h\ 06^m,\ 5^h\ 23^m.$

Sun's zenith distance $= 18°\ 16',\ 25°\ 50',\ 73°\ 59';$

the corresponding values of M being 1.517, 1.111, 3.501, the value of the barometer above given being unchanged through the day, $\beta = 7.35$ dm., and the consequent air masses being as follows: in the morning 11.15, at noon 8.17, and in the afternoon 25.73.

The heat in any ray from the center of the sun may be treated for our present purpose as being constant. The heat in this ray, as it would be observed before absorption at the upper surface of the earth's atmosphere and at the sun's mean distance, would be constant also, but would sensibly vary with the earth's distance from the sun, being greatest in winter, when the earth is in perihelion, and least in summer at aphelion. Let $\rho =$ the radius vector, unity being its value at the earth's mean distance, whence, to reduce any observation to what it would have been if taken when the earth was at its mean distance from the sun, we have only to divide it by ρ^2. We observe, however, that the present observations being for the purpose of comparing the heat in one ray with another and of determining their coefficients of absorption in the earth's atmosphere, changes of their relative heat, introduced by the variation of the earth's distance from the sun, are quantities of the second order, and all of them negligible. It is not necessary, then, to apply the correction just given to the present bolometric observations, though it cannot be omitted from any determination of the absolute amount of heat. Our data above obtained, then, are sufficient for determining the heat in this ray, and the coefficient of transmission; for, calling E the original rate of emission of solar energy, a the coefficient * of transmission of a ray from a zenith sun through the entire terrestrial atmosphere (whose pressure is equivalent to that of 7.6 dm. of mercury), and d, the galvanometer deflection produced by this heat at noon, $d_{,,}$ that in the

* It has been found desirable to modify the original notation of Pouillet, where J is the solar constant, p the coefficient of transmission, e the air-mass traversed by the ray, and t the temperature for this notation embodies ideas which have been so greatly changed that new symbols are more appropriate.

afternoon, and observing that $M_{,}\delta_{,}$ is the number of units of air mass interposed at noon ($= 8.17,$) and $M_{,,}\delta_{,,}$, the number of units of air mass interposed in the afternoon $= (25.73)$, we have evidently

$$E = \frac{d_{,}}{a^{\frac{M,\delta}{7.6}}} = \frac{d_{,,}}{a^{\frac{M,\delta}{7.6}}}$$

since the heat emitted from the sun is sensibly the same at all times of the day, whence

$$\log a = \frac{(\log d_{,,} - \log d_{,}) \times 7.6}{M_{,,}\delta_{,,} - M_{,}\delta_{,}}.$$

Substituting in the above equation the numerical values just given, we find, from a comparison of noon and afternoon values on April 29,

$$\log a = \frac{(\log d_{,,} - \log d_{,}) \times 7.6}{17.56},$$

and from the morning and noon comparison

$$\log a = \frac{(\log d_{,,} - \log d_{,}) \times 7.6}{2.98}.$$

Hence we obtain the following results for this day's observations, remarking that in such a climate as that of Allegheny considerable differences in the coefficients of the most refrangible rays will be found on different days, because these rays, as has been observed, almost wholly disappear in the early morning or late afternoon, and the probable error of their value is very great.

TABLE 3.

λ	.375	.40	45	50	.60	.70	80	.90	1.00
April 29, 1881:									
From morning and noon observations, a -	512	495	244	611	552	687	927	851	916
From afternoon and noon observations, a -	663	573	627	682	712	758	753	791	879
Average value of a	588	504	471	657	582	688	840	811	898

We now give in the following table a summary of all the early observations at Allegheny, which can be utilized for a determination of atmospheric transmission. The example just given in full, will serve as a type of the rest.

TABLE 4.

The observed galvanometer deflections are reduced to a scale on which the readings are proportional to the current passing through the galvanometer.

$d_{,}$ = galvanometer deflection with high sun.
$d_{,,}$ = galvanometer deflection with low sun.

λ	θ_μ 375	.400	450	500	.600	.700	.800	.900	1.000
1881.									
January 28 ... d	.....	101	.....	474	583	320	221	111	102
d	.....	43	.....	167	268	215	221	116	78
February 2 ...... d	21	80	215	280	367	293	175	.....	93
d	3	20	61	104	141	155	91	.....	17
February 17 d	25	62	120	202	250	227	188	.....	71
d	8	25	58	110	173	151	80	.....	39
April 22 d	19	40.5	154	206	262	226.5	177.5	124.5	98
d	4.5	17	63	119.5	171.5	149.5	122.5	89.5	81
April 29, a m d	.....	59	152	206	263	227	191	121	94
d	.....	11	124	189	258	237	187	122	90
April 29, p m d	.....	59	152	206	267	277	191	121	94
d	.....	32	103	121	188	198	140	80	66
April 29, a m ... d	13	29	113	151	235	205	139	100	80
d	10	22	65	126	156	197	155	93	86
April 29, p m ...... d	13	29	113	151	235	235	139	100	80
d	5	8	40	62	107	116	72	58	66
April 30 .... d	21	55	121	186	245	239	175	118	90
d	18	23	97	118	220	232	168	97	89
May 28 d	8	24	59	109	144	134	89	64	52
d	....	2	9	27	52	66	61	33	39

The next table gives the sun's position and the corresponding air-mass for each series in the previous table.

TABLE 5.

Date of observation		High sun.				Low sun.			
		Sun's hour angle	Sun's zenith distance	Barometer (d.)	Air mass (M. δ.)	Sun's hour angle	Sun's zenith distance	Barometer (d.)	Air mass (M. δ.)
1881		h. m.	° '	d. m.		h. m.	° '	d. m.	
January 28	M.	0 09	58 29	7.45	14.25	P.M.	2 57	71 28	7 14	22 24
February 2	M.	0 09	57 09	7.39	14.02	P.M.	2 49	70 45	7 39	22 71
February 17	M.	0 38	52 57	7.43	12 53	P.M.	2 56	66 01	7 12	18 25
April 22	M.	0 12	27 13	7 26	8 05	P.M.	4 36	66 22	7 26	18 22
April 23 a.m.	M.	0 11	27 19	7 40	8 07	A.M.	2 45	45 39	7 16	10 56
April 23 p.m.	M.	0 11	27 49	7 10	8 07	P.M.	1 26	63 57	7 10	16 85
April 29 a.m.	M.	0 06	25 50	7 25	8 17	A.M.	3 11	19 16	7 75	11 15
April 29	M.	0 09	25 50	7.35	8 17	P.M.	3 23	71 30	7 25	25 71
April 30	M.	0 01	25 31	7 41	8 21	A.M.	3 54	56 31	7 41	13 44
May 2	M.	0 11	19 06	7 02	7 75	P.M.	5 33	71 44	7 32	22 53

By combining the high and low sun observations of each day separately, the following coefficients of atmospheric transmission are obtained by means of the formula

$$\log a^\delta = \frac{\log d_{lt} - \log d_l}{\mathrm{M}_{lt}\delta_{lt} - \mathrm{M}_l\delta_l},$$

where a^δ is the coefficient of vertical transmission by air at a barometric pressure of one decimeter

TABLE 6.

λ =	.375	.400	.450	.500	.600	.700	.800	.900	1.000
1884.									
January 28		.910		.914	.957	.957	1 000	.976	.971
February 2	.735	.851	.865	.889	.914	.951	.927		.925
February 17	.857	.858	.884	.881	.893	.934	.856		.904
April 22	.898	.910	.915	.934	.959	.972	.984	.998	.988
April 23 a.m.		.917	.911	.961	.991	.967	.986	1 005	1 009
April 23 p.m.		.860	.955	.942	.964	.920	.964	.952	.950
April 29 a.m.	.916	.912	.941	.941	.972	.912	.990	.976	.989
April 29 p.m.	.917	.929	.984	.951	.956	.961	.963	.970	.985
April 30	.971	.987	.900	.937	.900	.979	.990	.961	.978
May 28		.851	.819	.982	.933	.953	.975	.976	1 001

Mean a^δ	.857	.022	.888	.905	.903	.911	.924	.906	.942	.908	.955	.605	.967	.908	.971	.904	.970	.907

Adopted a^δ	.881	.892	.900	.923	.942	.955	.965	.970	.971

Transmission for one atmosphere a	.392	.420	.480	.544	.640	.705	.763	.794	.799

It should be understood that, owing to the incessant changes of our atmosphere, years of observation might be spent without giving to this table all the exactness which is finally attainable. Later observations, carried on through the changing seasons of a whole year and with improved apparatus, appear to show that all these coefficients should be somewhat modified. They are here given as examples of the results first attained.

The noon observations on like rays, where made on successive days, through like air masses should give nearly like results, if the transmissibility of the atmosphere for heat were always the same for the same mass of air and the same ray. A comparison of this with the preceding tables shows, however, that the heat transmissibility must often change considerably from one day to another, even when the sky is clear on both. We are forced, (at least in these preliminary researches,) to make the ordinary assumption that the transmissibility is constant between noon and afternoon; but we recognize that the transmissibility does probably change even in these few hours, and that this assumption, though it is usual and here necessary, cannot be considered exact.

If we take the probable errors of these coefficients, we shall, in accordance with what has just been observed, find the largest probable error attached to the shortest wave-length. The probable errors derived from each column by the ordinary process are given against the values marked "mean a^δ." If these values be made the ordinates, and the wave-lengths the abscissæ, a smooth curve may be drawn through the points. A line drawn between the points representing the original and entirely uncorrected observations of a gives the smooth curve in Fig. 3. The very slightly different values given by the smooth curve are adopted as those of a, and a consideration of the manner in which these values have been obtained, of the probable errors, and of the illustration, will put the reader in full possession of all the means of forming a judgment on the trustworthiness of the results which the writer himself possesses. If he bear in mind that being obtained on only approximately homogeneous rays, there is reason why they should in theory (as is demonstrated later) indicate rather too large than too small a transmission. One remark may, however, be made in relation to the probable errors of the numbers corresponding to 0s.90 and 1s.00. There is here a very great interruption of the spectral energy (see λ .94 on the chart of the normal spectrum). In these early observations with a wide bolometer, the neighborhood of this "crevasse" in the curve was a source of slight irregularities, which appear in regard to these wave-lengths.

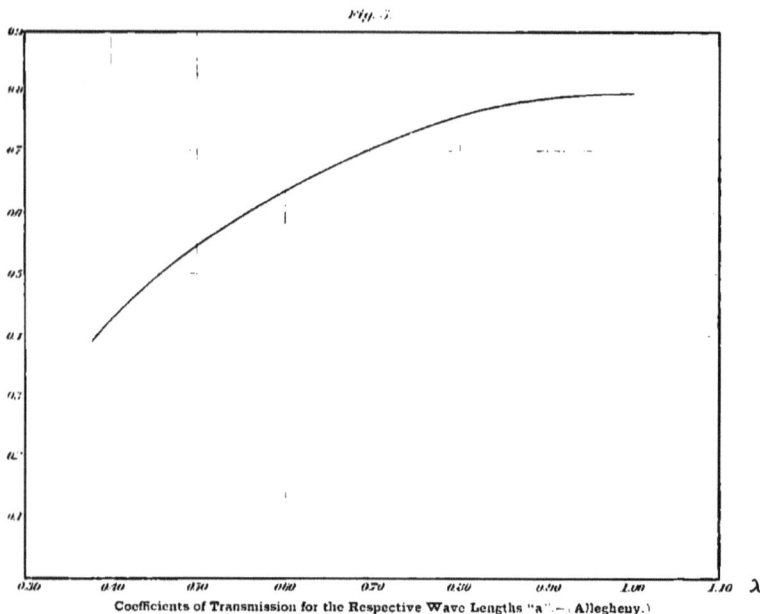

Fig. 5.

Coefficients of Transmission for the Respective Wave Lengths "a". — Allegheny.)

Up to the time of observation, it had been almost universally admitted by physicists that the infra-red heat was in general more absorbed by our atmosphere than the luminous. This is the testimony of many, and even at the date of writing these lines (October 1883) it may be considered to be still the generally received opinion, so far as the most recent and approved treatises on physics can be recognized as the exponents of scientific opinion on this point. As soon as accurate means had been devised for comparing the absorption in the infra-red with that in the luminous part of the spectrum, evidence began to accumulate that the latter was really the least transmissible. Considering the weight of authority against his own conclusion (that the infra-red heat within the range of his researches was more transmissible than the luminous), the writer felt bound to repeat his experiments in every manner and with every precaution. The reader's special attention is called to the nature and weight of the evidence given in the preceding table to the fact of the greatly increased transmission of heat-rays of as great a wave-length as $0^{\mu}.001$ over those in the luminous part of the spectrum. Indeed, it is here seen that except in the case of absorption bands each wave-length is, broadly speaking, more transmissible as it is found farther and farther in the infra-red.

We may add, that, besides the above days of exceptionally uniform atmospheric conditions, a great number were partially utilized when the series were so interrupted by the gathering of mist or clouds that they have not been cited here at all; but that in all cases the observations have been found to lead to the same result here given, and to warrant us in stating that, speaking without regard to local absorptions like those of the telluric lines, the coefficients of transmission *increase* with the wave-length from within the observed range of $\lambda = 0^{\mu}.375$, in the ultra-violet, to $\lambda = 1^{\mu}.00$ of the infra-red.

On many days, which do not appear above, when both morning and afternoon series could not be obtained, good noon series were observed. Those on which good noon series only were obtained cannot be used for finding coefficients of transmission, but may still be useful, if we like to compare the relative transmission of these rays in spring and in winter. For this purpose, all good noon observations have been reduced to a uniform battery current of 0.25 amperes, and the results, arranged in two sets, the first for winter and the second for spring measures, are as follows, the values given being deflections of the galvanometer in divisions of its arbitrary scale:

TABLE 7.

WINTER.

λ		.375	.400	.450	.500	.600	.700	.800	.900	1.000
1880.	h. m. h.m.									
December 18	1.00– 1.35		85		207	311	305	177	110	87
1881.										
January 28	11.35–12.25		95		159	353	300	207	175	55
February 2	11.50–12.20	32	74	201	271	297	274	104		87
February 3	11.15–11.35	33	118	243	156	302	311	184		139
February 5	12.10–12.55	21	52	182	206	300	207	157		67
February 17	11.05–1.10	22	60	116	225	252	220	182		69
February 22	12.00–12.50	25	50	137	240	282	201	151	87	101
Mean di.		31	84	199	204	328	259	172	111	91

STATE OF SKY DURING THE ABOVE EXPERIMENTS.

December 18.—Sky cleared unexpectedly to a good blue.

January 28.—Fair milky blue sky, continuing very fair to close of day.

February 2.—Sky hazy, improving towards noon; p. m. very thickly milky.

February 3.—Sky unusually clear and free from haze, except in measures on .700, .800, and 1.000 when thin cirrus had commenced to form.

February 5.—Sky hazy.

February 17.—Sky hazy, irregular, but becoming more uniform after noon, thicker at the last observations, p. m.

February 22.—Sky hazy, halo around the sun, occasional wisps of cirrus.

TABLE 8.

SPRING.

λ		.375	.400	.450	.500	.600	.700	.800	.900	1.000
1881.	h. m. h. m.									
March 10	11.30–12.15		43	12.	214	251	251	196	50	97
March 23	12.00–12.41	9	53	156	271	310	292	195	148	99
March 26	12.05–12.20	25	81	181	224	260	275	182	130	70
April 14	10.55–11.20	23	52		296	340	349	228		104
April 22	11.50–12.25	25	40	104	247	271	254	186	150	105
April 23	11.35–12.21	14	46	170	215	275	290	204	127	95
April 29	11.40–12.31	14	30	114	155	246	210	145	104	95
April 30	11.55–12.40	22	50	127	195	256	274	151	124	51
May 12	12.25–1.20		50	85	144	202	201	132	95	21
Mean di.		18	37	139	218	264	271	188	121	54

STATE OF THE SKY DURING THE ABOVE EXPERIMENTS.

March 10.—Sky a thickly milky blue, nearly uniform.

March 25.—Sky fair blue, with passing clouds.

March 28.—Sky fair, milky blue.

April 14.—Excellent blue sky; light clouds passing.

April 22.—Milky blue sky, uniform, but increasing slightly in thickness through the day.

April 23.—Sky rather thickly milky, only moderately good, but quite regular.

April 29.—Sky milky blue, with frequent clouds; apparently of about the same intrinsic value all day.

April 30.—Sky milky blue; nearly the same as on April 29; milkiness slightly increasing.

May 12.—Sky milky blue, with occasional clouds.

The last line of each table gives a summary, that for the winter deflections being the mean of seven series, and for the spring the mean of nine series. The absolute galvanometer deflections interest us little. What is most important is a comparison of their relative amounts.

The bolometer, like the thermo-pile, is an instrument intended primarily for differential work. Even if the absolute amounts of heat measured by it were open to question, this would be a matter of secondary importance were the *relative* results trustworthy. As a matter of fact, however, the absolute amounts of heat indicated by the bolometer, when the same instrument is used under the same conditions, are found to be much more exact than the writer (who did not have this end primarily in view in its construction) anticipated. It may therefore be interesting at this stage to apply the corrections for the selective absorption of the materials of the apparatus, so far as they are known, which will tend to give the values of these measures in terms whence the absolute amount of energy in each ray can be calculated with a certain approximation. We repeat, however, that it is the relative amounts of energy which the instrument is primarily designed to furnish. We will observe here that, in accordance with the general considerations already introduced (see p. 5 of introduction), we admit that what the sun sends us is, properly speaking, *energy*, conveyed in vibrations of certain wave-lengths, and that this energy, after falling on the mirror of our siderostat, the surface of our grating, and that of our bolometer strip or thermometer bulb, causes finally certain mechanical effects in our galvanometer or thermometer, which we take to be proportional to the heat in the ray, "heat" being the name we give to the solar energy, as interpreted to us by the above-mentioned media, each of which exercises some minute degree of selective absorption of its own. Thus, even if we suppose that our apparatus were placed at the upper limit of the atmosphere, receiving there the unmodified solar energy, the silver of the siderostat mirror would, for instance, of two rays of equal energy, one in the blue, the other in the infra-red, absorb, in the act of reflection, more of the former than of the latter, so that the two would be unequal after reflection from the siderostat, though equal before; and the heat in each, when it fell on the bolometer or thermometer, would not be proportional to the original energies. Another selective action takes place through the metal of the reflecting grating, and still another through the glass and mercury of the thermometer, or through the iron or platinum of the bolometer, or through the lampblack with which these are covered. If all the solar energy, of whatever wave-length, were treated indifferently by each of these substances (even if each absorbed some of it), the final result, in the reading of the thermometer or galvanometer, would evidently be proportional to the energy of the original ray, and the relative heats in any two rays would be strictly a measure of the relative energies originally sent in them earthward from the sun. Though the selective reflection which the ray has suffered in our catoptric apparatus be much less important than that which would take place with a lens and prism, yet every part has exercised some absorption of its own. We conclude, then, that if we could determine the selective absorption of each agent and make a correction for it, we should restore the exact proportions of the original energy existent in different wave-lengths. We cannot do this with absolute exactness, but in proportion as we succeed in doing so, will our final and corrected results be proportional to the original energy itself.

We have also to make corrections, first, for the overlapping portion of the weak second spectrum, which is found from experiment to have an intensity of ⅓₀ that of the first spectrum; second, for the diminution of heat in the diffraction spectrum, with increase of the angle of diffraction, which is here taken as proportional to secant r.

These last two corrections are instrumental and independent of the selective absorption. On applying them to the last table we get the following results:

TABLE 9.

λ	.375	.40	.45	.50	.60	.70	.80	.90	1.00
Correction I (subtractive)	0.	0.	0.606	0.	0	0	.0	.0	.0
Correction II (factor)	.005	2.067	1.116	1.118	1.201	1.207	1.152	1.206	1.117
Winter, 1881 (corrected for I and II)	86.1	152.0	95.8	175.7	126.7	317.8	201.5	1.3.6	92.7
Spring, 1881 (corrected for I and II)	74.1	117.9	124.8	115.7	265.9	432.5	317.7	135.9	50.6

We are now prepared to apply the corrections for selective absorption. The third correction is for the absorption by three surfaces of silver deduced from independent experiments at Allegheny not here given.

The fourth correction is for the absorption by the surface of speculum metal. There remains the possible selective absorption due to the bolometer strip itself, or to the lamp-black with which it is covered. It is usual to neglect the selective absorption of lamp-black upon the thermo-pile, and, indeed, there is so much doubt as to what this selective absorption is, and how it is exercised, that it is difficult to take it into account. Our own investigations upon it are not yet completed. We are led by them to think it probable that lamp-black is almost transparent to certain infrared rays, which, however, lie beyond the limits of the part of the spectrum we are now studying. Between $\lambda = 0^\mu.375$ and $\lambda = 1^\mu000$, it exercises a certain selective absorption which is however, treated here as negligible, since, on repeating our experiments, both with lamp-black and without, we do not find within the present limits any differences deserving remark. We must admit, however, that the ignorance which we, in common with all physicists, labor under concerning the selective absorption of this substance in the infra-red, is much to be regretted. Rigorously speaking, then, we ought perhaps to describe our present results as giving the energy (to use Mr. Lockyer's expression) "in terms of lamp-black absorption." Applying these corrections, we now have the following results:

TABLE 10.

λ	.375	.40	.45	.50	.60	.70	.80	.90	1.00
Correction III (factor)	2.000	1.923	1.877	1.635	1.570	1.104	1.168	1.159	1.150
Correction IV (factor)	1.031	1.039	1.051	1.061	1.059	1.105	1.195	1.176	1.306
Winter 1881 (corrected for I, II, III, and IV)	192.6	305.4	379.3	765.9	721.9	522.0	355.3	293.4	470.6
Spring 1881 (corrected for I, II, III, and IV)	111.9	295.1	429.7	389.6	621.0	272.5	375.3	355.0	254.6

We repeat that, in the degree in which we have above eliminated the selective absorption of the media of the apparatus, we are entitled to speak of the resultant values as proportional to the solar energy itself. We do not suppose ourselves to have accomplished so untried and difficult a task with exactness, but regard these curves as useful as a first approximation to the absolute energy curve.

The air-masses on the days included in this summary were as follows:

TABLE 11.

Date.	Mean hour angle.	Sun's zenith distance.	M. s.	Date.	Mean hour angle.	Sun's zenith distance.	M s.
	h. m.	° '			h. m.	° '	
1880.				1881.			
December 18	1.19	66 30	18.51	March 10	0.14	44 26	10.22
1881.				March 25	0.14	38 34	9.31
January 24	0.01	58 29	14.25	March 28	0.07	37 17	9.18
February 2	0.00	57 09	13.63	April 14	9.35	32 58	13.98
February 3	0.39	57 30	13.78	April 22	0.12	28 13	8.35
February 5	0.07	56 12	13.37	April 23	0.11	27 19	8.37
February 17	0.38	52 37	12.31	April 29	0.06	25 50	8.17
February 22	0.66	50 26	11.51	April 30	0.04	25 31	8.21
				May 12	0.54	24 09	8.15
Mean air-mass for winter			13.48	Mean air-mass for spring			9.53

If we take these same winter observations and select special observations made at the same altitude in spring, we obtain, by a process readily understood from what has preceded, the following results, in which we suppose *the same amount* of heat in spring and in winter, represented in each case by 1000, to fall upon our bolometer, the mean altitude of the sun at the time of observation being the same.

TABLE 12.

OBSERVATIONS AT THE SAME ALTITUDE OF THE SUN

λ	Winter. d_1	Spring. d_2	Reduced to sum 1000.	
			Winter	Spring
.375	192.6	71.5	49.6	30.2
.50	364.4	119.8	93.7	50.6
.15	579.3	275.6	149.2	116.5
.56	767.0	363.1	197.7	156.1
.60	724.9	439.0	186.6	185.7
.70	327.9	433.9	135.0	183.4
.80	338.3	298.5	87.1	126.2
.90	215.1	190.4	55.5	80.9
1.00	173.6	166.4	44.7	70.4
			1004.0	1000.0

It will be seen from tables 10 and 11 that, although the absorbing air-mass was during the winter nearly half as large again as in the spring, the heat received from the shorter wave-lengths was actually greater in the winter. (See also columns "Winter" and "Spring" of the table just given (12), where it is seen that for equal air-masses the actual infra-red deflections were not greatly different at the two seasons, *within the limit of these observations.**) It appears probable, then, that the transmissibility of the atmosphere for the light-producing radiations is relatively greater in winter than in spring. As this effect may be connected in some way with the unequal prevalence of atmospheric moisture at the two seasons, it may be well to state that the tension of aqueous vapor during the winter observations was in the neighborhood of 2 millimeters, in the the spring of 8 millimeters.

We now proceed to the calculation of the energy outside the atmosphere, for homogeneous rays, with the data which have been given. For this purpose we have used the formula—

$$\text{Log } E = \log d_1 - \frac{M_1 \, \beta_1}{7.6} \, \log a.$$

* This restriction must be noticed. The wide absorption bands subsequently discovered in the infra-red lie below the limits of these early researches, which cover but a small part of the great infra-red region. It would be premature, then, to base any general conclusions as to the effect of water-vapor upon the invisible heat region, in these early observations. It may be observed, in regard to the whole subject of winter and spring comparisons, that to obtain the same air-mass in winter we have to observe systematically at a different time of day than in spring, and we have already remarked that there is probably a systematic change in the quality of the absorption as the sun approaches the horizon. It is with the latter fact in view that I have written above: "As this effect *may* be connected with the unequal prevalence of atmospheric moisture," &c.

Where E is the energy in any ray outside the atmosphere (*i. e.*, before telluric absorption), d_0, the average galvanometer deflection at noon for the same ray, B_0, the barometer pressure in units of one decimeter, or the mass of air in the vertical column; $M_1 B_1$, the corresponding air-mass for the sun's zenith distance at noon, and a the adopted coefficient of transmission for the ray in question by an air mass of unity, represented by 7.6 dm. in the barometer.

TABLE 13.

WINTER.

$M_1 B_1 = 13.88.$

λ	$\text{Log } a \div \tau$	$\dfrac{M_1 B_1}{\text{Log } a \div \tau}$	Log d_0	Log E	E
375	— .0005	— .7420	2.2847	3.0273	1065
400	— .0496	— .6887	2.5604	3.2169	1774
450	— .0311	— .5746	2.7629	3.3375	2175
500	— .0018	— .4880	2.8857	3.3951	2485
600	— .0239	— .4305	2.8603	3.2198	1659
700	— .0200	— .2376	2.7226	3.0002	1000
800	— .0155	— .2151	2.5293	2.7444	555
900	— .0132	— .1892	2.7323	2.5165	729
1000	.0128	— .1777	2.2394	2.4173	261

TABLE 14.

SPRING.

$M_1 B_1 = 9.33.$

λ	$\text{Log } a \div \tau$	$\dfrac{M_1 B_1}{\text{Log } a \div \tau}$	Log d_0	Log E	E
375	— .0505	— .4992	2.0486	2.5478	353
400	— .0496	— .4628	2.5714	2.8346	683
450	— .0311	— .2902	2.6271	3.0111	1051
500	— .0018	— .2987	2.7255	3.0802	1203
600	— .0239	— .2116	2.7901	3.0347	1083
700	— .0200	— .1866	2.7423	2.9283	849
800	— .0155	— .1406	2.5701	2.7155	519
900	— .0132	— .1232	2.7366	2.4898	416
1000	.0128	— .1194	2.2704	2.4898	399

TABLE 15.

MEAN OF WINTER AND SPRING

$M_1 B_1 = 11.6.$

λ	$\text{Log } a \div \tau$	$\dfrac{M_1 B_1}{\text{Log } a \div \tau}$	Log d_0	Log E	E
375	— .0505	— .6206	2.1826	2.8032	636
400	.0496	— .5754	2.4763	3.0517	1127
450	.0311	— .4462	2.7002	3.1801	1515
500	— .0018	— .4057	2.8257	3.2200	1691
600	— .0239	— .3004	2.8201	3.1254	1311
700	.0200	— .2330	2.7526	2.9616	962
800	— .0155	— .1708	2.5566	2.7304	538
900	— .0132	— .1531	2.2555	2.4086	324
1000	.0128	— .1415	2.3088	2.1587	267

The following table has been prepared with the values observed in the spring of 1881, using mean coefficients of transmission, to show the relation between energy outside the atmosphere and that for high and low sun at Allegheny, the various actual absorbing air-masses at the low sun observations being reduced to a uniform value, double that at high sun.

TABLE 16.

λ	.275	.400	.450	.500	.600	.700	.800	.900	1.000
E energy before absorption	353	684	1031	1256	1088	819	539	316	309
d energy after absorption corrected high sun....	112	253	424	570	621	553	372	238	235
d energy after absorption corrected low sun ..	27	83	140	225	314	324	246	167	167

E can be computed from d_l and d_{ll} by the formula already given, and with these values the curves in Plate I have been plotted.

The middle curve (I) is that showing the distribution of the energy in the normal spectrum at high sun. Except for the heat below wave-length 1ᵘ.0, the area of the curve may be considered to represent the heat actually observed by the actinometers at noon, as presently given. Its maximum ordinate is near 0ᵘ.60 in the orange-yellow.

The lower curve (II) is that at low sun. Its area is proportional to the heat received when the sun shone through double the absorbing air-mass that it did at noon, and it will be seen that the maximum ordinate is near wave-length 0ᵘ.70, or near the extreme red.

The upper dotted curve is "the curve outside the atmosphere." Its area will give the heat, which would be observed if our apparatus were taken wholly above the absorbing air, and the distribution of this heat (energy) before absorption. Its maximum ordinate is near 0ᵘ.50 to 0ᵘ.55 in the green.

If we know the values in calories corresponding to the middle curve, we can now obtain the absolute heat before absorption, i. e., the *solar constant*.

It should be noticed that if we had attempted to deduce this latter value, by applying our logarithmic formulæ directly to ordinary actinometric observations (i. e., to observations where only the indiscriminate effect of all heat rays is noted by the thermometer) made at high and low sun, we should have obtained a quite different result. This has been the usual process, but it can never be a correct one; for, we repeat, these exponential formulæ are in theory only applicable to homogeneous rays.

The above values (in Table 16) are relative only. To obtain absolute ones we have now to combine this result with the actual measurements of solar radiation in calories, or other units furnished by actinometers under approximately the same conditions. We shall at the same time thus obtain a preliminary value for the solar constant. Taking the mean of our observations with the Violle and Crova actinometers on clearest days, we have 1.81 calories* observed at Allegheny in March, 1881. This is the absolute amount of heat represented by the area of a completed "high sun" curve.

To this result, the energy distributed through the whole spectrum has contributed, while our bolometer measurements in the diffraction spectrum end at wave-length 1ᵘ.00. Nevertheless, since we do in fact know from subsequent measures (to be given later) where the effective spectrum ends, we can by the aid of these later measures prolong the curves and obtain their relative areas with close approximation. In this way we determine, by measuring the charted areas, and making allowance for the (here) uncharted area below $\lambda = 1^u.0$:

    Area outside curve above $\lambda = 1^u.000$ 47.17
    Area outside curve below $\lambda = 1^u.000$ 26.19

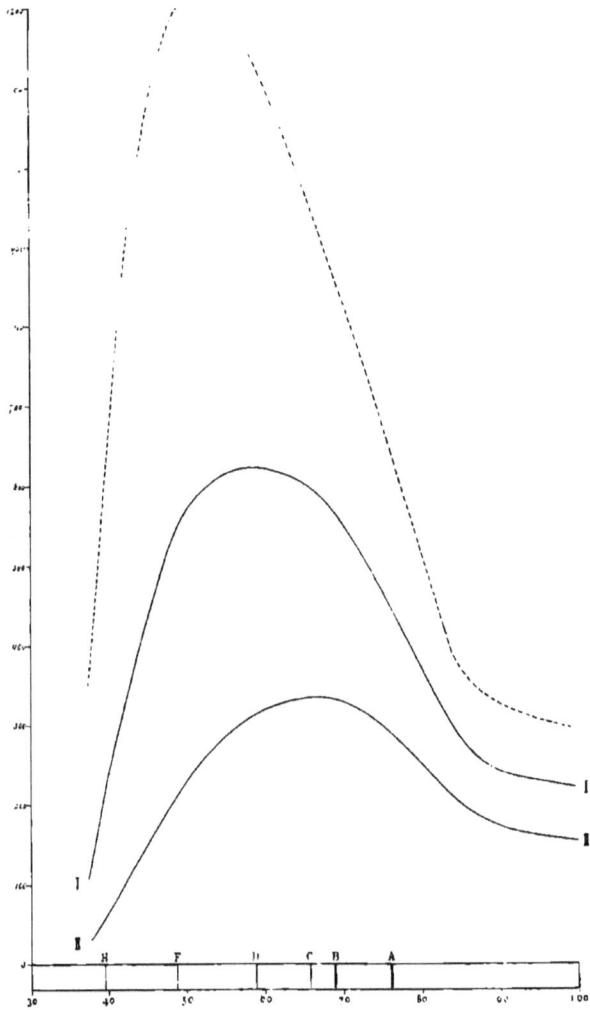

PLATE I.
RELATIVE ENERGY FOR HIGH AND LOW SUN.

·

We have, then, adopting 1.81 cal. as the solar radiation at Allegheny with clear sky, 1.81 cal. × 1.57 = **2.84** calories (*i.e.* calories per minute per square centimeter) as an approximate value of the solar constant.

In all these observations, the object has been to avoid the registering of small variations analogous to the Fraunhofer lines, and to give only the general distribution of the energy. The mapping of the interruptions of the energy caused by visible or invisible lines or bands forms a distinct research, and the results are given later in the present volume.

We find from these preliminary observations that the maximum energy in the normal spectrum of a high sun at the earth's surface is near the yellow, and that the position of the maximum of heat does not in fact differ widely from that of the maximum of light. It has been long known that certain ultra-violet and violet rays were much absorbed, but it has been supposed that the absorption increased also in the infra-red, so that the luminous part of the spectrum was, on the whole, the most transmissible.

But we see here, not only how enormous the absorption at the violet end really is, but that *the light rays have suffered a larger absorption before they reach us than the "heat" rays (i. e., than the extreme red and infra-red rays)*, a conclusion opposed to the present ordinary opinion, and, if true, of far-reaching importance. For if this "dark" heat escapes by radiation through our atmosphere more easily than the luminous heat enters, our view of the heat storing action of this atmosphere, and of the conditions of life on our planet, must be changed. Within the limits of the present charts, the "dark" heat apparently does so escape.

We can, from the data now gathered as to the rate of absorption for each ray, compute the value of the heat or energy before absorption (the solar constant) by a new process which is in strict accordance with theory. This preliminary value indicates that the true solar constant is larger than that commonly given.

The ratio of the dark to luminous heat has been so wholly changed by selective absorption that we must greatly modify our usual estimates, not only of the sun's heat radiation, but of his effective temperature. We infer also, that *the sun, to an eye without our atmosphere, would appear of a bluish tint.*

According, then, to the observations which have been detailed, the actual value of the solar heat is greater than has been supposed. The action of this heat on our atmosphere is also very different from that customarily asserted, and the extent of our misapprehension of the real circumstances which nature presents to us, may be said to be presented to us face to face, in our universal belief that the sun is white, while its real color may be unknown. Our present conditions of observation are, however, in many respects most deficient. They all rest on the assumption that the transmission of heat by like air masses remains the same throughout the day. There is too much reason to believe that it not only varies casually, but also changes systematically, both with the time of the day and with the seasons of the year. It will be seen, on consideration of the method by which the transmissibility for the high and low sun has been obtained, that we learn only the mean transmissibility of our atmosphere, and never its composition in other respects. The air, for instance, a little way above our heads, might have a different chemical constitution and different transmissibility from that in the lower stratum, without such means as the present giving us any hint of the fact. All these circumstances have an immediate bearing on the determination of the solar constant and the above value of 2.84 has never been regarded by me as more than a first approximation. Even as such, however, it shows the constant to be much larger than heretofore supposed, and the absorption of our atmosphere to be greater than has been imagined.

It must be constantly borne in mind that we assert that the formula of Pouillet has but a limited application, and is, as generally used, erroneous. We prove, later, by actual demonstration, that it always gives too small a value for the absorption, and hence the less absorption there is above us, the less important does the error of this formula become. Accordingly, other things

---

* The "caloric" here is the "small caloric" of Pouillet and later investigators. It is the amount of heat required to heat 1 gramme of water from 0° C. to 1° C., and is referred to the minute and square centimeter. 1 caloric = 42,200,000 ergs = 138 "actines" of Herschel = 0^{mm}.138 ice melted per square centimeter per minute.

being equal, the observations made even by the present process, if repeated on high mountains, may be expected to give more perfect values than those made at the sea level. We have, it is true, observed on approximately homogeneous rays, in obtaining our coefficients of transmission, yet we must remember that though we may allowably speak in common terms of the bolometer's hair-like strip as "linear," it is not absolutely so, and the true coefficients of transmission are, inferentially, smaller than those obtained even by its means. In other words, the amount absorbed by our atmosphere is not improbably greater even than the present observations make it.

We have reached in these preliminary investigations some conclusions quite at variance with accepted beliefs. We have found that the absorption of the heat, on the whole, diminishes as we go into the dark heat region, and that the "light" is more absorbed than the "heat," while it has been generally understood that the contrary is the case. It would appear from this that within the limits of the present observations ($0^\mu.4$ to $1^\mu.0$) the "dark" heat escapes more easily through our air than it enters as "light" heat, so that the familiar comparison of our atmosphere to the cover of a hot-bed does not here seem to be just, and so far as these preliminary observations extend, we do not find at all what the ordinary belief leads us to expect.

The construction of our ordinates for the curves outside the atmosphere, has shown us that the maximum point continually advances toward the blue; in other words, that the sun must really be of a bluish tint, so that we have never seen it as it is, and what we are accustomed to speak of as white light and "the sum of all radiations" is merely that remainder of rays, whether of "light" or "heat," which has filtered down to us. Our view of the solar light and heat, and of their effect on our atmosphere must be modified if such results are true, and it would be most desirable to prove their truth by some independent mode of observation, since we have nearly exhausted the capacity of our present means of research.

There does remain an entirely different method of observation, but one presenting peculiar difficulties. It is to ascend a very high mountain, and to compare observations made at its summit with others carried on at its base, so that we can not only estimate, but directly measure, the absorptions which the rays have actually undergone. The preparations for this form the subject of the next chapter.

CHAPTER II

JOURNEY TO MOUNT WHITNEY, 1881.

Toward the close of 1880 it had already become clear that the gain in our knowledge by repeating the observations then in progress at the Allegheny Observatory, at the base and at the summit of a lofty mountain, would justify the labor and expense of such an undertaking. There would have been little probability, however, of such a plan being carried out by the Observatory, were it not for the generosity of a citizen of Pittsburg, who placed at its disposal the considerable means demanded for the outfit of an expedition for this purpose.

By his own wish his name is not mentioned in this connection, but it is proper to acknowledge here, and first of all, the timely and indispensable aid which made the project a reality.

The expedition was, as at first designed, to be made wholly on the account of the Allegheny Observatory, whose trustees authorized me to use any of its apparatus for the purpose, it being understood that the special expenditures involved would be met from the source mentioned.

Upon the objects of the expedition and their bearings upon meteorology becoming known to the Chief Signal Officer of the United States Army, he consented to give it the advantage of his official direction and the aid of Signal Service observers, and upon the reasons which made the choice of its objective point in a remote part of the United States territory being approved by him, he contributed further material aid in transportation. The considerable expenses of reduction have been chiefly met from the private source just mentioned.

The principal conditions desirable in the mountain chosen should be—

(1) Clear air.

(2) Great altitude.

(3) Very abrupt rise, so that two contiguous stations may be found with very different altitudes.

(4) Southern latitude.

(5) A dry climate.

(1) The first and fifth conditions are almost inseparable. Such summits as Pike's Peak, and the neighboring summits in the Rockies, are rarely free from mist and cloud during the summer, and both from the nature of the observations and the fact that the stay must be brief, an almost absolutely pure and cloudless sky is indispensable.

(2) We ought if possible to leave at least a third of the atmosphere below us, which implies a height of at least 14,000 feet.

(3) An elevated plateau is unsuitable, for it is almost indispensable to have a relatively low station quite near the high one, and, if possible, in sight from it.

(4) We must, other things being equal, prefer a southern latitude which will enable us to view the sun nearly in the zenith. As no point east of the Rocky Mountains unites these requisites, inquiries were made at all sources, particularly of officers in various departments of the Government familiar with the Western Territories. I am specially indebted to Maj. J. W. Powell, the present head of the United States Geological Survey, and to the late Superintendent of the Coast Survey, to Capt. C. E. Dutton, of the United States Army, and to Mr. Clarence King (late in

35

charge of the Geological Survey), for valuable information. Among the points carefully considered were—

	Longitude	Latitude
Mount Nebo	111 15	39 50
San Bernadino	117	34 05
Pine Valley Mountains	113 15	37 25
Toyabe Range	117 05	39 20
Brian's Head	112 15	37 20

sites from 11,000 to 12,000 feet in height, each of which had its several advantages, but none of which met all the conditions. Finally, upon the advice of Mr. Clarence King, and with the concurrently favorable opinion of officers of the Coast Survey and others familiar with that region, Mount Whitney, in the Sierra Nevada Range of Southern California—approximate longitude, 118   30' (7 h. 54 m.); latitude, 36   35'—was found to be, on the whole, most desirable. Its height was known to be between 14,000 and 15,000 feet. Its eastern slopes are so precipitous that two stations can be found within 12 miles, visible from each other, and whose difference of elevation is 11,000 feet, and it rises from and overlooks one of the most desert regions of the continent, while its summit is almost perpetually clear during June, July, August, and September. It is, it is true, far from any railroad, in a wild region, and it had been ascended so rarely, and with such difficulty, that it was not certain that heavy instruments could be transported to the extreme summit. As there were neighboring mountains both in the Sierra Nevadas and Panamint Ranges, offering not greatly inferior advantages, to fall back on, and as it was certain that a very considerable altitude, at any rate, ought to be reached on Whitney, in spite of the imperfectness of our knowledge of the extreme summit, the site was submitted to the Chief Signal Officer and approved by him. Capt. O. E. Michaelis, of the Ordnance, temporarily on Signal Service duty, was ordered to establish a Signal Service station there, and Sergeants Dobbins and Nanry, observers of the Service, were detailed to join the expedition in San Francisco.

It was most desirable that we should reach the scene of operations so as to commence our observations in July, but delays occurred, in spite of our wishes, which, as it will be seen, frustrated this purpose. The first portion of the party, consisting of Captain Michaelis, Messrs. J. E. Keeler, of the Allegheny Observatory, W. C. Day, of the Johns Hopkins University, and the writer, left Allegheny, Pa., on the 7th of July, 1881. The instruments, weighing in their outer cases about 5,000 pounds, were to accompany us all the way; and I have to express the very great obligation of the whole expedition to Mr. Frank Thomson, vice president of the Pennsylvania Railroad, by whose kindness transportation was furnished to Chicago for a private car, which was occupied by us, with our instruments, and which, through his introduction to Messrs. S. H. H. Clark, general manager of the Union Pacific Railroad, Omaha, and A. N. Towne, general manager of the Central Pacific Railroad, San Francisco, was, by the courtesy of these gentlemen, continued in our use till we reached the Inyo Desert.

We reached San Francisco on the 22d. It was considered advisable, in the possible contingency of our being forced to choose our station in some point in the desert region east of the Sierra Nevada Range, that an escort should accompany us, and through this need an unforeseen delay of nine days occurred in San Francisco, a time which was shortened to us by the courtesy of General McDowell, commanding the department, and of Prof. George Davidson, of the Coast Survey, but which we could not but regret. On the 22d we left on the Southern Pacific road, the party having been joined by Mr. George F. Davidson, who accompanied it as a volunteer, by Sergeants Dobbins and Nanry of the Signal Service, by Mr. Frost, a carpenter, engaged in San Francisco to accompany us, and by the escort of Corporal Lamonette, of Company B, Eighth Infantry, and five enlisted men, who joined us at Benicia Barracks. We reached Caliente, where the writer, Mr. Keeler, and a part of the escort left the railroad, while Captain Michaelis and the rest of the party proceeded in the same train to Mojave, about 40 miles farther, where the instruments and provisions were to be taken across the desert by wagons. By riding day and night, I reached our station at the mountain foot (Lone Pine) on the evening of the 24th. The road lay along the Inyo Valley, a shadeless, waterless desert, on the west side of which (on our left) the Sierra Nevadas rose in constantly higher summits as we went northward, till we found ourselves looking up through the desert air, where the shade temperature was over 100° F., to the patches of snow on their summits, which told of the real altitude of the almost unknown upper regions to which we were finally bound. The mountains on the right, at first low and distant, drew closer and grew higher, making the valley character more and more distinct as it narrowed, while the desert over which we traveled constantly ascended, without losing its aspect of a narrow extended plain, shut in closer by mountains as we went northward. Near the upper extremity of Owen's Lake (a small dead sea), we got our first sight of Whitney, and in a few miles more reached the little hamlet of Lone Pine, built on a small patch of green, due to the moisture of a snow fed stream from the mountains which threads the valley, here about 6 miles wide, between the foot-hills, and almost perfectly flat. As we rode in, we noticed cellars belonging to houses shaken down in the last earthquake, which destroyed a large part of the inhabitants of the little place, for we were now in the earthquake country.

The outline of Whitney and the neighboring peaks seen from Lone Pine is very extraordinary, the serrated edge and the snow, justifying the name of the "Sierra Nevada." The air is so clear

Outline of Mt. Whitney Range.—As seen from Lone Pine.

that the appearance of nearness is most delusive. The mountains look like large rocks close by, covered with moss, on which patches of white are glistening, but only on looking through the telescope, which resolves the apparent moss into large forests of great trees, and the white patches into snow fields, can we realize the actual distance to the summits, which is about 12 miles, the interval,—"the foot-hills"—being an elevated desert table land, broken into low hills, extending back with a gradual rise to rather more than half this distance, where the eastern wall begins and attains most of its final altitude of over 14,000 feet in about 5 miles, reckoned on the level. I give in the form of a diary the events of each following working day, as far as the statement seems necessary.

July 25th.—With the aid of Mr. W. L. Hunter, of Lone Pine, to whom we were indebted for this and other kindnesses, we made a preliminary survey of the place and selected a site for our camp on the grounds of Mr. Begole. We passed in our reconnaissances a grave where seventeen persons, victims of the earthquake, are buried together. They were killed by the falling in on them of their "adobe" houses, and we felt our tents a safer shelter. Sergeants Dobbins and Nairy arrived.

26th.—Set up piers for the expected instruments.

27th.—Captain Michaelis arrived. Mr. W. Crapo, of Cerro Gordo, was engaged as guide, and arrangements were made for first mule-trains.

28th.—Private Nairy, two soldiers of the escort, and the carpenter were sent up the mountain with the guide, muleteers, and a small mule-train, carrying tent, provisions, and heliotropes. Their instructions were to establish a camp at the highest point, within reach of wood and water (a point on the other side of the ridge, to which Mr. Crapo undertook to guide them). After this the peak of Whitney was to be ascended from their camp, and heliotrope signals exchanged with our own station, from which, as I have said, the peak was visible.

From the 28th to the 1st was passed in enforced idleness, waiting for the instruments which were still on their slow way across the desert. Clouds hung over the mountains (for the only time during our whole stay), but no rain fell. One or two flashes came one morning from the peak of Whitney, showing that it had been reached by the party with the heliotropes, but no answer to our signals was returned. It may here be stated that owing to the difficulties of making any stay upon the peak without fire or shelter (which we had ourselves to experience later), and to the lack of men experienced in heliotrope signaling, communication with the camp on the other side of the ridge was kept up only by special messenger; although the conditions for heliotrope signaling between Lone Pine and the peak (were a station once established there) are excellent, the two places being full in view of each other, with almost constant sunshine.

29th.—The fencing in of a piece of ground in the village, 60 by 150 feet, was completed.

August 1.—The wagons arrived at noon. The instruments were at once unpacked, when it was found that the desert dust had penetrated every crevice and settled on every instrument, however carefully boxed. They were cleaned and observations begun immediately.

To Mr. W. C. Day was assigned the large actinometer.

To Mr. J. E. Keeler the spectro-bolometer.

To Mr. G. F. Davidson the comparator and the preliminary observations for time and latitude.[*]

Sergeant Dobbins was directed to make observations with the pyrheliometer and (subsequently) with the small actinometer, and to take the readings of the barometer and wet and dry bulb thermometers usual in the Signal Service.

The writer observed with each of these instruments and observers in turn, till it was certain that each understood what was novel in his work and had acquired fair expertness at it, but his chief time, whenever other duties admitted, was given, with the aid of Mr. Keeler, to the spectro-bolometer.

2d.—A small tent was set up for this instrument and the reflecting galvanometer, with a black cloth inner lining, to form a dark room for the latter. The heat was excessive without, and within it rose to a point beyond human endurance, while the light proved not to have been shut out even at the cost of the quite intolerable heat. After a day's trial this plan was then abandoned.

3d.—Set up a larger or "hospital" tent (about 14 feet square) and attempted, unsuccessfully, to construct a separate dark room within. In the adjacent hot box a thermometer in air, but under glass, rose to 235° F. At the same time packing for the mountain went on, and a mule train started for Mount Whitney this evening, carrying apparatus and quartermaster's stores. It was guided by Mr. Crapo, and accompanied by two soldiers of the escort.

4th.—The dark chamber was, through Mr. Keeler's ingenuity, completed so as to get the light excluded, without a heat such as to make observation impossible.

5th.—The systematic reading of the barometer and wet and dry bulb thermometers commenced to-day. Sergeant Dobbins was, however, directed by me to omit readings at those hours which interfered with his observations with the pyrheliometer or actinometer, it being impossible to spare a second observer for these latter.

6th.—Many clouds over the valley (for the first time) and a few drops of rain. The wind was violent all day; the tent was shaken so as to make it doubtful whether it could stand, and all the instruments in it were covered with sand and dust, while the lights for the galvanometer were extinguished by the penetrating gusts so as to make its use impossible.

7th (Sunday).—Still cloudy.

8th.—A slight earthquake shock at night. Commenced bolometer observations.

11th.—After two days' struggle with difficulties incident to the novel conditions, the first complete series of morning, noon, and evening bolometer observations was obtained. The very considerable changes of temperature in the tent through the day caused a troublesome drift of the galvanometer needle, but otherwise the result was satisfactory. On this as on previous days the other observations were successfully pursued in the prescribed manner and call for no remark here.

During all the previous days hiring of mule drivers and animals and the arrangement for the transportation of the somewhat elaborate apparatus, to the distant summit, had been a constant preoccupation, for the season was already far advanced, and we had originally hoped to have completed our chief observations in the valley and been at work by the first of August at the mountain station, which rose above us in constant view, apparently so near and really so distant. Every

---

[*] The observations for latitude had not been commenced when Mr. Davidson left for the mountain. We find from the Army map (Lieutenant Wheeler's expedition), longitude, Lone Pine, 118° 03′ 17″; latitude, 36° 36′.

effort was made to have the packing done so that the parts of each instrument should be kept together as nearly as possible, and in any case where these necessarily occupied two or more boxes, they were placed on the same mule, with strict orders that the arrangement should not be disturbed by the muleteers in the ascent. It was necessary, however, to send off separate trains, as the mules could be gathered for them, and hence the muleteers could only be in part overlooked.

Captain Michaelis, with Mr. Davidson, Sergeant Naury, and two soldiers, left to-day for the Mountain Camp with a train carrying part of the intruments and quartermaster's stores, intending to make the ascent through Cottonwood Cañon. The muleteers promised that this train should reach the Mountain Camp on the 14th and be back in Lone Pine by the 17th. I remained (hoping to complete the lower station bolometer observations) with Mr. Keeler. Mr. Day, Sergeant Dobbins, Corporal Langmette, and Private Kelly remained also in the camp.

13th.—My anxiety to know personally of the arrival of the instruments at the Mountain Camp induced me to leave to-day, though another day's observations was desirable. This I left to Mr. Keeler, with directions to pack as soon as it was made and to follow with Mr. Day on the 17th, or as soon as the expected mule-train had returned. I left myself in the afternoon with Mr. Crapo, reaching " Ridgers," an elevated ranch in the foot-hills, about eight in the evening.

14th.—After a night passed in the open air, I started southward with the guide, our object being to reach a cañon which would lead us over " the Great Divide " and thence to the mountain camp already established at the western base of Whitney Peak (on the other side of the range as seen from Lone Pine) where it was expected that the instruments already forwarded by the other route would be ready for work.

Our course first lay across a sloping table-land already elevated 800 or 1,000 feet above the valley, dotted with sage-brush, but still below the lowest edge of the timber-belt. After three hours' riding the trail began to ascend rapidly and the air to grow cooler, while we passed occasional dark stunted pines, which rose from the white gravel, like posts planted in it, there being no grass under them anywhere, nor any verdure, even when we had fairly entered the timber. The large dark trunks were so far apart, and formed such a contrast with the white ground beneath them, that the eye followed the color of the latter through the distant forests, which looked as though a fire passed through them, and presented a most desolate aspect from the absence of moisture and consequent verdure. The trail made sharp turns, plunged down into ravines into which descent on the saddle seemed at first impossible; and wormed its way between bowlders, and climbed over rocks and fallen timber, in such a manner as to give a formidable impression of the dangers our apparatus must have incurred in the ascent, though it had taken a somewhat easier and longer route than ours. All trace of the trail itself finally ceased in the bed of a water-course, seemingly barred to all passage by large bowlders and trees which had tumbled from above into the channel, up which, however, with the occasional aid of the ax, we slowly forced a road, reaching at nightfall a small meadow whose altitude must have been 8,000 or 9,000 feet, watered by the stream whose bed we had been following. The temperature had now fallen greatly, and ice formed thickly during the night, which we passed like the preceding one, under an unclouded sky, whose stars seemed perceptibly brighter than we see them in the clearest night from lower stations.

15th.—In the morning there could be no question of the change in the blue of the heavens, which was darker and more violet colored than that at Lone Pine; itself purer than that seen except at very rare intervals at Allegheny. We had now parted from any signs of a trail, and a long and tedious ascent was followed by a sharp descent, during which we lost nearly all the elevation gained since sunrise, and this brought us into " Diaz meadows," up from which, after another formidable climb, we got over the " Divide." Then came a descent of about 2,000 feet, and then another mountain to be climbed whose slope was so steep that occasionally the insecurely poised bowlders which covered it, as we stepped from one to another, rolled from under our feet, and went leaping downward. The lost labor consumed in this incessant alternation of ascent and descent is enormous, and (as I found later), by the construction of a mule-path along another and direct route (which we took in finally descending again to Lone Pine), may be almost wholly avoided.

The distant scenery had been much more monotonous than might have been expected from this account of the route, but now one more descent brought us into a great cañon running westward, with a magnificent view of the vast amphitheater of precipices, behind Sheep Mountain.

I noticed now, that, though long tanned by the hot sun of the valley before starting, the skin of both my face and hands was beginning for the first time to burn badly; a striking effect in this cold air, and which could not be attributed, as it has been in the case of some Alpine and other climbers, to reflection from snow, for we had, as yet, seen none but at a distance.

16th.—After another night like the last, we climbed, with some hours' work, several thousand feet up the cañon sides over the roughest country we had found yet. From this, as from other eminences, we could see the smoke of forest fires at one or two very distant points, fires which the guide said would grow more numerous later in the season; a sight which added to my anxiety to get to the mountain work. We now came down into Whitney Cañon by a descent which was actually worse than anything that had preceded, but which finally brought us in view of Whitney Peak, for the first time since leaving Lone Pine. It was still high above us, but looking most delusively near. It was difficult, indeed, even with all one's experience of the deception as to distances common here, to believe that the peak was even a mile away, or that the little patch of white on its flank was more than a few yards in diameter. The summit was really, however, six or eight hours further, and the white patch was a snow field which fed the considerable mountain torrent now falling past us.

The rest was easy. We ascended by the stream past little meadows and small lakes filled with the clearest ice-cold water.

A little further we found the woods burning over many acres, the fire having been apparently wantonly set by some sheep-herders, who are the great destroyers of the timber in this upper region where the few spots of herbage are found. Shortly, we finally rose above the entire timber belt, and at five o'clock we reached camp at an elevation of about 12,000 feet, for which an excellent site had been chosen by Mr. Crapo. It was beautifully placed on a nearly circular and well-watered meadow about 200 yards in diameter, while an amphitheater of very precipitous cliffs from 1,000 to 1,500 feet, forming the base and flank of Whitney Peak, rose immediately from its northern and eastern sides and was continued by others more remote on the south.* the only distant view from the camp being toward the west through the long valley along which we had ascended, and looking back through which we saw a horizon of mountain summits. Here I learned the dismal news, that the mule train, which was to have delivered the freight here on the 14th, and which we had been looking to meet on its return, had not yet arrived at the camp, and that only a single instrument (an actinometer) had arrived in condition for work, of those sent by the preceding train; not that the others were broken, but that the mule-drivers had left the boxes and parcels along the route, so that nothing was complete.

Captain Michaelis had, I learned, ascended the peak that day with Sergeant Nanry, and both returned soon after I entered camp. Captain Michaelis reported the ascent very trying. The sergeant, indeed, was sick in consequence.

I pass over days spent in anxious waiting. We were cut off completely from communication with Lone Pine except by special messenger. By scaling the mountain wall on the east of us, it is possible, it seems, to descend on foot through Lone Pine Cañon, direct to Lone Pine itself (as we proved by our own subsequent experience), in a day. An Indian guide, sent by this nearly unknown route to Lone Pine, returned the next day, bringing letters, but no news of the mule train, which was as completely lost to the knowledge of those to whom it was coming, as a ship at sea could be.

Most of us, while waiting for the instruments, had occasion to note, without their help, that the solar radiation was wholly altered in character from that in the valley. I, for instance, have alluded already to the fact that my hands and face were considerably more burned on the way up, through the cool air, than in the hotter desert below. On the day following my arrival in camp, my hands presented the appearance of as severe burns as though they had been held in an actual fire, and my face was hardly recognizable. Others suffered less, but all of us, with skins thoroughly tanned and indurated by weeks in the desert, were more or less burned.

* See view in Frontispiece of the Mountain Camp at the base of Whitney Peak. Longitude of camp, from Army map, 118° 18′ 30″; approximate latitude by sextant (Davidson), 36° 34′.

19th.—The sky to-day, as always, is of the most deep violet blue, such as we never, under any circumstances, see near the sea level. It is absolutely cloudless, and there is only a slight orange tint about the horizon at sunset. Carrying a screen in the hand between the eye and the sun, till the eye is shaded from the direct rays, it can follow this blue up to the edge of the solar disk without finding in it any loss of this deep violet or any milkiness as it approaches the limb. It is an incomparably beautiful sky for the observer's purposes, such as I have not seen equaled in the Rocky Mountains, in Egypt, or on Mount Etna. It had been part of my object to make an effort to see the solar corona by directly cutting off the sun's light by a very distant cliff, though I was aware that Bond had tried a similar experiment unsuccessfully in the Alps, and though I had myself been foiled in a similar attempt on Mount Etna. On the south of the camp was a range of cliffs, running nearly east and west, and whose almost perpendicular wall rose from 1,000 to 1,200 feet. They appeared to be within a pistol shot of the camp. I left it for them at about eleven o'clock, but reached them at nearly half-past twelve, after an hour and a half of hard scrambling. I found that I could choose a position on the north of the cliff, along whose edge the sun was moving nearly horizontally; so that the shadow was fixed as regards the observer, and so sharp that, though I must have been over one-quarter of a mile from the portion of the cliff casting it, I could, without moving from my place, and by only a slight motion of the head, put the eye in or out of view of the sun's north limb. The rocks were, in these circumstances, lined with a brilliant silver edge, due to diffraction. This I had anticipated, but now I saw what could not be seen by screening the sun with a near object, that the sky really did not maintain the same violet blue up to the sun, but that a fine corona was seen about it of about 1° diameter, nearly uniform, though it was sensibly brighter through the diameter of 1½. Upon bringing to bear upon it an excellent portable telescope, magnifying about thirty times, I found it was composed of motes in the sunbeam, between the diffracting edge and the observer's eye. This result, if disappointing, is also interesting in another point of view, as showing that the dust-shell, which, as I have elsewhere stated, encircles our planet, exists at an altitude of at least 15,000 feet, and under favorable conditions for the purity of the atmosphere. The result is not without importance in its bearing upon our conclusions as to atmospheric absorption.[a]

20th.—A portion of the mule train came in about noon.

21st.—Captain Michaelis started at 5 a. m. for Lone Pine, to engage more mules and to arrange the trains. Actinometer readings and routine observations with the thermometer were all that we could do as yet.

22d.—I had been ill since my arrival in camp, and on this day first ascended the mountain. I started at 9 a. m., and, being somewhat weak, occupied over four hours in the ascent, while it might be made by an active person in less than three hours, though not without difficulty, the actual height above the camp being something like 5,000 feet, while the difference between exertion at this altitude and at the sea-level is extreme. The Peak would be wholly inaccessible (from the precipices on its side, which rise in steps of several hundred feet) were it not that the earthquakes have rent these into fissures, and that through these narrow cracks bowlders and rocks from above have poured down in past times, in a rocky river, forming a "*couloir*," as it is called in the Alps, the rocks being still poised so that the surface ones can easily be started downward. Through the nearest of these couloirs, called by the guide "The Devil's Ladder," I commenced the ascent, the stones occasionally rolling away and bounding down hundreds of feet below me. After one-half or three quarters of an hour in this interminable couloir, I got on to the mountain slope, still extremely steep, the surface presenting an appearance as though stones, from the size of a foot ball to that of a grand piano, had been hailed down on it and covered it to an unknown depth. After nearly three hours' time I came to the snow-field, which I have mentioned as having been seen from a distance. It was about one-quarter of a mile in length. At the summit were some Indian and Mexican laborers, who had been brought up to improve the path up the "Devil's Ladder," lying and smoking in the sun. The view from the summit was of a horizon of tumbled mountains on the north, west, and south, not continuously

---

[a] Prof. Clarence King, late head of the Geological Survey, whose familiarity with these regions and whose competence as a geologist are well known, informs me that he believes this dust above the Sierra Nevadas has been borne across the Pacific and owes its origin to the "Loess" of China.

white as in the Alps, for though at a more than Alpine height, I saw only scattered snow-fields here and there. The air was cold, but not very chilly, and the sky of a deeper violet overhead than in the camp below. On the east side, the mountains descended in a series of precipices between 3,000 and 4,000 feet to a little lake surrounded by a snow-field. The eye could follow the course of the stream running from it a little way down a cañon, with tremendous vertical walls, which lead in the direction of Lone Pine. Lone Pine itself, was descried as a little spot of green on the brown floor of the desert. Opposite, on the other side of the Inyo Valley, was a range of mountains nearly 10,000 feet in height, and beyond, to the south, the Panamint and other ranges. Between us and them was a reddish sea of desert dust, 4,000 or 5,000 feet above the valley floor, and almost covering the lower summits of the mountains. Through this dust ocean, we at Lone Pine must have been observing; yet the sky even there is, as I have said, of unaccustomed purity, and probably we observe under still worse conditions habitually when at home.

The top of Whitney is an area of perhaps three to four acres, nearly level, or with a slight downward slope toward the west. Stone for the erection of permanent buildings is here in unlimited quantity. We look immediately down on one of the driest regions of the globe, from an altitude of nearly three miles, in a sky of exquisite purity, and this station, once reached, is everything that I could have hoped to find it, and more; but existence is only possible on the summit with permanent shelter, for though at the moment I viewed it it was calm, yet the wind and cold would be fatal to life at other times, without house and fire. The nearest wood is over 3,000 feet below it. It became evident to me that we must forego, at this late season, further hope of making regular observations on the Peak, and confine ourselves to those at Mountain Camp, for it was evident that without mules to carry up wood and shelter no continued observations can be possible. In descending I noticed here and there parts of great tree-trunks, some 8 or 10 feet long, evidently very old, lying on the naked bowlders, without the slightest trace of vegetation within a mile or any sign to show how they came there. I afterward found these isolated trunks elsewhere, and it seems clear that they are relics of a remote day, when the forest grew 2,000 feet higher than it does at present, the pitch, saturating the wood, and the excessive dryness of the region, having preserved them here for an almost indefinite period. They are a most striking and curious evidence of a condition of things which once existed, and which exists no longer, the change being evidently due to a corresponding climatic alteration. What has caused this change it does not, perhaps, lie within my province to inquire, but I cannot doubt that the changes in those conditions of the atmosphere's transmissibility for heat, which we have climbed into this altitude to study, are connected with the answer to the riddle. I staid but a few minutes at the summit, took a final look at the snow fields about us, and down into the torrid regions of the desert, far below; and then descended to the camp, which I reached at about four in the afternoon.

25th.—Captain Michaelis, Messrs. Keeler and Day, accompanied by a train of twenty mules, arrived bringing the long looked-for instruments. By evening the siderostat was mounted. Through the kind assistance of Captain Michaelis the hospital tent was set up and two piers completed by the end of the next day.

Mr. Davidson left us on the 27th. Sergeant Naury was instructed in the use of the Regnault hygrometer, which was placed in his charge, together with the pyrheliometer. A considerable portion of our apparatus had been constructed with a special view of observing the solar corona here, if possible, without an eclipse. We were engaged for the following three days with this, and just as a possibility of success seemed near, a most disheartening accident robbed us of further hope in this direction. As the attempt was unsuccessful, I will not enlarge upon it, nor describe the intended means.

We turned to the spectro-bolometer, which, however, in spite of all our exertions, was not got to work until the 31st.

On the 2d of September Captain Michaelis went up the mountain, with Sergeant Naury, and Coles and Johnson as guides, carrying a tent and intending to stay three days for observations. Early the next morning, all the party made their appearance in camp, reporting that they had passed a sleepless night, without shelter or warmth, the wind being so high that they could not pitch the tent, while the quarter-cord of wood, carried up with great difficulty, had been all burned in a vain effort to keep warm.

September 4.—Sky, for the first time smoky, apparently from forest fires. Messrs. Keeler and Johnson ascended the mountain with the barometer.

5th.—I ascended the peak a second time. Keeler and Johnson had succeeded in passing a night there, though not a pleasant one, and had secured valuable observations, though it was evident that the tri-hourly series, being conducted at Lone Pine, would not have a counterpart on the summit. The afternoon was clear, though forest fires were numerous about the horizon. Looking north, the great masses of Mount Tyndall and Mount Williamson were prominent objects. Sheep Mountain (which has sometimes been called Mount Whitney) is to the south of us, and an almost numberless multitude of majestic, but still nameless, summits fills the western horizon. With a little Casella theodolite the following bearings were taken by Mr. Keeler close to the rude pile of stone (the "monument") which is on the extreme summit at the easternmost verge of Whitney Peak.

The bearing of Williamson by compass, 25° 45′ W., was made 0° 0′ by theodolite, and the following others taken by Mr. Keeler and the writer:

Williamson	0 00
Tyndall, eastern peak	321 17
Tyndall, western peak	304 16
Kawkah, highest or southern peak	256 11
Oleanche Peak	167 49
Point of Sheep Rock	155 04
Telescope Peak	125 40
"The Monument" (Whitney Peak)	34 05

6th.—A few cirrus clouds in the air, the first seen.

7th.—Our bolometer measurements had been made with the grating, and the glass and quartz prisms, and now having made the discovery of the great band "Ω" in the extreme red spectrum, we commenced to explore that interesting region with the rock salt prism also. It was thought safe in this dry air to leave the rock salt prism on the apparatus over night rather than disturb the adjustment by putting it away. The night proved to be phenomenally damp for that locality, and the prism was so injured that we could not do much with it the next day, a most unfortunate accident, as it prevented exploration of regions in the possibly existent solar heat spectrum, beyond those just discovered with the glass prism. The 5 inch equatorial, loaned us by Professor Pickering of Harvard College Observatory, had been installed largely through the aid of Mr. Crapo, the guide (who by profession is a surveyor), but when we came to take it out from its boxes upon the mountain, for the first time, it was not possible to use it to advantage, owing to the fact that none of the eye-pieces fitted the provisional tube (which had been made to carry the spectroscope). By holding the eye-pieces in the hand, we could at least determine what the quality of the atmosphere as shown by the easier usual star tests was, and the result was very satisfactory.

By the 8th, forest fires had multiplied in frequency, and the air was evidently not so pure as a week ago. We had, by hard struggling, and in spite of adverse circumstances, secured, however, what seemed most essential to our purpose, and though we had not done all we had hoped to do, we had done more than at one time seemed possible. In view of the fact that the sky, for our purpose, had commenced to deteriorate, I decided to descend. We worked till the afternoon of September 9th and then by hard labor at night, and all the next day into the evening, we got our instruments packed. Upon Sunday, Captain Michaelis, Mr. Keeler, Mr. Day, and myself, with Johnson as guide, started early in the morning on foot, to reach Lone Pine by the direct descent down Lone Pine Cañon—an almost unknown route. This day will always live in my memory, though I cannot describe the grandeur of the scenery nor its extraordinary character, here. Much of the route, we found, could only be followed by frequent actual climbing downward. We first ascended for over two hours, past snow-cliffs and along the frozen lakes in the northern shadow of Whitney Peak, and then passing through a defile in the rocks, so narrow that only one person could traverse it at a time, we suddenly found ourselves on the other side of the ridge, which had hidden the eastern view from us for weeks—so suddenly that we were startled as we looked down as through a window from our wintry height, to the desert, and the bright green of its oases far below, in a climate where it was still summer. We climbed down, until after many thousand feet, we reached the first of the little deeply blue lakes we had seen from the peak, and then, following the ice stream which flowed from this, we passed through a deep gorge, to other lakes and snow-fields below, and so on down all day, until we left snow behind us, and, till looking up the long distance through which we had come, we could see only the top of Whitney at the end of the vista. In the latter part of the day we traveled for over two hours through burnt or burning forests, always keeping on or near the bed of the stream, and amidst scenery which I remember nothing to equal.

Captain Michaelis and Johnson had pressed on to Lone Pine, while I, with Messrs. Keeler and Day, was walking more leisurely. As it grew dark we reached the desert. Shortly after night, Mr. Day, who had sprained an ankle in the descent, found himself unable to proceed further. The night air was that of the desert, cool but not chilly. We were still some hours from our destination. Giving our coats to Mr. Day, we left him for a night in the open air, which at this season involved no special hardship, and pushed on to Lone Pine, promising to send out for him in the morning. We reached there just at midnight, after seventeen hours of steady and violent exertion, and Mr. Day, to our agreeable surprise, got in on his own feet before sunrise.

There is little to add. We packed the remaining instruments at Lone Pine and made our way back across the desert: the original party leaving San Francisco on the 22d of September and arriving in Pittsburg on the 28th.

I hope I have made plain my own belief that Mount Whitney is an excellent station for the purpose for which it was chosen. The great drawback in our case was the inability to remain at the very summit, for to do this requires a permanent shelter, but a railroad will shortly run through Inyo Valley, and from this, by the aid of an easily constructed mule-path, the ascent of the very highest peak can be made in a day, while the telegraph will put it in direct communication with Washington. I do not think the Italian Government, in its observatory on Etna, the French, in that of the Puy de Dome, or any other nation at any other occupied station, has a finer site for such a purpose than the United States possess in Whitney and its neighboring peaks, and it is most earnestly to be hoped that something more than a mere ordinary meteorological station will be finally erected here, and that the almost unequaled advantages of this site will be developed by the Government.

CHAPTER III

ACTINOMETRY.

HISTORICAL INTRODUCTION.

I have already remarked that, while the determination of the amount of heat the sun sends the earth is equally important to Astronomical Physics and to Meteorology, the problem is one whose exact solution is not, at present, in our power. Fifty years ago, Herschel and Pouillet believed that they had fixed this value with precision. Later observers have successively employed improved methods of observation and inference, with a tendency in their results to higher and higher values; yet we are apt to look on the latest found as though they were final ones, not only noting, perhaps, the warning given by these constant and progressive increments, that no determination that has ever been made is probably to be considered as more than an approximation to the truth, which may, if we judge from the amount of these discrepancies, be very different from any.

The solution of the problem involves two chief difficulties, the first formidable, the second perhaps insuperable.

We have first to determine the amount of solar heat which the earth actually receives at the sea-level by observing the amount which falls in a given time on a given surface. This, at least, might seem to be easily ascertainable by direct experiment; but the difficulties, even here, are so great that the most competent observers differ by nearly a third of the whole amount in question, even as to what is directly measured. The variations in the heat-transmitting power of our atmosphere, even on clear days, are so surprising and anomalous that we can hardly adopt such assumptions in reducing our observations as make the method of least squares useful in other branches of physics; for when it may happen (as in the case of Forbes, cited later) that a single day's observation so outweighs years of previous work that they are to be set aside as of comparatively no value, our ordinary methods evidently fail us. No one, who has not personally carried on a long series of these observations, can have any idea of the difficulty of the conditions or their variety. We are as though at the bottom of a turbid and agitated sea, and trying thence to obtain an idea of what goes on in an upper region of light and calm. Were we indeed at the bottom of such a sea, it is obvious that if it grew momentarily clear above us, we should get in that moment a higher idea of the light outside than by any amount of previous direct observation, and further, that thus knowing that the best moments for observation were coincident with the highest observed values, we should justly deem these highest values our most trustworthy ones.

Here, then, is another respect in which we must depart widely from ordinary usage, which in almost every other branch of physics and astronomy obliges us to consider the mean of a large number of observations as the most probable value. In solar actinometry, the mean of all our observations is *never* really the most probable, and the true value is always, and necessarily, higher than this mean. This statement may appear strange, even paradoxical, to the reader unfamiliar with this particular class of observations. It is one of great importance, and whose meaning should be fully understood. We infer from it that if it were possible to make an actinometer free from purely instrumental error, that the highest observation of solar heat by it would always be the most trustworthy, and would in fact outweigh (in our imaginary case) an unlimited number of

15

lower ones. Connected with this same apparent paradox is the fact that when we begin to improve our actual instrument, and to allow for minute errors in its registration of some definite amount of radiant heat, we find that these errors tend to lie all in one direction. In other words, the corrections which we introduce for them will not have, on the whole, the negative sign as often as the positive, but however far we may push our investigations, the corrections tend to assume the positive sign.

In most physical observations, while we know that we cannot reach absolute exactness, and that the complexity of nature would oblige us to introduce minute corrections, and minuter, without end, ere we could reach to absolute truth, we yet know that we can, after having pushed precision to its practically attainable limit, rest assured that these neglected minutest corrections will on the whole balance each other. In *solar* actinometry this is not the case. When we have pushed precision to its practically attainable limit, where corrections become so minute that they are no longer individually manageable, we have reason to believe that the sum of those whose individual consideration we must forego is not negligible, for these corrections represent the loss and gain of heat in these measurements; of heat which is lost in unnumbered undetected ways, and gained in almost none.

But secondly the observed amount, even if its true value were found within near limits (as it yet conceivably may be), only represents that residual heat which has come down to the observer after a very large absorption by our atmosphere. He cannot *directly* observe the heat before this atmospheric absorption, and the absolute necessity of adopting some hypothesis as to its action introduces the second difficulty I have mentioned as perhaps insuperable. To our predecessors, Herschel and Pouillet, this difficulty scarcely presented itself as being one at all. They had inherited a formula representing a primitive hypothesis, a kind of scientific dogma, which was accepted on trust and used without question; and their successors down to the present day have, with less justification, employed nearly the same rule, which is, it must be admitted, so easily followed that it would be most convenient to us if nature would but follow it also; but, as the writer has already endeavored to show, the actual processes by which the solar heat is absorbed are almost infinitely more complex than this hypothesis makes them. What is novel in the present investigations is the attempt to accept, as far as our still imperfect knowledge admits, the difficult conditions nature actually imposes, and to discard what is called the exponential formula of Pouillet, even in the modified shape in which it has been employed by recent investigators of repute. We shall thus reach results which cannot possibly have the exactness which previous observers have attributed to their own, but which will lie between limits of error which seem determinable. The width of these limits is but a statement in other terms of the great extent of our ignorance on a matter where we have supposed ourselves, till of late, to know nearly all that there was to be learned. The most probable result between these limits, then, will be found to be materially greater than that of previous observers; but, though our estimate of the actual amount of heat which the sun sends the earth is thus increased, our conclusions will be that the effect of its direct radiation is far smaller than has been supposed. In other words, though our estimate of the heat received by direct solar radiation is increased, we also find that the actual *temperature* of the earth's surface on which organic life depends, *is maintained in very slight degree by the direct solar rays*, and in very large degree by some processes in our atmosphere intimately connected with that complex absorption which the old formula ignores.

I do not propose to give a full history of solar actinometry, or to give any complete list even of the notable contributors to it, but some brief mention of the following names (which I place in chronological order) is necessary to my purpose: *

1760.—Date of the completed posthumous edition of Bouguer's works, and the first enunciation

---

The reader desiring to learn more of the history of the subject is referred to the excellent little treatise by M. Radau called "Actinométrie"; he may also consult "La chaleur solaire," by M. Mouchot, and the numerous memoirs referred to in Houzeau's "Bibliographie de l'astronomie," as well as the theoretical investigations of Clausius (Pogg. Annal., vol. CXXI., p. 330 *et seq.*), of Lord Rayleigh (London, Edinburgh & Dublin Phil. Mag., Feb., 1871, *et seq.*). The well-known researches of Tyndall on the blue color of the sky have an importance in this connection, as from the fact of selective reflection or diffusion we may infer that corresponding parts have disappeared from the direct beam by a *selective* process of which Pouillet's formula takes no account.

of the formula adopted by Herschel, Pouillet, and their successors. This is stated by Bouguer in these words:

"When the thicknesses" (of the absorbing medium) "increase by equal quantities, light diminishes according to the terms of a geometric progression."

Bouguer points out that the diminution can be graphically represented by the logarithmic curve, where the abscissæ are proportional to the thickness of the absorbing medium and the ordinates to the amount of light or heat received, so that on the other hand, knowing the amount of heat received and the amount of the absorbing medium, we can determine what the heat was before absorption.*

This conclusion was remarkably in advance of the physical assumptions made before Bouguer. It embodied all the facts known at the time he wrote, and it is to his credit that he perceived as much as he did. A study of his original investigations enhances our opinion of him as a skillful and conscientious observer.

1760.—In the same year with the posthumous publication of Bouguer appeared the "Photometria" of Lambert, a book which I have not been able to consult directly, but which is understood to be a work of merit, based in many respects upon Bouguer's previous investigations. Lambert's work is remarkable for the clear description of several methods or results which have, in later times, been rediscovered or reapplied by others. He was aware that the true measure of radiant heat was the initial velocity of heating resulting from it. He applied this method to the determination of the permeability of successive glass plates to the solar rays, with a result which remarkably anticipates the law (usually attributed to De la Roche or to Melloni) that the "facility of transmission" through successive plates is variable, and "continually increases with the number already passed through."† (Bouguer's formula would make the "facility of transmission," *i. e.* the common ratio of the geometric progression, a *constant.*)

1819.—Leslie also points out "that the *initial* change of the thermometer is, in every case, the only certain and accurate measure of heat."

1825.—Sir John Herschel devises his actinometer and introduces the method of exposing alternately in sun and shade.

1838.—Date of the appearance in the "Comptes Rendus" of Pouillet's celebrated "Memoire sur la chaleur solaire, sur les pouvoirs rayonnants et absorbents de l'air et sur la temperature de la space." Pouillet reaches the following conclusions, which have obtained almost universal currency, and are even yet found in our text-books. Adopting as his unit the quantity of heat which the sun sends normally upon the surface of one square cm. exposed at the surface of the earth's atmosphere during one minute ("the small calorie"), he finds (1) that the observed heat at the sea-level from a vertical sun is about 1.1 calories, (2) that the amount of heat transmitted vertically by our atmosphere at the sea level is not quite ? of the whole, and hence (3) that ? of 1.1 = 1 ? calories, or, to give his exact value, that 1.7633 calories is the solar constant, or the amount of heat at the upper limit of the atmosphere. This value corresponds to an amount of heat which would melt a stratum of ice 31 meters thick over the whole earth annually. He then goes on to inquire whether the earth receives heat from any considerable source besides the sun, and concludes that *because the amount of solar heat just given does not account for the earth's actual surface temperature* it must receive from some other source almost as much heat as from the sun itself, reaching the remarkable result that this other source is radiation from the stars, whose united action may be represented by the radiation of an enveloping shell whose *mean* temperature is — 142° C. This — 142° C. then is, according to Pouillet, the "temperature of space," which seems to him nearly as important in warming the earth as the radiations of the sun.

As very few who quote Pouillet's value of the "temperature of space" have any knowledge of the way he derived it (his celebrated memoir being more often referred to than read), I may explain here that he determines by a most original, ingenious, and plausible, though not absolutely satisfactory train of reasoning, that the amount of heat required to maintain the surface of the earth at its known mean temperature, is that which would melt a stratum of ice equal to 57

---

* Bouguer, "Traité d'optique sur la gradation de la lumiere," Paris, 1760.
† Quoted by Forbes.

meters thick annually, *and because he has already found that 31 meters of this only is represented by the sun,* he is compelled to look for some other cause for heat to melt the remaining 26 meters, and he finds it by assigning, as we have just seen, to "space" the temperature of —142° C. Pouillet then did not determine the temperature of space by any direct experiment, as he is often supposed to have done. His so-called experiments on the temperature of space were inquiries to see what temperature should be assigned to it to meet the supposed necessity of melting 26 meters of ice annually by the heat of the stars in addition to that of the sun, since 31+26, in all 57 meters, was the amount, according to him, to be accounted for. If his methods of measurement of the direct solar heat had been correct, it will be seen from the Mount Whitney experiments that he would have found a quantity nearly representing his whole 57 meters from the sun alone, and in this case the temperature of space assigned by his theory would apparently have been the absolute zero. It is a legitimate inference from Pouillet's own theory, then, that in proportion as our measurements of heat from the sun give larger values does the necessity for assigning any sensible value to the "temperature of space" disappear.

Because we have pointed out changes which the progress of science has introduced in Pouillet's conclusions, we must not be understood to speak otherwise than with admiration of the work of this celebrated physicist. The memoir cited contains, in a highly condensed form, the result of great and conscientious labor. Owing to this condensation, the author's meaning is not, in all cases, as clear as might be desirable, but this arises from the great extent and painstaking character of his researches, and his limited space for the presentation of them.

1835–1838.—At about this time Melloni observed that in the case of glass plates or like media like proportions were *not* transmitted by like strata, and the cause of this was pointed out by Biot. An analogous observation had been (as we have mentioned) made in the last century by Lambert, but was forgotten; and even the observation of Melloni, important as it is in its application to our subject, has been slighted or altogether overlooked by nearly all subsequent investigators. A notable exception, however, is the work of Principal Forbes, who, even before Pouillet (in 1832), with the aid of Professor Käntz, and using Herschel's actinometer, made a series of observations at Brienz and on the Faulhorn. It is significant of the peculiar difficulties of such work as we now consider that he was led to throw away very numerous antecedent observations in favor of those of a single day, the 25th of September, 1832, and that he appears to have been occupied during a considerable part of ten years in reducing these observations of one day, which with a few confirmatory ones made on the 13th and 14th of August, 1841, were published in 1842 in the Philosophical Transactions.

This memoir is a model in many respects to be followed even at the present time. One of its most novel features is the application of Melloni's observation and Biot's conclusion to the determination of the solar constant, for Forbes sees clearly that if like masses do not absorb like proportions the old rule is useless. He therefore discards the exponential formula, and projecting his observations graphically, shows that, as a matter of fact, the absorptions cannot be represented by a simple geometric progression, and draws an empirical interpolating curve, by the aid of which he determines that the solar constant is 388.4 actines, or, in our notation, 2.85 calories, the heat at the sea-level being given by him as 1.52 calories, so that it will be seen that Forbes's observation of the heat received at the sea-level is only slightly in excess of Pouillet's. The great difference in the value of the solar constant comes in part from his conclusion that all equal air-masses have like absorptions, and in part from his discarding Pouillet's formula. He also finds by photometric researches that the amount of light reflected from the atmosphere is equal to that directly received from the sun. Forbes points out that the mass of air traversed ceases to be sensibly proportional to the secant of the sun's zenith distance as we approach the horizon, and he gives the correct value derived from La Place's, which we have ourselves employed. He found that the absorption for a like mass of air was the same whether that mass was of the quality of that upon the mountain or in the valley, an observation which our great confidence in Forbes as an observer leads us to admit may have been noted by him in some exceptional case, but which does not agree with our observations or with those of others.

1847.—Sir John Herschel publishes, in the Cape of Good Hope Observations, his experiments, made in 1836, to determine the solar constant by the heating of water, whence he obtained 191.4 actines (1.39 calories) for radiation at sea level, and for the heat outside the atmosphere, 2.09 calories.

1872.—M. Soret, of Geneva,[*] remarks that methods for measuring the solar intensity may be classed as static or dynamic. "In the static, two bodies are placed in identical conditions, except that one receives the solar rays and the other does not. The final difference of temperature taken up by these two bodies will give a measure of the intensity of the radiation." He gives his reasons for preferring the static method.

M. Soret concludes that in winter, when the air is dry, the radiation is more intense for the same air mass than in summer, and that in general, other things being equal, we obtain a higher value when the air is drier than when it is humid, even when it may appear more transparent to the eye. He further observes that the dust, germs, or water vapor particles, in the air, while partly intercepting all radiations, must particularly affect the most refrangible ones. He confirms this result by interposing a definite thickness of water between the sun and his thermometer bulb. He very justly observes that under such conditions Pouillet's formula ($t = Ap^x$) cannot be absolutely exact; nevertheless, he deems it practically sufficient for observations made in the course of a day at one station under like atmospheric conditions. By observations on Mont Blanc, he determines that the upper strata of the atmosphere are less absorbent than the lower for like masses, and concludes that the formula, which may be tolerated in the case of observations at the same station, proves quite insufficient where those at two different stations are to be compared. He proposes for observations at two different stations to write Pouillet's formula

$$t = Ap^{e-m}$$

thus introducing the barometric pressure under the second power. The observations of M. Soret are most instructive to the student, though it is not easy to express their result with certainty in our notation, since he does not give us the amount of the solar radiation, but only one of its effects, which is to raise the thermometer employed by him at the maximum about 16° above its surroundings at the sea-level, or about 19½° on the summit of Mont Blanc. That we do not have the heat outside the atmosphere given us in set terms by M. Soret appears to be due to his perception of the fact that there possibly may be certain kinds of rays quite absorbed, even before they reach the summit of the highest mountain, and hence his distrust of the best formula he can frame, which must rest on the results of observation of only such rays as have actually reached the observer. This wise reserve enhances our opinion of the value of the memoir.

1874–1879.—M. Violle, in various communications to the "Comptes Rendus," and especially in two in the "Annales de Chimie et de Physique" for 1877 and 1879, has given a new value of the solar constant. He believes that the relation between the heat before and after absorption is exactly expressible by the formula

$$I = 2.54 \times 0.946^{\frac{H(Z-z)f_z}{760}}$$

which is a modification of Pouillet's, I here being the amount of heat which reaches the soil, 2.54, the solar constant, H, the barometer, $(Z - z) f_z$ a quantity proportional to the vapor mass traversed.

Although M. Violle alludes to the fact that the radiations of the sun are not homogeneous, he gives, as it will be seen, little weight to this consideration in the above formula, of whose sufficiency he feels certain, remarking of it, indeed, that the true law of absorption is always one and the same, and always represented by it without any doubt. In his observations he employs a modification of the globe actinometer, his use of it being to see how much the thermometer will rise above the temperature of the surrounding inclosure. In such a use this instrument belongs then to the static method, but from this observed static excess M. Violle proposes to calculate by a very ingenious method the initial rate of heating of his thermometer bulb. This initial rate, then, on which everything depends, is not obtained by direct observation, but by a process of deduction from an observation of the static kind.

Knowing the initial rate of heating, we next require to know the exact mass of the thing

* Association Française pour l'avancement de science. Congrès de Bordeaux.

heated in order to determine the sun's effect. The thing heated here is not a large mass whose weight is readily determinable, but the minute amount of mercury contained in a very small thermometer bulb. It is easy to make a relatively considerable error in this minute determination, and as here again the whole value of the solar constant depends upon its accuracy, this is a difficulty special to this instrument, and fuller evidence would be desirable of the means by which the very large value of this important constant was obtained, and of the validity of the method for finding the initial rate, which the data given do not enable us to verify. M. Violle's methods are very noteworthy, the labor he has spent in observation (in the course of which he ascended Mont Blanc), has given general currency to his results.

1876.—We cannot omit mention even in this brief sketch of the extremely original methods and very assiduous observation of Mr. J. Ericsson, of New York. His observations, made by methods entirely his own, give results for the heat received at the sea-level not very different from those of Pouillet. Mr. Ericsson labors, as far as possible, to take nothing on trust, but to give us in every respect the results of observation only. Since it is impossible, however, for him to actually observe above the atmosphere, he is obliged to employ some hypothesis to find the heat before absorption. In this case the assumption is that a certain law of progression in the diminution of the absorption, observed at the earth's surface as we approach the zenith (where the atmospheric depth is unity, would also hold if we could rise above the earth's surface. Mr. Ericsson's most interesting investigations may be found in the volume published by him for private circulation in 1878 ("Contributions to the Centennial Exhibition"), and the reader who has not access to this will also find the more important ones in the early volumes of Nature, where he is advised to consult them.

M. Crova, of Montpellier, has dealt with our subject in many most valuable memoirs. In one published in 1876* we have an example of excellent observations, of great care in the conclusion, and of a reserve of judgment which recognizes the really great difficulties of the problem and the really wide limits of probable error. M. Crova recognizes the importance of the effect of the complex character of the solar radiation and the insufficiency of Pouillet's formula, but as he is unable to discriminate between these radiations, he resorts to a method of allowing for their complexity, which is a further improvement on that of Forbes. He also shows that our value of the solar constant is, as a matter of observation, greater, as the air masses by which we determine it are less. He points out some very useful precautions to take in preparing a thermometer bulb, and he devises a very convenient and rapid acting actinometer for his observations. (We are, ourselves, under the impression that his own rating of the constant of this actinometer would be raised with advantage if account were taken of some of the corrections which we have pointed out in a separate chapter.) M. Crova's observations appear to us to be among the most trustworthy made in recent times. He selects two days as especially good, January 8, 1875, when he obtains as the solar constant 1.898, and January 4, 1876, when he obtains the solar constant 2.323, or nearly 25 per cent. more than in the other case; and I may remark that I have myself met similar discrepancies under similarly favorable conditions. Remembering M. Crova's high qualities as an observer, we see that these discrepancies must be chiefly due to the different absorptive powers of the atmosphere on the two clearest and most similar days which he could find, and they present the reader an idea of the real difficulties of our task. In what other kind of observation than this could so great differences be expected as the final results of the greatest skill and the most favorable conditions! They are peculiar to our subject, and they should teach us all caution as to forming too absolute conclusions.

---

*Mesure de l'intensité calorifique des radiations solaires.

CHAPTER IV

PYRHELIOMETRIC OBSERVATIONS.

Perhaps the best known instrument for measuring solar heat is one which ought to be considered, among those still in use, the worst in principle, and in practice one of the most untrustworthy. I refer to the pyrheliometer of Pouillet, which was the best attainable fifty years ago, and which is to be found described in every text-book on the subject. While considering the use of this instrument as in our present knowledge full of objections, I have, however, employed it as an adjunct to others, in order to connect these late observations with the earlier ones made by its inventor, and because the pyrheliometer is so generally known. I mention it first in the order of observation, but the reader who desires to see its historical place is referred to the chapter on actinometry.

The instrument, in the form in which it was used on the Mount Whitney expedition, is essentially that described by Pouillet. It consists of a shallow, cylindrical box of thin copper, electroplated with silver, and blackened on its front surface, having a diameter of 113 millimeters, and exposing to the sun a surface of 100 square centimeters. It is 11.2 millimeters thick and holds when filled, after the thermometer is in and plugged, 104.2 grammes of water. (On Mount Whitney, after August 31, 1881, it was temporarily fitted with a cork and held nearly 108 grammes of water.) It has a stem, insulated from heat conduction by wood. Deducting the weight of the wood, which is here treated as absorbing no heat, we have—

	Grammes.
Weight of box, 172.2 grammes; water equivalent	16.4
Weight of brass piece holding thermometer, 8.4 grammes; water equivalent	0.8
Thermometer bulb and mercury, water equivalent	0.5
Total water equivalent of vessel and immersed portion of thermometer	17.7

The vessel can be rotated around the axis of the cylinder, by which motion the water is mixed and kept at a nearly uniform temperature.

"The observation is made in the following manner: *The water in the vessel being nearly of the surrounding temperature,*[*] the pyrheliometer is held in the shade, but very near the place where it is to receive the sun; it is placed so that it looks towards the same extent of sky, and there, for four minutes, its warming or its cooling is noted from minute to minute; during the following minute it is placed behind a screen, and then adjusted so that on removing the screen at the end of the minute, which will be the fifth, the solar rays strike it perpendicularly. Then, during five minutes, under the action of the sun, its warming, which becomes very rapid, is observed from minute to minute, and care is taken to keep the water incessantly agitated; at the end of the fifth minute the screen is replaced, the apparatus withdrawn into its first position, and for five minutes more its cooling is observed."

According to Pouillet ("Comptes Rendus," July 9, 1838) let R be the warming which the pyrheliometer undergoes during the five minutes of the solar action, r and r' the coolings during the

---

[*] A consideration of importance.

five-minute intervals preceding and following. The elevation of temperature t, which would take place if there were no radiation, is approximately (when the loss of heat by convection is slight)

$$t = R + \frac{r + r'}{2}. *$$

Let d be the diameter of the vessel expressed in centimeters, p the weight of the water which it contains expressed in grammes, p' the weight of the vessel and the immersed portion of the thermometer reduced to a specific heat of unity. The elevation of temperature t corresponds, then, to an amount of heat $t (p + p')$.

This heat having fallen in five minutes on a surface $\frac{\pi d^2}{4}$, each unit of surface has received $\frac{4 (p + p') t}{\pi d^2}$ during the five minutes, and in one minute $\frac{4 (p + p') t}{5 \pi d^2}$.

In the pyrheliometer used $\frac{\pi d^2}{4} = 100$ square centimeters, hence

Calories per minute per square centimeter $= \frac{t}{500} (p + p') t.$

If Pouillet's expression (which we have just stated) were correct, it would be quite immaterial whether the day was calm or whether a strong wind blew on the instrument, provided that it blew uniformly. Pouillet himself, however, has devised a special form of the instrument, covered with a lens to meet the case of the wind, but has to correct this special instrument by one of the primitive form we are discussing. Now, the above formula, simple as it seems, is theoretically wrong (as I will not stop to demonstrate), and the instrument is untrustworthy even in a uniform wind, as may be inferred from the following experiments. It is because we can so rarely command the condition of a perfect calm that the error becomes of so large importance, though even in a perfect calm the above equations seem to me to be still untrustworthy.

A second practical defect has been pointed out by others, and is, indeed, very apparent, even to casual observation. When we take the instrument out of the sun and place it in the shade, instead of falling, the temperature continues to rise for a perceptible fraction of a minute; and when we take it out of the shade and put it in the sun the temperature continues to fall for a fraction of a minute. If the water be most thoroughly agitated, as in our own experiments, it is certain that this effect will be much reduced; but it will be still apparent if the readings are taken at short intervals, as the reader may see on examining the illustrative observation in the foot-note.[†] A portion of the water, in fact, clings to the copper, and is not removed by agitation, however violent. It forms a non-conducting film against the copper surface, so that the actual conductivity is totally different from what we might expect, and nothing less than a system of internal brushes, which should scrape away this persistent film, would suffice to remove it. A very great improvement, therefore, has been made by Professor Tyndall in the substitution of mercury, a liquid which

---

* Pouillet says, "it is easy to see that the elevation of temperature t produced by the sun is $t = R + \frac{r + r'}{2}$." It seems to me that it is *not* easy to see, and that this formula does not, in fact, represent any accepted law of cooling.

†(March 2ᵈ, 1881; Station, Allegheny.)—The water in the pyrheliometer was incessantly agitated during the following experiment. In spite of this the result was somewhat as though the water were a poorly conducting solid, the temperature remaining constant, or even rising, during a considerable part of a minute, after the sun's rays were cut off, and usually falling for a considerable time after its exposure to the sun, as will be seen clearly from these figures taken from direct observation.

TABLE 17.

Transferred from shade to sun at 1ʰ 55ᵐ 00ˢ.		Transferred from sun to shade at 2ʰ 00ᵐ 00ˢ.		Transferred from shade to sun at 2ʰ 00ᵐ 00ˢ.		Transferred from sun to shade at 2ʰ 10ᵐ 00ˢ.	
Time.	Temp.	Time.	Temp.	Time.	Temp.	Time.	Temp.
h. m. s.	°C.	h. m. s.	°C.	h. m. s.	°C.	h. m. s.	°C.
1 55 00	18 94	2 00 00	20 08	2 05 00	17.70	2 10 00	19.00
1 55 15	18 98	2 00 15	20 08	2 05 15	17.40	2 10 15	19 05
1 55 30	18 82	2 00 30	20 06	2 05 30	17.38	2 10 30	19.01
1 55 45	18 95	....	......	2 05 45	17.22	....	......

These readings are typical of a great number observed and not here given.

does not wet the surface, and I desired to take a mercury pyrheliometer of his pattern with me, but could not obtain one till the return of the expedition. I have used, then, Pouillet's apparatus in its original form, for the reasons stated, although it has other errors besides those mentioned, and have compared its indications subsequently with those of a mercury pyrheliometer, and introduced a correction for the difference; but our "corrected calories" are corrected to this extent only. While in deference to the long-established repute of Pouillet's instrument, we give a full series of observations made by it, we do not attach great weight to these values as absolute determinations, although they may often be found convenient as relative ones. We have accordingly omitted observations representing much labor, where the series were not so complete as was desirable, for, even in this clear sky, wind or other causes frequently spoils a series, and perhaps a whole day's work, as if both morning and evening series be complete, they are useless without the noon one. We give an example of but a single series in full (that for noon of August 14, made at Lone Pine), though the values here given for other series are reduced from similar complete minute readings.

In our own reductions of the pyrheliometer we have (for the altitudes actually observed) treated the length of the path of the ray (M) as proportional to the secant of the zenith distance, and taking the barometer expressed in decimeters as (β), $M\beta$ represents the absorbing air-mass. According to Pouillet's symbols, A represents the heat outside the atmosphere (i.e., before absorption), p the coefficient of transmission (the proportion t transmitted by his unit stratum, the zenith depth), e the number of unit strata traversed; and it is (erroneously) assumed by Pouillet that the same *proportion* is transmitted by one stratum as by another, or that p is a constant, so that if t be the observed temperature $t = Ap^e$.

Our unit stratum is that which would support one decimeter of mercury. We designate the heat outside the atmosphere by E. The coefficient of transmission for a unit stratum, which Pouillet writes p, we call a, where unity is the absorbing mass overhead at sea-level or that supporting 7.6 decimeters mercury. Where it is expressed in terms of strata 1 decimeter mercury, this becomes a^s. We use Pouillet's formula of reduction, then, under a slightly different form, and writing c for the observed heat in calories, we have $C = E a^{\frac{M\beta}{.76}}$, a formula which we have given our reasons elsewhere for believing incomplete.

COMPARISON OF WATER WITH MERCURY PYRHELIOMETER BY SIMULTANEOUS OBSERVATIONS.

The instruments used in this comparison were, the water pyrheliometer (No. 1), used on the expedition and already described, and a mercury pyrheliometer (No. 2) made of cast-iron, nickel plated, and coated with lamp-black on its front surface. It exposed a surface of 20.268 sq. cm. to the sun. It was mounted as an altazimuth, the thermometer-stem projecting from the horizontal axis. Both instruments were protected from wind and variation of radiation from surrounding objects by cylindrical screens.

The water equivalent of pyrheliometer No. 1 on this occasion was $p + p' = 97 + 17.7 = 114.7$ grammes, and the number of calories per minute per square centimeter was $\frac{1}{500}(p+p')\,t = .2294\,t$, which is the formula used in reduction of series A and B.

For pyrheliometer No. 2:	Grammes
The water equivalent of the iron flask	9.84
The water equivalent of the inclosed part of thermometer	0.40
	$p = 10.24$

In series A':
$$p + p' = 9.84 + 10.24 = 20.08 \text{ grammes.}$$

In series B':
$$p + p' = 6.45 + 10.24 = 16.69 \text{ grammes.}$$

The formula becomes:

$$\text{Series A', No. of calories} = \frac{1}{101.34} \times (20.08)\,t = .1982\,t.$$

$$\text{Series B', No. of calories} = \frac{1}{101.34} \times (16.69)\,t = .1647\,t.$$

TABLE 18.

[Date, October 22, 1881. Station, Allegheny. Sky, very milky with haze and clouds near horizon. Wind, gentle breeze. Instrument, water pyrheliometer (No. 1). Protected from wind by a tin-plate cylinder open at both ends and covered with cotton. Its condition is therefore similar to that of the mercury pyrheliometer. Charge of water = 97 c. c. Observer, F. W. V. Observations synchronous with series A'.]

SERIES A.

No.	Interval.	Noon.	Fall.	Rise.
	Min.			
1	5	11ʰ 50ᵐ to 11ʰ 55ᵐ fall in shade	0°. 23	
2	5	11 55 to 12 00 rise in sun		2°. 24
3	5	12 00 to 12 05 fall in shade	0 .45	
4	5	12 05 to 12 10 rise in sun		2 .85
5	5	12 10 to 12 15 fall in shade	0 .55	
6	5	12 15 to 12 20 rise in sun		2 .85
7	5	12 20 to 12 25 fall in shade	0 .67	
8	5	12 25 to 12 30 rise in sun		2 .52
9	5	12 30 to 12 35 fall in shade	1 .05	

From No. 2: $t = 3.28 + \dfrac{.23 + .45}{2} = 3 .62.$ Calories per minute per square cm 0. 830

From No. 4: $t = 2.85 + \dfrac{.45 + .55}{2} = 3 .35.$ Calories per minute per square cm 0. 768

From No. 6: $t = 2.85 + \dfrac{.55 + .67}{2} = 3 .46.$ Calories per minute per square cm 0. 794

From No. 8: $t = 2.52 + \dfrac{.67 + 1.05}{2} = 3 .38.$ Calories per minute per square cm 0. 775

Mean 0. 792

TABLE 19.

[October 22, 1881 (continued). Sky, very milky with haze and clouds near horizon. Wind, gentle breeze. Instrument, case of mercury pyrheliometer (No. 2) filled with water. Protected from wind by pasteboard cylinder open at both ends and covered with cotton. Charge of water = 9.84 c. c. Observer, J. E. K. Observations synchronous with series A.]

SERIES A'.

No.	Interval.	Noon.	Fall.	Rise.
	Min.	After about 15 minutes' exposure to sun—		
1	5	11ʰ 50ᵐ to 11ʰ 55ᵐ fall in sun	2° 62	
2	5	11 55 to 12 00 rise in sun		1°. 37
3	5	12 00 to 12 05 fall in shade	2 .12	
4	5	12 05 to 12 10 rise in sun		2 .32
5	5	12 10 to 12 15 fall in shade	1 .56	
6	5	12 15 to 12 20 rise in sun		1 .79
7	5	12 20 to 12 25 fall in shade	2 .62	
8	5	12 25 to 12 30 rise in sun		1 .74
9	5	12 30 to 12 35 fall in shade	2 .97	

Uncertain.

From No. 2: $t = 1.37 + \dfrac{2.62 + 2.12}{2} = 3 .74.$ Calories per minute per square cm 0. 741

From No. 4: $t = 2.32 + \dfrac{2.12 + 1.56}{2} = 4 .16.$ Calories per minute per square cm 0. 824

From No. 6: $t = 1.79 + \dfrac{1.56 + 2.42}{2} = 3 .78.$ Calories per minute per square cm 0. 749

From No. 8: $t = 1.74 + \dfrac{2.42 + 2.97}{2} = 4 .41.$ Calories per minute per square cm 0. 880

Mean 0. 798

TABLE 20.

[October 22, 1881 (continued). Sky, very milky but somewhat clearer than at noon. Wind, gentle breeze. Instrument, water pyrheliometer (No. 1). Charge of water, 97 c. c. Observer, F. W. V. Synchronous with series B′.]

SERIES B

No.	Interval.	P. M.	Fall.	Rise.
	Min.			
1	5	1ʰ 00ᵐ to 1ʰ 05ᵐ fall in shade	0.51	
2	5	1 05 to 1 10 rise in sun		2.78
3	5	1 10 to 1 15 fall in shade	0.48	
4	5	1 15 to 1 20 rise in sun		2.80
5	5	1 20 to 1 25 fall in shade	0.58	
6	5	1 25 to 1 30 rise in sun		1.95
7	5	1 30 to 1 35 fall in shade	0.93	

From No. 2: $t = 2.7\mathrm{s} + \dfrac{.51 + .4\mathrm{s}}{2} = 3.23$. Calories per minute per square cm 0.755

From No. 4: $t = 2.80 + \dfrac{.4\mathrm{s} + .5\mathrm{s}}{2} = 3.33$. Calories per minute per square cm .... = 0.764

From No. 6: $t = 1.93 + \dfrac{.5\mathrm{s} + .93}{2} = 2.69$. Calories per minute per square cm 0.647

Mean.. 0.712

TABLE 21.

[October 22, 1881 (continued). Sky, very milky but somewhat clearer than at noon. Wind, gentle breeze. Instrument, mercury pyrheliometer (No. 2) filled with mercury according to its customary use. Charge of mercury, 1918 grammes. Observer, J. E. K. The water used in previous experiments had been emptied out and the case dried by heating it before it was filled with mercury. The case having been heated to expel water the temperature at the outset was considerably above the shade reading. Synchronous with Series B.]

SERIES B′

No.	Interval.	P. M.	Fall.	Rise.
	Min.			
1	5	1ʰ 00ᵐ to 1ʰ 05ᵐ fall in shade	4.47	
2	5	1 05 to 1 10 rise in sun		1.26
3	5	1 10 to 1 15 fall in shade	3.99	
4	5	1 15 to 1 20 rise in sun		2.56
5	5	1 20 to 1 25 fall in shade	2.75	
6	5	1 25 to 1 30 rise in sun		1.39
7	5	1 30 to 1 35 fall in shade	2.85	

From No. 2: $t = 1.26 + \dfrac{4.47 + 3.99}{2} = 5.49$. Calories per minute per square cm 0.904

From No. 4: $t = 2.56 + \dfrac{3.99 + 2.75}{2} = 6.33$. Calories per minute per square cm 1.046

From No. 6: $t = 1.39 + \dfrac{2.75 + 2.85}{2} = 4.76$. Calories per minute per square cm 0.774

Mean.. 0.908

TABLE 22.

From the above simultaneous observations we have the efficiency of pyrheliometer No. 2 in terms of No. 1.

No. 2 filled with water. Series A and A′.	No. 2 filled with mercury. Series B and B′
Ratio $\dfrac{\text{No. 2}}{\text{No. 1}} = \dfrac{.741}{.830} = .89$	Ratio $\dfrac{\text{No. 2}}{\text{No. 1}} = \dfrac{.904}{.755} = 1.20$
Ratio $\dfrac{\text{No. 2}}{\text{No. 1}} = \dfrac{.824}{.768} = 1.07$	Ratio $\dfrac{\text{No. 2}}{\text{No. 1}} = \dfrac{1.046}{.764} = 1.37$
Ratio $\dfrac{\text{No. 2}}{\text{No. 1}} = \dfrac{.749}{.794} = .94$	Ratio $\dfrac{\text{No. 2}}{\text{No. 1}} = \dfrac{.774}{.617} = 1.25$
Ratio $\dfrac{\text{No. 2}}{\text{No. 1}} = \dfrac{.880}{.775} = 1.14$	Mean efficiency of No. 2 filled with mercury. 1.27.
Mean efficiency of No. 2 filled with water. 1.01.	

Accordingly we adopt the multiplying factor 1.27 for the reduction of a result obtained with a water pyrheliometer to what it would have been if obtained with an instrument employing mercury as its liquid.

EXAMPLE OF A PYRHELIOMETER SERIES IN FULL.

TABLE 23.

[Date, August 14, 1881. Station Lone Pine. Observer, A. C. D. Sky, clear. Wind, fresh to gentle variable. Instrument, pyrheliometer No. 1. Charge of water, 85.26 grammes.]

Time.	Reading of thermometer.	Change per minute.	Exposure.	Time.	Reading of thermometer.	Change per minute.	Exposure.
11 30 A. M.	29.20	....	Shade	12 01 P. M.	39.82	+ .82	Sun.
31	30.26	+1.06	Sun.	02	40.58	+ .76	Do.
32	31.52	+1.26	Do.	03	41.60	+1.02	Do.
33	32.90	+1.38	Do.	04	42.23	+ .63	Do.
34	33.95	+1.05	Do.	05	43.10	+ .87	Do.
35	35.00	+1.05	Do.	06	42.50	— .60	Shade.
36	34.80	— .20	Shade	07	41.90	— .60	Do.
37	34.62	— .18	Do.	08	41.49	— .41	Do.
38	34.40	— .22	Do.	09	40.99	— .50	Do.
39	34.38	— .02	Do.	10	40.50	— .49	Do.
40	34.22	— .16	Do.	11	41.32	+ .82	Sun.
41	35.30	+1.08	Sun.	12	42.00	+ .68	Do.
42	36.21	+ .91	Do.	13	42.49	— .49	Do.
43	37.62	+1.41	Do.	14	43.00	+ .51	Do.
44	38.83	+1.81	Do.	15	43.74	+ .74	Do.
45	39.80	+1.97	Do.	16	43.30	— .44	Shade.
46	38.52	— .48	Shade.	17	42.80	— .50	Do.
47	38.18	— .34	Do.	18	42.43	— .37	Do.
48	37.90	— .28	Do.	19	42.10	— .33	Do.
49	37.58	— .32	Do.	20	41.70	— .40	Do.
50	37.20	— .38	Do.	21	42.12	+ .42	Sun.
51	37.99	+ .79	Sun	22	43.00	+ .88	Do.
52	36.85	+1.86	Do	23	43.82	+ .82	Do.
53	39.70	+ .85	Do.	24	44.40	+ .58	Do.
54	40.58	+ .88	Do.	25	45.00	+ .60	Do.
55	41.30	+ .72	Do.	26	44.24	— .76	Shade.
56	40.83	— .47	Shade.	27	43.59	— .65	Do.
57	40.32	— .51	Do.	28	43.20	— .39	Do.
58	39.98	— .34	Do.	29	42.64	— .56	Do.
59	39.59	— .39	Do.	30	42.00	— .64	Do.
12 00 M	39.20	— .39	Do.				

[For reduction of these observations see Table 33.]

OBSERVATIONS WITH THE PYRHELIOMETER MADE AT LONE PINE.

TABLE 24.

[Date, August 11, 1881. Station, Lone Pine. Observer, A. C. D. Sky, clear. Wind, gentle. Charge of water, 81.6 c. c. Barometer (B_i) = 6.64. Length of path of ray (M_i) = 2.281. M_i, B_i, = 15.15.]

No.	Interval.		A. M.	Fall.	Rise.
	Min.				
1	5	7 05 to 7 10	fall in shade	1.51	
2	5	7 10 to 7 15	rise in sun		3 .31
3	5	7 15 to 7 20	fall in shade	1.55	
4	5	7 20 to 7 25	rise in sun		3 .34
5	5	7 25 to 7 30	fall in shade	1.68	
6	5	7 30 to 7 35	rise in sun		3 .13
7	5	7 35 to 7 40	fall in shade	1.73	
8	5	7 40 to 7 45	rise in sun		3 .40
9	5	7 45 to 7 50	fall in shade	1.60	
10	5	7 50 to 7 55	rise in sun		3 .18
11	5	7 55 to 7 60	fall in shade	1.97	

Note.—Very clear sky and no wind. The above series considered by the observer an excellent one.

From No. 2: $t = 3.31 + \dfrac{1.54 + 1.55}{2} = 4.86$. Calories per minute per square cm 0.985

From No. 4: $t = 3.31 + \dfrac{1.55 + 1.68}{2} = 4.96$. Calories per minute per square cm 1.005

From No. 6: $t = 3.43 + \dfrac{1.68 + 1.73}{2} = 5.84$. Calories per minute per square cm 0.980

From No. 8: $t = 3.40 + \dfrac{1.73 + 1.60}{2} = 5.07$. Calories per minute per square cm 1.027

From No. 10: $t = 3.18 + \dfrac{1.60 + 1.97}{2} = 4.97$. Calories per minute per square cm 1.007

Mean 1.001

Reduced to standard mercury pyrheliometer (1.001 × 1.27) 1.271

TABLE 25.

[Date August 1), 1881. Station, Lone Pine. Observer, A. C. D. Sky, clear. Wind, gentle. Charge of water, 50 s. c. (put in at 10 a. m.). The excessive temperature of the air put a stop to the readings at 11° 50' by sending the mercury above the scale. Barometer (θ) 6.64. Length of path of ray (M) 1.075. (M 4.) 7.16.]

No.	Interval	Noon.	Fall.	Rise.
	Min.			
1	5	11° 35' to 11° 40' fall in shade	0 .50	......
2	5	11 40 to 11 45 rise in sun	......	1 .50
3	5	11 45 to 11 50 fall in shade	1 .25	......

From No. 2: $t = 4.60 + \dfrac{.50 + 1.25}{2} = 5.43$. Calories per minute per square cm 1.253

Reduced to standard mercury pyrheliometer 1.501

TABLE 26.

[Date August 11, 1881. Station, Lone Pine. Observer A. C. D. Sky, clear. Wind, fresh. Charge of water, 90.6 c. c. Barometer (θ) 6.62. Length of path of ray (M) 2.103. (M 4.) 11.53.]

No.	Interval	P. M.	Fall.	Rise.
	Min.			
1	5	4 05' to 4 10' fall in shade	1 .00	......
2	5	4 10 to 4 15 rise in sun	......	1 .80
3	5	4 15 to 4 20 fall in shade	1 .26	......
4	5	4 20 to 4 25 rise in sun	......	3 .15
5	5	4 25 to 4 50 fall in shade	1 .86	......
6	5	4 45 to 4 50 fall in shade	1 .44	......
7	5	4 50 to 4 55 rise in sun	......	3 .05
8	5	4 55 to 5 00 fall in shade	2 .24	......

From No. 2: $t = 3.86 + \dfrac{1.00 + 1.26}{2} = 4.99$. Calories per minute per square cm 1.111

From No. 4: $t = 3.15 + \dfrac{1.26 + 1.86}{2} = 4.71$. Calories per minute per square cm 1.048

From No. 7: $t = 3.05 + \dfrac{1.44 + 2.24}{2} = 4.89$. Calories per minute per square cm 1.089

Mean 1.083

Reduced to standard mercury pyrheliometer 1.375

TABLE 27.

[Date August 12, 1881. Station, Lone Pine. Observer A. C. D. Sky, fair with drifting cirro-cumulus clouds. Wind, gentle. Charge of water, 91.3 c. c. (Put in August 11, at 8 p. m.) Barometer (θ) 6.64. Length of path of ray (M) 2.184. (M 8.) 14.51.]

No.	Interval	A. M.	Fall.	Rise.
	Min.			
1	5	7 15' to 7 20' fall in shade	0 .94	......
2	5	7 20 to 7 25 rise in sun	......	3 .60
3	5	7 25 to 7 30 fall in shade	0 .81	......
4	5	7 30 to 7 35 rise in sun	......	3 .41
5	5	7 35 to 7 40 fall in shade	1 .21	......
6	5	7 40 to 7 45 rise in sun	......	3 .41
7	5	7 45 to 7 50 fall in shade	1 .40	......
8	5	7 50 to 7 55 rise in sun	......	3 .21
9	5	7 55 to 7 60 fall in shade	1 .53	......

From No. 2: $t = 3.60 + \dfrac{.94 + .81}{2} = 4.46$. Calories per minute per square cm 0.972

From No. 4: $t = 3.41 + \dfrac{.81 + 1.21}{2} = 4.42$. Calories per minute per square cm 0.963

From No. 6: $t = 3.41 + \dfrac{1.21 + 1.40}{2} = 4.72$. Calories per minute per square cm 1.029

From No. 8: $t = 3.21 + \dfrac{1.40 + 1.53}{2} = 4.68$. Calories per minute per square cm 1.020

Mean 0.996

Reduced to standard mercury pyrheliometer 1.265

TABLE 28.

[Date, August 12, 1881.　Station, Lone Pine.　Observer, A. C. D.　Sky, fair.　Wind, fresh to brisk.　Charge of water, 91.3 c. c.　Barometer (B.) 6.64.　Length of path of ray (M.) 1.078.　(M, B.) 7.16.]

[Owing to clouds no observations were taken from 11.30 to 12.]

No.	Interval.	Noon.	Fall.	Rise.
	Min.			
1	5	12ʰ 05ᵐ to 12ʰ 10ᵐ fall in shade	0°.67	
2	5	12 10 to 12 15 rise in sun		3°.86
3	5	12 15 to 12 20 fall in shade	1 .80	
4	5	12 20 to 12 25 rise in sun		3 .38
5	5	12 25 to 12 30 fall in shade	2 .22	

From No. 2: $t = 3.86 + \dfrac{.67 + 1.80}{2} = 5.10$.　Calories per minute per square cm 1.112

From No. 4: $t = 3.38 + \dfrac{1.80 + 2.22}{2} = 5.39$.　Calories per minute per square cm 1.175

Mean .... : 1.144
Reduced to standard mercury pyrheliometer 1.453

TABLE 29.

[Date, August 12, 1881.　Station, Lone Pine.　Observer, A. C. D.　Sky, fair.　Wind, fresh breeze.　Charge of water, 91.3 c. c.　Barometer (B.) 6.62.　Length of path of ray (M.) 2.302.　(M, B.) 15.25.]

[Clouds prevented further observation.]

No.	Interval.	P. M.	Fall.	Rise.
	Min.			
1	5	4ʰ 30ᵐ to 4ʰ 35ᵐ fall in shade	1 .03	
2	5	4 35 to 4 40 rise in sun		3°.10
3	5	4 40 to 4 45 fall in shade	1 .55	

From No. 2: $t = 3.10 + \dfrac{1.03 + 1.55}{2} = 4.39$.　Calories per minute per square cm 1.022

Reduced to standard mercury pyrheliometer 1.298

The charge of water was measured with a 20 c. c. pipette up to and including August 12, 1881.　After this date the water was weighed.

TABLE 30.

[Date, August 13, 1881.　Station, Lone Pine.　Observer, A. C. D.　Sky, clear.　Wind, very gentle.　Charge of water, 88.6 grammes.　Barometer (B.) 6.62.　Length of path of ray (M.) 2.302.　(M, B.) 15.25.]

No.	Interval.	A. M.	Fall.	Rise.
	Min.			
1	5	7ʰ 05ᵐ to 7ʰ 10 fall in shade	1 .19	
2	5	7 10 to 7 15 rise in sun		3 .54
3	5	7 15 to 7 20 fall in shade	1 .50	
4	5	7 20 to 7 25 rise in sun		3 .39
5	5	7 25 to 7 30 fall in shade	1 .45	
6	5	7 30 to 7 35 rise in sun		3 .31
7	5	7 35 to 7 40 fall in shade	1 .46	
8	5	7 40 to 7 45 rise in sun		3 .40
9	5	7 45 to 7 50 fall in shade	1 .69	
10	5	7 50 to 7 55 rise in sun		3 .62
11	5	7 55 to 8 00 fall in shade	1 .80	

From No. 2: $t = 3.54 + \dfrac{1.19 + 1.50}{2} = 4.89$.　Calories per minute per square cm 1.040

From No. 4: $t = 3.39 + \dfrac{1.50 + 1.45}{2} = 4.78$.　Calories per minute per square cm 1.016

From No. 6: $t = 3.31 + \dfrac{1.45 + 1.46}{2} = 4.77$.　Calories per minute per square cm 1.014

From No. 8: $t = 3.40 + \dfrac{1.46 + 1.69}{2} = 4.98$.　Calories per minute per square cm 1.059

From No. 10: $t = 3.62 + \dfrac{1.69 + 1.80}{2} = 4.84$.　Calories per minute per square cm 1.029

Mean 1.032
Reduced to standard mercury pyrheliometer 1.311

TABLE 31.

[Date, August 13, 1881. Station, Lone Pine. Observer, A. C. D. Sky, clear. Wind very brisk at times reaching a gale. Charge of water, 88.6 grammes. Barometer (b) 6.62. Length of path of ray (M) 1.67. (M, B.)—7.44.]

[Observations taken in rear of building, to shield the instrument from the wind as much as possible.]

No.	Interval.	Noon.	Fall.	Rise.
	Min.			
1	5	11h 45m to 11h 50m fall in shade	0.00	
2	5	11 50 to 11 55 rise in sun		1.63
3	5	11 55 to 12 00 fall in shade	1.44	
4	5	12 00 to 12 05 rise in sun		1.94
5	5	12 05 to 12 10 fall in shade	1.99	
6	5	12 25 to 12 30 fall in shade	1.70	
7	5	12 30 to 12 35 rise in sun		1.74
8	5	12 35 to 12 40 fall in shade	1.74	

From No. 2: $t = 1.63 + \dfrac{.60 + 1.41}{2} = 5.65$. Calories per minute per square cm 1.201

From No. 4: $t = 3.84 + \dfrac{1.41 + 1.98}{2} = 5.52$. Calories per minute per square cm 1.195

From No. 7: $t = 1.45 + \dfrac{1.70 + 1.74}{2} = 5.87$. Calories per minute per square cm 1.248

Mean 1.215
Reduced to standard mercury pyrheliometer.................................. 1.543
(No evening observation was obtained on this day.)

TABLE 32.

[Date, August 14, 1881. Station, Lone Pine. Observer, A. C. D. Sky, clear. Wind, calm. Charge of water, 85.3 grammes. Barometer (b) 6.62. Length of path of ray (M) 2.302. (M, B.)—15.25.]

No.	Interval.	A. M.	Fall.	Rise.
	Min.			
1	5	7h 05m to 7h 10m fall in shade	1.00	
2	5	7 10 to 7 15 rise in sun		3.91
3	5	7 15 to 7 20 fall in shade	1.16	
4	5	7 20 to 7 25 rise in sun		3.82
5	5	7 25 to 7 30 fall in shade	1.56	
6	5	7 30 to 7 35 rise in sun		3.64
7	5	7 35 to 7 40 fall in shade	1.80	
8	5	7 40 to 7 45 rise in sun		3.51
9	5	7 45 to 7 50 fall in shade	1.95	
10	5	7 50 to 7 55 rise in sun		3.37
11	5	7 55 to 8 00 fall in shade	2.27	

From No. 2: $t = 3.91 + \dfrac{1.00 + 1.16}{2} = 5.14$. Calories per minute per square cm 1.020

From No. 4: $t = 3.82 + \dfrac{1.16 + 1.56}{2} = 5.38$. Calories per minute per square cm 1.038

From No. 6: $t = 3.64 + \dfrac{1.56 + 1.80}{2} = 5.32$. Calories per minute per square cm 1.066

From No. 8: $t = 3.51 + \dfrac{1.80 + 1.95}{2} = 5.39$. Calories per minute per square cm 1.110

From No. 10: $t = 3.37 + \dfrac{1.95 + 2.27}{2} = 5.48$. Calories per minute per square cm 1.129

Mean 1.098
Reduced to standard mercury pyrheliometer (1.098 × 1.27) 1.394

TABLE 33.

Date August 14, 1881. Station, Lone Pine. Observer A. C. D. Sky, clear. Wind, fresh; variable. Charge of water, 85.3 grammes. Barometer (θ_o) 6.63. Length of path of ray (M_o) = 1.082. (M, θ_o)=7.18.

No.	Interval.	Noon	Fall.	Rise.
	Min.			
1	5	11h 35m to 11h 40m fall in shade	0.78	
2	5	11 40 to 11 45 rise in sun		4.78
3	5	11 45 to 11 50 fall in shade	1.80	
4	5	11 50 to 11 55 rise in sun		4.10
5	5	11 55 to 12 00 fall in shade	2.10	
6	5	12 00 to 12 05 rise in sun		3.90
7	5	12 05 to 12 10 fall in shade	2.60	
8	5	12 10 to 12 15 rise in sun		3.24
9	5	12 15 to 12 20 fall in shade	2.04	
10	5	12 20 to 12 25 rise in shade		3.30
11	5	12 25 to 12 30 fall in shade	3.00	

From No. 2: $t = 4.78 + \dfrac{.78 + 1.80}{2} = 6.07$. Calories per minute per square cm 1.250

From No. 4: $t = 4.10 + \dfrac{1.80 + 2.10}{2} = 6.05$. Calories per minute per square cm 1.246

From No. 6: $t = 3.90 + \dfrac{2.10 + 2.60}{2} = 6.25$. Calories per minute per square cm 1.287

From No. 8: $t = 3.24 + \dfrac{2.60 + 2.04}{2} = 5.56$. Calories per minute per square cm 1.145

From No. 10: $t = 3.30 + \dfrac{2.04 + 3.00}{2} = 5.82$. Calories per minute per square cm 1.199

Mean 1.225
Reduced to standard mercury pyrheliometer 1.556

TABLE 34.

Date, August 14, 1881. Station, Lone Pine. Observer, A. C. D. Sky, clear. Wind, gentle to fresh. Charge of water, 85.3 grammes. Barometer (θ_o) = 6.64. Length of path of ray (M_o) = 2.148. (M, θ_o) 14.26.

No.	Interval.	P. M.	Fall.	Rise.
	Min.			
1	5	4h 05m to 4h 10m fall in shade	1.42	
2	5	4 10 to 4 15 rise in sun		1.46
3	5	4 15 to 4 20 fall in shade	1.66	
4	5	4 20 to 4 25 rise in sun		3.41
5	5	4 25 to 4 30 fall in shade	1.92	
6	5	4 30 to 4 35 rise in sun		3.10
7	5	4 35 to 4 40 fall in shade	2.55	
8	5	4 40 to 4 45 rise in sun		2.47
9	5	4 45 to 4 50 fall in shade	2.40	
10	5	4 50 to 4 55 rise in sun		2.58
11	5	4 55 to 5 00 fall in shade	2.42	

From No. 2: $t = 3.46 + \dfrac{1.42 + 1.66}{2} = 5.00$. Calories per minute per square cm 1.030

From No. 4: $t = 3.44 + \dfrac{1.66 + 1.92}{2} = 5.23$. Calories per minute per square cm 1.077

From No. 6: $t = 3.10 + \dfrac{1.92 + 2.55}{2} = 5.34$. Calories per minute per square cm 1.100

From No. 8: $t = 2.47 + \dfrac{2.55 + 2.40}{2} = 4.94$. Calories per minute per square cm 1.018

From No. 10: $t = 2.28 + \dfrac{2.40 + 2.42}{2} = 4.69$. Calories per minute per square cm 0.966

Mean 1.038
Reduced to standard mercury pyrheliometer 1.318

OBSERVATIONS WITH THE PYRHELIOMETER MADE AT MOUNTAIN CAMP, MOUNT WHITNEY.

TABLE 35.

Date August 29 1881. Station, Mountain Camp. Observer, J. J. N. Sky deep blue. Wind, light breeze. Charge of water, 90 grammes. Barometer (d.) 4.98. Length of path of ray (M.) 1.955, (M. θ) 9.71.

No.	Interval.	A. M.	Fall	Rise
	Min	After 15 minutes' exposure to sun—		
1	5	8 00 to 8 05 fall in shade		2 61
2	5	8 05 to 8 10 rise in sun		2 57
3	5	8 10 to 8 15 fall in shade	2 00	

From No. 2: $t = 2.57 + \frac{2.61 + 2.00}{2} = 4.89$. Calories per minute per square cm.......... 1.053

Reduced to standard mercury pyrheliometer.. 1.337

TABLE 36.

Date, August 29, 1881. Station, Mountain Camp. Observer, J. J. N. Sky deep blue. Wind light breeze. Charge of water, 91.3 grammes. Barometer (d.) 4.98. Length of path of ray (M.) 1.125, (M. θ) 5.60.

No.	Interval.	Noon.	Fall	Rise
	Min	After 15 minutes' exposure to sun—		
1	5	11 40 to 11 45 fall in shade	3 57	3 88
2	5	11 45 to 11 50 rise in sun		
3	5	11 50 to 11 55 fall in shade	3 21	
4	5	11 55 to 12 00 rise in sun		3 48
5	5	12 00 to 12 05 fall in shade	4 04	
6	5	12 05 to 12 10 rise in sun		3 66
7	5	12 10 to 12 15 fall in shade	3 92	

From No. 2: $t = 3.88 + \frac{3.57 + 3.21}{2} = 7.27$. Calories per minute per square cm.......... 1.554

From No. 4: $t = 3.48 + \frac{3.21 + 4.04}{2} = 7.11$. Calories per minute per square cm.......... 1.520

From No. 6: $t = 3.66 + \frac{4.04 + 3.92}{2} = 7.49$. Calories per minute per square cm.......... 1.567

Mean 1.547
Reduced to standard mercury pyrheliometer............................... 1.930

TABLE 37.

Date August 29 1881. Station, Mountain Camp. Observer, J. J. N. Sky deep blue. Wind light breeze. Charge of water, 90.5 grammes. Barometer (d.) 4.99. Length of path of ray (M.) 2.791, (M. θ) 13.90.

No.	Interval	P. M.	Fall	Rise
	Min	After 15 minutes' exposure to sun—		
1	5	4 25 to 4 30 fall in shade	2 68	
2	5	4 30 to 4 35 rise in sun		2 59
3	5	4 35 to 4 40 fall in shade	2 61	
4	5	4 40 to 4 45 rise in sun		3 12
5	5	4 45 to 4 50 fall in shade	2 74	
6	5	4 50 to 4 55 rise in sun		3 24
7	5	4 55 to 5 00 fall in shade	4 12	

From No. 2: $t = 3.59 + \frac{2.68 + 2.61}{2} = 6.24$. Calories per minute per square cm.......... 1.348

From No. 4: $t = 3.32 + \frac{2.61 + 2.74}{2} = 6.00$. Calories per minute per square cm.......... 1.296

From No. 6: $t = 3.24 + \frac{2.74 + 3.12}{2} = 6.17$. Calories per minute per square cm.......... 1.333

Mean 1.326
Reduced to standard mercury pyrheliometer 1.684

Owing to the high cliffs on the east, and the consequently late sunrise at the camp, the morning observations here were necessarily taken later than those at Lone Pine, while the evening ones were at the same hour

TABLE 38.

[Date, August —, 1881. Station, Mountain Camp. Observer, J. J. N. Sky, deep blue. Wind, calm. Charge of water 91.1 grammes. Barometer (B.)=5.00. Length of path of ray (M.)=1.786. (M., β.)=805.]

No.	Interval.	A. M.	Fall.	Rise.
	Min.	After 15 minutes' exposure to sun—		
1	5	8h 05m to 8h 10m fall in shade	1.77	
2	5	8 10 to 8 15 rise in sun		3.67
3	5	8 15 to 8 20 fall in shade	1.86	
4	5	8 20 to 8 25 rise in sun		3.76
5	5	8 25 to 8 30 fall in shade	2.02	
6	5	8 30 to 8 35 rise in sun		3.28
7	5	8 35 to 8 40 fall in shade	2.24	

From No. 2: $t = 3.67 + \frac{1.77 + 1.86}{2} - 5.39$. Calories per minute per square cm 1.195

From No. 4: $t = 3.76 + \frac{1.86 + 2.02}{2} - 5.70$. Calories per minute per square cm 1.241

From No. 6: $t = 3.28 + \frac{2.02 + 2.24}{2} - 5.41$. Calories per minute per square cm 1.177

Mean 1.204
Reduced to standard mercury pyrheliometer 1.529

TABLE 39.

[Date, August 30, 1881. Station, Mountain Camp. Observer, J. J. N. Sky, deep blue. Wind, light breeze. Charge of water, 90.2 grammes. Barometer (B.) 5.02. Length of path of ray (M.) 1.130. (M. β.) 5.67.]

No.	Interval.	Noon.	Fall.	Rise.
	Min.	After 15 minutes' exposure to sun—		
1	5	11h 40m to 11h 45m fall in shade	2.40	
2	5	11 45 to 11 50 rise in sun		4.38
3	5	11 50 to 11 55 fall in shade	2.78	
4	5	11 55 to 12 00 rise in sun		3.21
5	5	12 00 to 12 05 fall in shade	2.65	
6	5	12 05 to 12 10 rise in sun		3.90
7	5	12 10 to 12 15 fall in shade	2.30	

Fresh breeze.

From No. 2: $t = 4.38 + \frac{2.40 + 2.78}{2} - 6.97$. Calories per minute per square cm 1.504

From No. 4: $t = 3.21 + \frac{2.78 + 2.65}{2} - 5.93$. Calories per minute per square cm 1.240

From No. 6: $t = 3.90 + \frac{2.65 + 2.30}{2} - 5.54$. Calories per minute per square cm 1.195

Mean 1.326
Reduced to standard mercury pyrheliometer 1.684

TABLE 40.

[Date, August 30, 1881. Station, Mountain Camp. Observer, J. J. N. Sky, deep blue. Wind, light breeze. Charge of water, 89.6 grammes. Barometer (B.) 5.01. Length of path of ray (M.) 3.073. (M. β.) 18.21.]

No.	Interval	P. M.	Fall.	Rise.
	Min.	After 15 minutes' exposure to sun—		
1	5	4h 25m to 4h 30m fall in shade	1.71	
2	5	4 30 to 4 35 rise in sun		3.84
3	5	4 35 to 4 40 fall in shade	2.73	
4	5	4 40 to 4 45 rise in sun		3.93
5	5	4 45 to 4 50 fall in shade	2.24	
6	5	4 50 to 4 55 rise in sun		3.92
7	5	4 55 to 5 00 fall in shade	2.41	

From No. 2: $t = 3.84 + \frac{1.71 + 2.73}{2} - 6.06$. Calories per minute per square cm 1.301

From No. 4: $t = 3.93 + \frac{2.73 + 2.24}{2} - 5.42$. Calories per minute per square cm 1.163

From No. 6: $t = 3.92 + \frac{2.24 + 2.41}{2} - 5.55$. Calories per minute per square cm 1.191

Mean 1.218
Reduced to standard mercury pyrheliometer 1.547

TABLE 11.

[Date, August 31, 1881. Station, Mountain Camp. Observer, J. J. N. Sky, deep blue. Wind, light breeze. Charge of water, 100.7 grammes. Barometer, 24.50. Length of path of ray $\cdot M = 1.061$ M' ...]

No.	Interval.	A. M.	Fall	Rise
	Min.	After — minutes exposure to sun—		
1	5	8.20 to 8.25 fall in shade		
2	5	8.25 to 8.30 rise in sun		
3	5	8.30 to 8.35 fall in shade		
4	5	8.35 to 8.40 rise in sun		
5	5	8.40 to 8.45 fall in shade		
6	5	8.45 to 8.50 rise in sun		
7	5	8.50 to 8.55 fall in shade		

From No. 2: $t = 2.91 + \dfrac{2.42 + 1.98}{2} = 5.41.$ Calories per minute per square cm... 1.217

From No. 4: $t = 2.75 + \dfrac{1.98 + 2.68}{2} = 5.08.$ Calories per minute per square cm... 1.203

From No. 6: $t = 2.81 + \dfrac{2.68 + 2.18}{2} = 5.26.$ Calories per minute per square cm... 1.216

Mean... 1.222
Reduced to standard mercury pyrheliometer... 1.552

TABLE 12.

[Date, August 31, 1881. Station, Mountain Camp. Observer, J. J. N. Sky, deep blue. Wind, light breeze. Charge of water, 100.7 grammes. Barometer, ... Length of path of ray $M = 1.136$ M' ...]

No.	Interval.	Noon.	Fall	Rise
	Min.	After 18 minutes exposure to sun—		
1	5	11.40 to 11.45 fall in shade		
2	5	11.45 to 11.50 rise in sun		
3	5	11.50 to 11.55 fall in shade		
4	5	11.55 to 12.00 rise in sun		
5	5	12.00 to 12.05 fall in shade		
6	5	12.05 to 12.10 rise in sun		
7	5	12.10 to 12.15 fall in shade		

From No. 2: $t = 3.50 + \dfrac{2.35 + 3.55}{2} = 6.37.$ Calories per minute per square cm... 1.500

From No. 4: $t = 3.53 + \dfrac{3.35 + 2.30}{2} = 6.36.$ Calories per minute per square cm... 1.198

From No. 6: $t = 2.87 + \dfrac{2.30 + 2.60}{2} = 5.32.$ Calories per minute per square cm... 1.253

Mean... 1.417
Reduced to standard mercury pyrheliometer... 1.799

TABLE 13.

[Date, August 31, 1881. Station, Mountain Camp. Observer, J. J. N. Sky, deep blue. Wind, light breeze. Charge of water, 107.5 grammes. Barometer, 24.50. Length of path of ray $M = 2.925$ M' ...]

No.	Interval.	P. M.	Fall	Rise
	Min.	After 8 minutes exposure to sun—		
1	5	4.25 to 4.30 fall in shade		
2	5	4.30 to 4.35 rise in sun		
3	5	4.35 to 4.40 fall in shade		
4	5	4.40 to 4.45 rise in sun		
5	5	4.45 to 4.50 fall in shade		
6	5	4.50 to 4.55 rise in sun		
7	5	4.55 to 5.00 fall in shade		

From No. 2: $t = 2.75 + \dfrac{1.92 + 2.37}{2} = 4.93.$ Calories per minute per square cm... 1.231

From No. 4: $t = 2.41 + \dfrac{2.37 + 1.96}{2} = 4.58.$ Calories per minute per square cm... 1.147

From No. 6: $t = 2.21 + \dfrac{1.96 + 2.21}{2} = 4.34.$ Calories per minute per square cm... 1.087

Mean... 1.176
Reduced to standard mercury pyrheliometer... 1.468

TABLE 14.

[Date, September 1, 1881. Station, Mountain Camp. Observer, J. J. N. Sky, deep blue. Wind, light breeze. Charge of water, 107.7 grammes. Barometer (β_1) 5.09. Length of path of ray (M_1) 1.836. ($M_1\beta_1$) 9.18.]

No.	Interval.	A. M.	Fall.	Rise.
	Min.	After 15 minutes' exposure to sun—		
1	5	8 00 to 8 05 fall in shade	0 83	......
2	5	8 05 to 8 10 rise in sun....................		1 86
3	5	8 10 to 8 15 fall in shade.................	1 25	......
4	5	8 15 to 8 20 rise in sun..................		3 85
5	5	8 20 to 8 25 fall in shade................	1 37	......
6	5	8 25 to 8 30 rise in sun.................		3 60
7	5	8 30 to 8 35 fall in shade...............	1 88	......

From No. 2; $t = 1.60 + \dfrac{.83 + 1.25}{2} = 5$ .64. Calories per minute per square cm........ 1.119

From No. 4; $t = 3.85 + \dfrac{1.25 + 1.37}{2} = 5$ .16. Calories per minute per square cm........ 1.290

From No. 6; $t = 3.60 + \dfrac{1.37 + 1.66}{2} = 5$ .21. Calories per minute per square cm........ 1.302

Mean 1.304
Reduced to standard mercury pyrheliometer............................ 1.494

TABLE 15.

[Date, September 1, 1881. Station Mountain Camp. Observer, J. J. N. Sky, deep blue. Wind, fresh breeze. Charge of water, 107.7 grammes. Barometer (β_1) 5.09. Length of path of ray (M_2) 132. ($M_2\beta_1$) 5.67.]

No.	Interval.	Noon.	Fall.	Rise.
	Min.	After 15 minutes' exposure to sun—		
1	5	11 40 to 11 45 fall in shade	1" 85	......
2	5	11 45 to 11 50 rise in sun		3 68
3	5	11 50 to 11 55 fall in shade...............	1 71	......
4	5	11 55 to 12 00 rise in sun................		3 95
5	5	12 00 to 12 05 fall in shade..............	2 74	......
6	5	12 05 to 12 10 rise in sun...............		3 25
7	5	12 10 to 12 15 fall in shade.............	2 98	......

From No. 2; $t = 3.68 + \dfrac{1.85 + 1.71}{2} = 5$ .46. Calories per minute per square cm........ 1.365

From No. 4; $t = 3.93 + \dfrac{1.71 + 2.74}{2} = 6$ .16. Calories per minute per square cm........ 1.540

From No. 6; $t = 3.24 + \dfrac{2.74 + 2.98}{2} = 5$ .74. Calories per minute per square cm........ 1.435

Mean... 1.447
Reduced to standard mercury pyrheliometer.............................. 1.838

TABLE 16.

[Date, September 1, 1881. Station, Mountain Camp. Observer, J. J. N. Sky, deep blue. Wind, fresh breeze. Charge of water, 107.7 grammes. Barometer (β_1) 4.99. Length of path of ray (M_3) 2.959. ($M_3\beta_1$) 14.77.]

No.	Interval.	P. M.	Fall.	Rise.
	Min.	After 15 minutes' exposure to sun—		
1	5	4 25 to 4 30 fall in shade...........	1 80	......
2	5	4 30 to 4 35 rise in sun............		3 09
3	5	4 35 to 4 40 fall in shade..........	1 90	......
4	5	4 40 to 4 45 rise in sun...........		2 85
5	5	4 45 to 4 50 fall in shade.........	1 60	......
6	5	4 50 to 4 55 rise in sun..........		2 60
7	5	4 55 to 5 00 fall in shade........	2 33	......

From No. 2; $t = 3.09 + \dfrac{1.80 + 1.90}{2} = 4$ .94. Calories per minute per square cm........ 1.239

From No. 4; $t = 2.85 + \dfrac{1.90 + 1.60}{2} = 4$ .70. Calories per minute per square cm........ 1.179

From No. 6; $t = 2.60 + \dfrac{1.60 + 2.33}{2} = 4$ .56. Calories per minute per square cm........ 1.140

Mean... 1.186
Reduced to standard mercury pyrheliometer................................ 1.519

WHITNEY PEAK.

TABLE 47.

Date: September 5, 1881. Station, peak of Whitney. Observer, L. E. K. Sky, deep blue but somewhat smoky from forest fires. Wind, fresh, variable. Charge of water, 1061 grammes. Barometer (t°) 1.16. Length of path of ray (M₀) 1.197. (M₀ 6) 5.57. Strong gust of wind.]

No.	Interval	Noon	Fall	Rise
	Min.	After 5 minutes' exposure to sun—		
1	5	12° 45′ to 12° 50′ fall in shade	1 .80	
2	5	12 50 to 12 55 rise in sun		3 .50
3	5	12 55 to 1 00 fall in shade	2 .80	
4	5	1 00 to 1 05 rise in sun		3 .55
5	5	1 05 to 1 10 fall in shade	2 .24	
6	5	1 10 to 1 15 rise in sun		3 .36
7	5	1 15 to 1 20 fall in shade	2 .72	

From No. 2: $t = 3.70 + \dfrac{1.80 + 2.80}{2} = 5.80.$ Calories per minute per square cm 1.436

From No. 4: $t = 3.55 + \dfrac{2.80 + 2.24}{2} = 6.07.$ Calories per minute per square cm 1.503

From No. 6: $t = 3.36 + \dfrac{2.24 + 2.72}{2} = 5.84.$ Calories per minute per square cm 1.446

Mean 1.462

Reduced to standard mercury pyrheliometer... 1.857

TABLE 48.

Date, September 8, 1881. Station peak of Whitney. Observer, O. E. M. Sky, very hazy. Wind, . Charge of water, 1061 grammes. Barometer (t°) 1.16. Length of path of ray (M₀) 1.216. (M₀ 6) 5.54.]

No.	Interval	P. M.	Fall	Rise
	Min.			
1	5	12° 50′ 30″ to 12° 56′ 30″ fall in shade	0 .58	
2	5	12 56 30 to 1 01 30 rise in sun		3 .24
3	5	1 01 30 to 1 06 30 fall in shade	1 .45	
4	5	1 06 30 to 1 11 30 rise in sun		3 .89
5	5	1 11 30 to 1 16 30 fall in shade	1 .39	
6	5	1 16 30 to 1 21 30 rise in sun		4 .04
7	5	1 21 30 to 1 26 30 fall in shade	0 .50	
8	5	1 26 30 to 1 31 30 rise in sun		3 .70
9	5	1 31 30 to 1 36 30 fall in shade	1 .50	
10	5	1 36 30 to 1 41 30 rise in sun		3 .02
11	5	1 41 30 to 1 46 30 fall in shade	1 .62	

From No. 2: $t = 3.23 + \dfrac{0.58 + 1.45}{2} = 4.40.$ Calories per minute per square cm 1.045

From No. 4: $t = 3.89 + \dfrac{1.45 + 1.39}{2} = 5.05.$ Calories per minute per square cm 1.245

From No. 6: $t = 4.04 + \dfrac{1.39 + 0.70}{2} = 5.00.$ Calories per minute per square cm 0.989

From No. 8: $t = 3.70 + \dfrac{0.70 + 1.50}{2} = 4.80.$ Calories per minute per square cm 1.158

From No. 10: $t = 3.02 + \dfrac{1.50 + 1.62}{2} = 4.58.$ Calories per minute per square cm 1.131

Mean 1.114

Reduced to standard mercury pyrheliometer... 1.415

DISCUSSION OF PYRHELIOMETER OBSERVATIONS.

We observe considerable variations in the measured heat between one day and another, even when the sky appears equally clear. In fact, with this instrument only a quite absolute calm is suitable for observation, and as we approach this rarely attained condition the readings will rise, so that these may be considered as usually too low, owing to this cause alone. We have just obtained by our comparison with the mercury pyrheliometer the multiplying factor 1.27, which may be considered to introduce an approximate correction for the non conductivity of the water, already referred to; but there are several small corrections which Pouillet omits, and which, though not properly negligible, we omit also here, though they are given in full in connection with the acti. nometer. It is, in fact, a waste of labor to attempt the serious task of determining these special values for the pyrheliometer, and for the purpose of improving so unsatisfactory an instrument; but we remark that the omitted corrections are in general of the positive sign, so that the value of the solar constant, which we now proceed to deduce by Pouillet's method, would be still greater if these were introduced. The values we use are taken from the preceding tables, where they are expressed in calories, one calorie being the amount of heat required to warm one gramme of water from 0° to 1 Centigrade, and the solar constant being expressed by the number of calories per minute given by the sun's rays before absorption falling normally on 1 sq. cm. Thus, on August 14th, we obtained 1.394 calories as the heating effect of the sun per minute in the morning at Lone Pine, where its rays fall normally on a surface 1 cm. square, a value which we consider as below the truth.

Let M_{t}, z_{t} represent the absorbing air-mass at the noon observation, M_{tt}, z_{tt} that at the morning or evening observations, or their mean. Let C_{t} denote the corrected value in calories found by the noon observation and C_{tt} that by the morning or evening observations, or their mean.

Then E being the heat before absorption, i. e. the solar constant, a the coefficient of transmission through the entire atmosphere (such as would support 7.6 dm. of mercury), $a^{\frac{z}{7.6}}$ is the coefficient of transmission.

$$a = \left(\frac{C_{tt}}{C_{t}}\right)^{\frac{7.6}{z_{t}\beta - z_{tt}\beta_{tt}}} \qquad E = \frac{C_{t}}{z_{t}\beta / 7.6}$$

Thus, on August 14—

$$M_{t}, z_{t} = 7.18 \qquad M_{tt}, z_{tt} \text{ (morn.)} = 15.25 \qquad M_{tt}, z_{tt} \text{ (eve.)} = 14.20$$

$$C_{t} = 1.558 \qquad C_{tt} \text{ (morn.)} = 1.394 \qquad C_{tt} \text{ (eve.)} = 1.318$$

and if we determine a from noon and morning observations, we have

$$a = 0.9005 \qquad E = 1.789$$

If from noon and evening observations—

$$a = 0.8342 \qquad E = 1.783$$

We should, by Pouillet's theory, find the same values for a and E under either circumstance.

TABLE 49.

Reduction of Lone Pine pyrheliometer observations by Pouillet's formula.

Computation of a. Computer, A. B. S

Data.	Morning and noon.			Evening and noon.		
	August 12.	August 13.	August 14.	August 11.	August 12.	August 14.
Log 7.6	0.8808	0.8808	0.8808	0.8808	0.8808	0.8808
M, β	14.51	15.25	15.25	14.54	15.25	14.20
$M \circ$	7.16	7.14	7.18	7.16	7.16	7.18
$M \circ - M \circ$	7.35	8.11	8.07	7.17	8.09	7.02
Log $M \circ - M \circ$	0.8663	0.9090	0.9069	0.8555	0.9079	0.8461
Log a	9.9115	9.9718	9.9739	9.9253	9.9729	9.9645
a	1.611	0.957	0.942	1.060	0.940	1.083
C	1.295	1.511	1.394	1.375	1.298	1.318
C_{t}	1.451	1.513	1.558	1.594	1.481	1.558
Log C_{tt}	0.1623	0.1176	0.1445	0.1364	0.1133	0.1199
Log C_{t}	0.1623	0.1804	0.1926	0.2017	0.1653	0.1926
Log $C_{tt} - \log C_{t}$	−0.0062	0.0768	0.0483	0.0654	0.0480	0.0727
Log a	0.0022	0.0663	−0.0455	0.0672	0.0464	0.0787
Tab Log a	9.9178	9.9677	9.9545	9.9328	9.9538	9.9243
a	0.8663	0.9584	0.9005	0.8565	0.8993	0.8342

$$a = \frac{7.6}{M \circ - M \circ}$$

TABLE 50.

[Computation of F. Computer, A. B. S.]

Data	August 11		August 12		August 13		August 14	
	Mean of morning and evening	Noon	Mean of morning and evening	Noon	Mean of morning and evening	Noon	Mean of morning and evening	Noon
Log C	0.181	0.2017	0.1675	0.1623	0.117	0.1581	0.1521	0.1926
Log $M\delta$	1.182	0.8519	1.1726	0.8518	1.1811	0.8517	1.162	0.8511
Log $7G$	0.8808	0.8809	0.8808	0.8809	0.8808	0.8807	0.8808	0.8809
Log $\frac{7}{G}M\delta$	0.2751	9.9741	0.2318	9.9741	0.2025	9.9740	0.2974	9.9751
$\frac{M\delta}{7b}$	1.885	0.9421	1.706	0.9421	0.9985	1.955	0.947
Log a^{7b}	0.167	0.0502	−0.1080	0.0610	0.131	−0.0625	0.1263	−0.1587
Log E	0.2650	0.2050	0.2515	0.2513	0.2307	0.2507	0.2525	0.2211
E	1.841	1.811	1.675	1.814	1.751	1.751	1.789	1.751

Comparing the morning and noon and evening and noon observations at Lone Pine, we have then the following table:

TABLE 51.

LONE PINE

Dates	a		F
	Morning and noon	Evening and noon	Noon
August 11		0.8567	1.841
August 12	0.8465	0.8567	1.611
August 13	0.8584	...	1.751
August 14	0.8605	0.8512	1.781

The mean transmissibility at Lone Pine is then here found by Pouillet's method to be about 87 per cent. That determined by him near the sea-level was about 80 per cent. The limited time at our disposal for observation renders conclusions from our present series less trustworthy than from his fuller ones. It seems clear, however, that at an altitude of almost 1,000 meters, where our observations were taken, the air, *weight for weight, is much more transparent to the heat rays* (diathermanous) *than at the sea-level.*

Even with this instrument, then, we see that the *quality* of the absorbing medium as well as its density changes as we ascend. The mean value of the solar constant from noon observations (calories 1.760) is remarkably near Pouillet's (1.764). The only significance of this appears to be that like methods bring about like results. The results themselves are, as we believe, in both cases widely wrong.

TABLE 52.

Reduction of Mountain Camp observations ; pyrheliometer.

Computation of a

Date	Morning and noon omitted.				Evening and noon			
	August 29.	August 30.	August 31.	September 1.	August 29.	August 30.	August 31.	September 1.
$M\delta$					12.97	13.21	14.15	11.77
$M\delta_0$					5.87	5.16	5.16	5.67
$M\delta - M\delta_0$					8.21	8.54	8.99	9.10
a_0					0.91	0.89	0.85	0.81
C					1.681	1.547	1.408	1.529
F					1.000	1.651	1.408	1.851
Log C					0.265	0.1895	0.1967	0.1846
Log C_0					0.2784	0.2151	0.2351	0.2647
Log $C - $Log C_0					0.0726	0.0655	0.0886	0.0897
Log a					0.1183	0.1254	0.1753	0.1801
Tab log a					9.970	9.976	9.917	9.901
a					0.835	0.845	0.848	0.859

TABLE 53.

Computation of E.

Data.	August 29.		August 30.		August 31.		September 1.	
	Evening.	Noon.	Evening.	Noon.	Evening.	Noon.	Evening.	Noon.
Ma.	13.95	5.60	15.21	5.67	14.65	5.66	14.77	5.67
$M\delta$								
7.8	1.830	0.757	2.010	0.746	1.930	0.745	1.910	0.746
$Log.\ t'$	0.2361	0.2969	0.1995	0.2303	0.1867	0.2373	0.1846	0.2643
$Log.\ a$	−0.0661	−0.0661	−0.0291	−0.0291	−0.0753	0.0753	−0.0886	−0.0886
$v4$								
$Log.\ a^{7s}$	0.1210	0.0187	0.0591	−0.0219	0.1473	−0.0561	0.1331	0.0512
$Log.\ E$	0.3173	0.3476	0.2486	0.2482	0.3120	0.3114	0.3147	0.3155
E	2.225	2.226	1.772	1.771	2.054	2.049	2.064	2.068

TABLE 54.

We now give a similar table for the observations at the Mountain Camp.

Dates	a		E
	Morning and noon.	Evening and noon.	Noon.
August 29 ...	0.4454	0.8768	2.226
August 30 ...	0.5995	0.9015	1.772
August 31 ...	0.6543	0.8909	2.049
September 1 ...	0.8575	0.8529	2.068

We have remarked before that owing to the high cliffs on the east which concealed the sun till its altitude was high, morning observations there are in general, less trustworthy than the evening ones, and this is specially seen in the present anomalous values of a from the morning series. The noon and evening ones give $a = .872$ or very near that found at Lone Pine. Omitting the morning values, we have the mean Solar Constant 2.029, a greater value than that at Lone Pine.

In accordance with the results of previous observers, then, and of our own with other instruments, *we find a larger value of the Solar Constant as we deduce it from observations through a smaller air mass.*

Our observations on the mountain and at Lone Pine not being synchronous, we can best compare summaries of the results at the higher station with those at the lower.

TABLE 55.

Summary of pyrheliometric results.

AT LONE PINE

Date.	Air-mass.			Date	Uncorrected calories.		
	A. M.	M.	P. M.		A. M	M.	P. M
August 11	15.15	7.16	14.33	August 11	1.004	1.251	1.083
August 12	14.31	7.16	15.25	August 12	.986	1.141	1.022
August 13	15.25	7.11	....	August 13	1.032	1.215	....
August 14	15.25	7.18	14.29	August 14	1.108	1.225	1.038
Mean M s	15.01	7.16	14.30	Means	1.032	1.209	1.048
				Corrected calories..	1.311	1.535	1.331

AT MOUNTAIN CAMP

August 29	9.74	5.60	15.93	August 29	1.053	1.567	1.326
August 30	8.95	5.67	15.21	August 30	1.204	1.326	1.212
August 31	8.32	5.64	14.85	August 31	1.222	1.417	1.158
September 1 ...	9.18	5.67	14.77	September 1	1.334	1.417	1.196
Mean M s	9.05	5.65	14.64	Means	1.204	1.49	1.224
				Corrected calories..	1.529	1.898	1.554

Applying Pouillet's formula to these results, we have, comparing like hours of observation on the mountain and at Lone Pine:

cal.

Mountain Camp, morning: $c = 1.529$ ⎫ whence $a = .823$; $E = 1.924$.
Lone Pine, morning: $c = 1.311$ ⎭

Mountain Camp, noon: $c = 1.828$ ⎫ whence $a = .415$; $E = 1.993$.
Lone Pine, noon: $c = 1.535$ ⎭

Mountain Camp, evening: $c = 1.554$ ⎫ whence $a = 1.639 \times 10^{2}$; $E = .004$.
Lone Pine, evening: $c = 1.331$ ⎭

These results seem to us most instructive in regard to the defects of Pouillet's formula for our present purpose. Its use here independently of its other errors tacitly assumes that a given air-mass always exercises the same absorption whatever the constituents of the air may be. The results of this erroneous assumption do not appear notable when we compare high and low sun observations at the same station, as Pouillet himself did, for the error affects each in turn; but when we compare an air mass taken on the mountain with an equal air-mass taken in the valley, the consequences of the error may become salient. Thus, in the evening observations we have the air masses at Lone Pine and at Mountain Camp almost identical, they being in the one case such as would support dm. 14.59 of mercury, and in the other 14.61. We should then have the absorptions also almost identical if the assumption were correct, but nothing of the kind happens, and we get, in fact, the monstrous results given in the last instance. It does not seem necessary to give here further illustration of the untrustworthiness of the formula in other ways (see discussion in article on spectro-bolometer) to determine us to depend as little as possible on results obtained with this instrument and in this manner. We are forced to conclude that the instrument gives the results, even of direct observation, much too small, and, as we demonstrate later, the formula itself will infallibly deduce too small values for the Solar Constant even from correct observations. We will pass on, then, to the consideration of more trustworthy instruments and methods.

CHAPTER V.

USE OF GLOBE ACTINOMETER.

When it became necessary to determine on the use of some form of actinometer in conjunction with the spectro-bolometer, the instruments of M. Violle, M. Crova, and the conjugate bulbs of M. Marie Davy were selected. I take this opportunity of acknowledging the kindness of all these gentlemen, who were good enough to undertake to see that I had suitable copies made of their respective instruments; but that of M. Marie Davy was broken in transit, and that of M. Crova, most unfortunately, did not arrive in time for the expedition. Only that of M. Violle

Fig 5

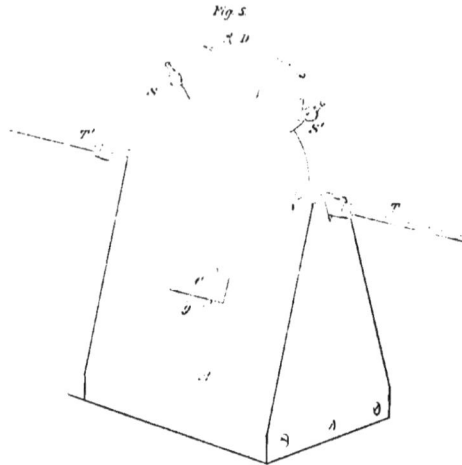

Small Actinometer, Exterior

came before the expedition started. I therefore ordered the construction of two small globe actinometers in Pittsburg.

The use of the globe actinometer presupposes the knowledge of the fact that the sun's temperature, whatever it may be, is at any rate so far higher than that of the inclosed shell of water or ice to which the sun thermometer radiates, that the excess of the latter under solar radiation is sensibly the same under all conditions of actual observation. According to Mr. Ericsson's most careful determinations, the excess of such a thermometer is the same whether it be radiating to an inclosed shell of melting ice or to one of red hot iron; and though all are not agreed that the excess is so absolutely independent of the surroundings, there is no doubt that the temperature of excess may be here treated as independent of the temperature of the water used in our actual experiment.

I had experimented in all the time at my command with the large globe actinometer of M. Violle, on the outside of which I had a gnomon placed so that it might be accurately directed to the sun without preliminary exposure, and after the expedition started had the two smaller actinometers mounted altazimuthly (an equatorial mounting would be probably better) in an improvised wooden support, which proved more manageable than the ring. The large actinometer of M. Violle may be found described in the Annales de Chimie et de Physique for 1876, vol. 10, pages 15 and 16, and in numerous other places, so that it is not necessary to redescribe it here. The two smaller actinometers which I employed with it are represented in Figs. 5 and 6, where D is the diaphragm

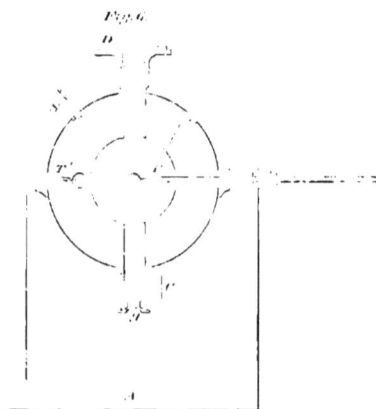

Fig. 6.

Smaller Actinometer.—Interior.

plate admitting the solar rays; S' the inlet for the water between the globes; S the exit; T the thermometer which registers the water temperature; T' the thermometer which registers the excess caused by the sun over the surroundings; C a counterpoise; G a very slightly ground glass plate which prevents the entrance of air currents from below and receives the shadow cast by the bulb of the central thermometer; A is the temporary mounting arranged on the expedition. With either one of these two forms of the globe actinometer we have no glass or absorbing material as a cover, and consequently do not attempt to observe in vacuo but, by a means to be directly explained, obtain in theory the same results as though we did so. We may use them in various ways. According to M. Violle's method the instrument is exposed to the sun's radiation for a certain time, usually fifteen or twenty minutes. The internal thermometer T rises at first rapidly, then more slowly, till it sensibly attains its temperature of equilibrium, where it is radiating as much to the surrounding globe as it is receiving from the sun. As the thermometer rises, the temperature is to be read from minute to minute until it becomes stationary, then the solar radiation is cut off and the thermometer allowed to cool during a like time till it has sensibly regained the temperature of the globe around it. When the excess of temperature of the sun thermometer over that of its inclosure is extremely small, the loss of heat is sensibly proportional to the temperature, so that at the first instant of its heating (when the temperature of excess is 0) there is in theory all gain and no loss, whether from radiation or convection, and if this *initial* rate of heating could be determined we should have the same result whether our thermometer was in air or in vacuo.

The excess which the thermometer may finally reach is, in this point of view, immaterial, for when we know the area our thermometer bulb exposes to the sun (call this S) and the specific heat or water equivalent of its bulb (M), we need only the initial rate (V) to determine the solar radiation in calories per minute, for this $= \frac{VM}{S}$. This initial rate, however, is evanescent, and cannot be directly observed, but it may be determined within very narrow limits by actual observation of the rate during the first quarter or half minute, during which the mean rate can be but very little inferior to the initial one, and by also experimentally determining another value, which must necessarily be a very little in excess of it (which is easily done), and between these closely contiguous values the true one must lie. These methods are those of direct experiment. The following ingenious method, due to M. Violle, may be rather said to be that of a mathematician.

If we represent the rise of the thermometer graphically, and if we admit the supposition that, under the varied circumstances which affect it, its rise is represented rigorously by some simple curve (e. g., a logarithmic curve), it is evident that, by watching the thermometer long enough, we could obtain the equation of this curve, and then by differentiating this equation obtain the initial rate.

It would seem, however, that any law which unites every part of a known curve to every other in a simple and rigorous geometrical relation, is one which nature rarely exactly follows. The reading of the thermometer, for instance, at the end of the fifth or tenth minute depends upon the clouds that have passed, or the breeze that has blown, since it began to rise; and, so long as it is true that even under the most favorable conditions the reading is affected from moment to moment by numberless minute and casual circumstances, it is evident that this reading can be exactly connected with the initial rate by no known law. M. Violle's formula, however, virtually assume that the curve representing this is actually such a logarithmic one, or, at any rate, a curve all whose points are interdependent and connected by some simple law. He calculates the initial rates on this assumption, and in doing so appears to us to reason correctly and elegantly as a mathematician, but on premises which the physicist may perhaps be permitted to question.

Careful experiment at Allegheny, which the reader will find later in detail, has shown that the initial rate determined by this process is always somewhat *too small*. We take (out of nearly a hundred examples we might cite) the observations with the small actinometer under the most favorable conditions from $11^h 30^m$ to 12^h on August 25, 1881, the observations themselves being given in table 58. We draw, in our own investigations, upon as large a scale as practicable (e. g., 1 in = 1 or 1 minute), the curve representing the actual observations. In the heating curve the abscissæ may be proportional to times and the ordinates to the observed temperatures of excess, and we have then a curve bearing most resemblance to the logarithmic one. Taking in this curve of observation three suitable points with equidistant abscissæ (those corresponding, for instance, to the excesses of the thermometer at the beginning, middle, and end of the time), we next pass a logarithmic curve through these points, and determine its axis of X. If the curve of observation is indeed represented by a logarithmic curve, the subtangents of the actual curve of observation on this axis of X will be sensibly constant. As a matter of fact they *systematically* decrease toward the initial point, so that the actual rate of rise is greater than the rate derived from the formula.

All the observers were exercised at every opportunity, for some weeks before the actual records began, in acquiring expertness in the use of the instrument and accuracy in the reading of the thermometers. No less than 180 minute readings, each taken on a thermometer, reading direct to 0.1° C. and by estimation to 0.01° C., were made daily on each instrument independently of the readings of the water thermometer. The results of this direct observation, whether made by the instrument and thermometers used by M. Violle or the instruments made here, with thermometers by other makers, in the clearest days and in one of the driest climates in the world, were never as large as M. Violle has found, though, from the favorable conditions, we should expect that they would have been larger. I cannot undertake to, at present, explain this. I have, however, found what appears to be a possible cause of the discrepancy. On the return from the expedition our own thermometers gave a less water equivalent in proportion to their size than M. Violle finds for his, and this led me to search critically in his writings for some statement as to the size of his bulbs and the mass of mercury and glass in them. He says in a footnote: "M, the water equivalent, was carefully determined. It was measured indirectly by experiments on cooling, and directly by a thermometer in every way resembling that which had been used, broken at the stem. In both cases M was found equal to 0.222 grammes," and he found $\frac{M}{S}$ 0.365.

We give M. Violle's own remarks:

"Soit, a un instant donné, U la vitesse d'échauffement du thermomètre pour l'excès actuel θ, U' la vitesse de refroidissement que l'on observerait a ce même excès θ si l'on interceptait l'action de la source ; la somme $U+U'$ représente effectivement la vitesse d'échauffement du thermomètre corrigée du refroidissement qui correspond a la même température ; elle exprime donc l'action constante du Soleil dégagée des effets du refroidissement. Par conséquent, si après avoir observé pendant quelques minutes l'échauffement du thermomètre exposé a la radiation solaire on intercepte cette radiation, et qu'on observe alors le refroidissement, on trouvera que, en effet, a chaque valeur de θ répond une même valeur constante de $U+U'$, quoique séparément U et U' changent avec θ. En multipliant cette somme constante par la valeur en eau M de la portion du thermomètre qui s'échauffe et en la divisant par la section de cette boule, on aura la quantité de chaleur solaire qui tombe en une minute sur 1 centimètre carré de surface normale aux rayons, c'est-à-dire la *mesure absolue* de la quantité de chaleur reçue par notre globe au point et a l'instant considérés.

Voici, par exemple, les observations faites dans la matinée du 16 août 1875 au sommet du Mont-Blanc (les temps t sont comptés a partir de l'ouverture ou de la fermeture de tube d'admission) :

Temps t.	Échauffements θ.	Refroidissements θ'.	$\theta + \theta'$.
m.			
0	0.0	18.0	18.0
5	14.9	3.0	17.9
10	17.6	0.6	18.2
15	17.9	0.1	18.0
20	18.0	0.0	18.0

La somme $\theta + \theta'$ est constante et égale a θ_0 ; l'échauffement et le refroidissement se sont faits avec la même vitesse. Les températures d'échauffement sont données par la formule

$$\theta = \theta_0 (1 - e^{-at}) = 18 (1 - e^{-0.30t})$$

et les températures de refroidissement par la formule

$$\theta' = \theta_0 e^{-at} = 18 e^{-0.30t}$$

Les vitesses d'échauffement et de refroidissement sont donc respectivement

$$U = m\theta_0 \quad a\theta_0 = 0.552 \quad 0.30\theta$$

et

$$U' = m\theta_0 \quad 0.30\theta.$$

d'où il suit que, pour une même température $\theta = \theta_0$, on aura

$$U + U' = m\theta_0 \quad 0.552.$$

Un avantage précieux de la méthode est l'emploi de toutes les données au calcul de la quantité cherchée : il y a là un contrôle rigoureux dont l'importance ne saurait échapper.

La somme $U + U'$ représente l'action du Soleil ; multipliée par la valeur en eau M de la portion du thermomètre qui s'échauffe (1) et divisée par la surface d'un grand cercle, elle constituera l'expression numérique de la quantité absolue de chaleur I reçue a 10h 22m du matin, le 16 août 1875, au sommet du Mont-Blanc, $I = 2.332$ (unité de chaleur étant la quantité de chaleur nécessaire pour élever de 1 degré la température de 1 gramme d'eau.

(1) M avait été soigneusement déterminée ; on l'avait mesurée indirectement par les expériences de refroidissement, et directement sur un thermomètre tout semblable a celui qui avait servi, rompu a la naissance de la tige ; or on a trouvé dans les deux cas $M = 0.222$ et l'on avait $\frac{M}{S} = 0.365$."

The above, given as a footnote (1), contains all the information he affords us about the determination of the fundamental value on which his solar constant rests. S (the area of a great circle of his thermometer bulb) we see $\frac{222}{365}$ 0.608, whence we find the diameter of his bulb

$$D = 2 \sqrt{\frac{.608}{\pi}} \quad 0.88 \text{ cm}$$

the volume of such a sphere being

$$\frac{(0.88)^3 \times 7}{6} = 0.357 \text{ c. cm}$$

If the bulb was spheroidal rather than spherical the result is sensibly the same, and we see on what a very minute quantity (about one-third of a cubic centimeter) the final determination reposes, and how easily a relatively large error might be made in the determination of so small a quantity.

0.357 cubic centimeter, then, is the volume of M. Violle's bulb, and the water equivalent of such a sphere, were it composed entirely of mercury at a specific gravity of 13.6 and specific heat of .0333, as determined by Regnault, would be but .162 instead of .222. The glass in these thermometer bulbs is of the thinnest description, but were the thermometer bulb of solid glass with a specific gravity of 3 and of a higher specific heat than any which we can find assigned to glass by Dulong and Petit or by Regnault, the water equivalent would still be less than the .222 assigned to it by M. Violle, which apparently can be represented by no combination of glass and mercury. It would appear, since M. Violle's value of the solar constant (2.54) is proportional to this seemingly inadmissible water equivalent, that it is to be desired that we should have an explanation of these apparent discrepancies.

Had it been known when we commenced our observations that the initial rate was smaller than that given by M. Violle's formula, we should have saved ourselves great labor and attained greater accuracy by direct observations of this initial rate. As we acquired this knowledge only by experience, when it was too late to alter the method of observation already commenced, we continued taking minute readings on all three actinometers, and have subsequently applied corrections to them by means of careful determinations given hereafter. The three actinometers are thus discriminated. The large actinometer with 30.5 cm. globe is called No. 1. It is supplied with two thermometers by Baudin of Paris, divided to fifths of degrees, and read by estimation to fiftieths. The other thermometers are divided to tenths of degrees and read by estimation to hundredths. "Baudin 8739," the water thermometer, has a bulb 0.991 cm. in diameter. The bulb of "Baudin 8737," the sun thermometer, is 0.963 cm. in diameter. The water equivalent of this thermometer will be given later.

Actinometer No. 2, having a globe 15 cm. in diameter, is mounted so as to have an altazimuth motion, and is supplied with a water thermometer by Grunow, the diameter of whose bulb is 0.836 cm, and with a sun thermometer by Green, No. 4571, having a bulb diameter of 1.247 cm.

The third actinometer is of the same size and mounting as No. 2. In it have been used a water thermometer by Grunow, similar to that for No. 2, and a sun thermometer by Green, No 4572, whose bulb had a diameter of 1.207 cm. In all our observations the water has been constantly agitated. We at first did this by forcing the water in at one tube and out at the other, so as to keep up a uniform flow and temperature, but substituted for this another method of agitating the water in the actinometer itself.

In the following tables we have used the same symbols as M. Violle, for the reader's convenience. Thus, θ_0 indicates the final temperature of the sun thermometer in excess of its surroundings. m is a constant corresponding to the reciprocal of the subtangent of M. Violle's assumed logarithmic curve; accordingly $m \theta_0$ is the initial rate of warming of the particular thermometer in question when the sun's rays fall upon it. If we suppose this rate to be exact and to continue unchanged for a minute, we evidently have the effect of the solar radiation for one minute, on the supposition that all the heat has been retained by the instrument and none of it dissipated through radiation, through air currents, or in any other way. To this latter part of the assumption we may safely agree. From this rate per minute, the area of the exposed surface in centimeters and the water equivalent of the heated bulb in grammes, we can evidently obtain the solar radiation in calories, referred to the gramme, centimeter and minute. We give in tables 56 and 57 an example of an ordinary observation in detail.

[footnote] *I. e., the final temperature corresponding to complete equilibrium of the heat received from the sun with that radiated from the bulb.

TABLE 56.

Date, August 25, 1881.   Station, Mountain Camp, Mount Whitney, California.   Actinometer No. 2, aperture, medium.   Sun thermometer.
Green 672.   Water thermometer, Green 1.   Observer, J. J. N.   Local time used.   Condition of sky clear.   Wind light.

Time.	Thermometers		Adjustment.	Remarks
	Water.	Sun.		
h m				
11 30 a m	20 10	20 15	Shade	
11 31 a m		21 90	Sun	
11 32 a m		20 80	.. do	
11 33 a m		29 51	.. do	
11 34 a m	20 51	31 10	do	Agitated water
11 35 a m		32 50	do	
11 36 a m		33 53	do	
11 37 a m		34 42	do	
11 38 a m		34 91	do	
11 39 a m	20 15	35 35	do	Do
11 40 a m		35 74	.. do	
11 41 a m		35 95	.. do	
11 42 a m		36 15	do	
11 43 a m	20 02	36 27	.. do	Do
11 44 a m		36 43	do	
11 45 a m		36 50	do	
11 46 a m		3 22	Shade	
11 47 a m	20 00	36 95	.. do	Do.
11 48 a m		28 21	.. do	
11 49 a m		26 63	.. do	
12 50 a m		25 55	.. do	
11 51 a m	20 18	24 36	do	Do
11 52 a m		23 54	.. do	
11 53 a m		23 03	.. do	
11 54 a m		22 42	do	
11 55 a m	20 18	22 00	.. do	Do.
11 56 a m		21 68	do	
11 57 a m		21 55	.. do	Fresh wind variable
11 58 a m		21 12	do	
11 59 a m	20 03	20 86	.. do	Agitated water.
12 00 noon		20 58	.. do	

TABLE 57.

Date, August 25, 1881.   Station, Mountain Camp, Mount Whitney, California.   Actinometer No. 2, aperture, medium.   Sun thermometer,
Green 672.   Water thermometer, Green 1.   Observer, J. J. N.   Local time used.   Condition of sky clear.   Wind light.

Time.	Thermometers		Adjustment.	Remarks
	Water.	Sun.		
h m				
12 00 noon	20 05	20 58	Shade	
12 01 p m		24 11	Sun	
12 02 p m		27 07	.. do	
12 03 p m		29 55	do	
12 04 p m	20 10	31 10	.. do	Agitated water
12 05 p m		32 40	do	
12 06 p m		33 45	.. do	
12 07 p m		34 21	do	Light wind.
12 08 p m	20 08	34 85	do	Agitated water
12 09 p m		35 30	do	
12 10 p m		35 68	.. do	
12 11 p m		35 92	do	
12 12 p m		36 08	.. do	
12 13 p m	20 00	36 27	.. do	Do.
12 14 p m		36 55	.. do	
12 15 p m		36 48	.. do	
12 16 p m		33 42	Shade	
12 17 p m		30 50	do	
12 18 p m	20 00	28 37	.. do	Do
12 19 p m		26 70	.. do	
12 20 p m		25 45	.. do	
12 21 p m		24 56	.. do	
12 22 p m		24 55	.. do	
12 23 p m	20 03	23 87	.. do	Do
12 24 p m		22 21	do	
12 25 p m		21 85	do	
12 26 p m		21 60	do	Fresh wind, variable.
12 27 p m	20 07	21 52	do	Agitated water
12 28 p m		21 08	do	
12 29 p m		20 90	.. do	
12 30 p m	20 00	20 75	.. do	Fresh wind variable

After correcting the very minute accidental irregularities of the *water* thermometer above given by a smooth curve, we obtain the following table, where the temperatures of excess of the *sun* thermometer are those directly representing the difference between the latter and the temperature of the inclosed water-shell, to whose lamp-blacked copper surface the solar thermometer radiates.

TABLE 58.

Time, August 25, 1st, A. M.	Adopted water thermometer	Excess of sun thermometer	Time August 25, 1st, P. M.	Adopted water thermometer	Excess of sun thermometer
h. m.			*h. m.*		
11 30	20 13	0 05	12 40	20 10	0 72
11 31	....	3 81	12 41	....	1 63
11 32	....	6 30	12 42	....	7 01
11 33	....	9 21	12 43	....	9 32
11 34	....	11 00	12 44	....	11 04
11 35	....	12 40	12 45	....	12 45
11 36	....	13 41	12 46	....	13 29
11 37	....	14 22	12 47	....	14 18
11 38	20 19	14 82	12 48	20 05	14 80
11 39	....	15 31	12 49	....	15 25
11 40	....	15 61	12 50	....	15 65
11 41	....	15 88	12 51	....	15 87
11 42	....	16 04	12 52	....	16 03
11 43	....	16 18	12 53	....	16 22
11 44	....	16 31	12 54	....	16 33
11 45	20 08	16 12	12 55	20 04	16 45
11 46	....	17 14	12 56	....	13 28
11 47	....	10 28	12 57	....	10 46
11 48	....	8 30	12 58	....	8 33
11 49	....	6 55	12 59	....	6 66
11 50	....	5 30	13 00	....	5 39
11 51	....	4 38	13 01	....	4 32
11 52	....	3 46	13 02	....	3 51
11 53	20 07	2 88	13 03	20 03	2 81
11 54	....	2 35	13 04	....	2 30
11 55	....	1 90	13 05	....	1 90
11 56	....	1 61	13 06	....	1 57
11 57	....	1 26	13 07	....	1 29
11 58	....	1 03	13 08	....	1 05
11 59	....	0 80	13 09	....	0 87
12 00	20 06	0 72	13 10	20 02	0 71

M. Violle's formula may be written, $mt \log e = \log \theta_0 - \log \theta'$, where θ_0 is the sum of the corresponding ordinates in the heating and cooling curves (which he assumes to be identical curves, though referred to different axes); θ_0, then, should be a constant and proportional to the final excess. We do not find it to be so in our own practice; and as the best we can do, we always determine it from the mean of eight observations taken on four corresponding points of the heating and cooling curves; m is the reciprocal of the subtangent, on the assumption that the curve is a logarithmic one, e is the Napierian base. It is our uniform experience that m, which should be on this assumption a constant, varies, and varies systematically, according as it is determined by the comparison of θ_0 with an ordinate which represents the excess near the commencement of exposure or later. We have scarcely met an exception in scores of examples, of which the following, cited at random, may be considered typical. We have just given the original observations, with readings taken, like all those on the expedition, from minute to minute (though to give all the observations in such detail here would be impossible in the space at command).

We now proceed to give the reduction by M. Violle's method, applied successively to the fifth, the tenth, and the fifteenth minute of heating or cooling.

TABLE 59.

(Transcript from original reduction-book.)

[Date, August 25, 1881. Actinometer No 3. Station, Mountain Camp. Aperture, medium. State of sky, clear. Sun thermometer, "Green 1872. Wind, light. Water thermometer, "Grunow I. Barometer, 5″s 02. Observer, J. J. N.].

Time	Water thermometer	Sun thermometer	Difference	Exposure
h. m.				
11 30 a. m.	20 10	20 15	0 05	Shade.
11 35 a m.	20 10	32 50	12 40	Sun.
11 40 a m.	20 09	35 72	15 63	Sun.
11 45 a m.	20 08	36 50	16 42	Sun.
11 50 a m.	20 08	25 38	5 30	Shade.
11 55 a m.	20 07	22 00	1 93	Shade.
12 00 m.	20 06	20 78	0 72	Shade.

Mean value of m from 3 points of curve $= .222$

What we see of m in this particular example we see in every other, that under the circumstances of actual observation it is not a constant either in the heating or the cooling curve, and that apart from little irregularities of observation, it is largest when taken near the first moment of heating or of cooling. The mean of the three values just given is $m = .222$; hence the initial rate derived in this particular instance for $m\,\theta_0 = .222 \times 17.22 = 3.82$, a value which must be too small. We have, from the fact that the observations were originally taken by M. Violle's method, been led to reduce them by his formula, and to afterward introduce a correction. Wholly independent observations taken by such means as to show by direct experiment the initial rate within very narrow limits, give a confirmatory result, and these will be found more particularly mentioned under the head of "Correction A."

When the globe is entirely closed to extraneous radiations from the sun or any other source, and the thermometer is heated only by radiation from the walls of its inclosure, m is much more nearly constant, as we see by the experiments detailed in the Appendix.

CHAPTER VI.

DETERMINATION OF WATER EQUIVALENTS OF THERMOMETER BULBS.

Since all measurements of solar radiation by the globe actinometers depend upon the thermic capacity or water equivalent of the bulb of the thermometer employed (a constant on which the value of the *solar constant* immediately depends), it becomes necessary to determine this with all possible precision. The liability to relatively large error in the determination of such minute quantities is great, and this liability must be admitted to be a serious objection to an otherwise excellent instrument. If the determination is made by ordinary calorimetrical processes (which are usually ill adapted to this special case), we ought to check them by some wholly independent means.

Two such methods have been devised by the writer for finding the separate water equivalents of the mercury and glass in the thermometer bulb without destroying the instrument.

In each of these two methods the weights of mercury and of glass in the thermometer bulb are indirectly determined by balancing the entire thermometer upon a fulcrum applied at selected points on the stem. We have used these results only as a check on the adopted value, which is that derived from the third method, where the joint specific heat of mercury and glass in the bulb has been directly measured in a small calorimeter. In view of the great importance of this constant, we give a description of the application of each of these methods in detail.

FIRST METHOD.

Determination of the water equivalent of the bulb of thermometer ("Baudin 8737") by measuring its dimensions and balancing it on a fulcrum in air.

The external dimensions of the stem are first measured, and by these and its calibration[1] the center of gravity of the stem alone is determined. If this point be now made the fulcrum, it is obvious that the glass stem in no way affects the balancing of the bulb and the mercury in the stem, which may be done by a known weight at a known distance.

Then let the following data be obtained:

a = total weight of thermometer.
b = external volume of thermometer.
g = external volume of bulb.
m = specific gravity of mercury.
k = distance from center of bulb to fulcrum.
l = distance from center of mercury in stem to fulcrum.
u = product of known balancing weight by its distance from fulcrum.
x = total weight of mercury.
cx = volume of mercury in stem (determined from coefficient of expansion of mercury).
dcx = volume of empty part of stem (determined from coefficient of expansion of mercury).
cmx = weight of mercury in stem.

---

[1] Its calibration, with sufficient accuracy for this purpose, presupposes a preliminary approximate knowledge only of the quantity of mercury, as the whole internal capacity of the stem is excessively small. We can indeed so form our equations as to dispense even with this approximate knowledge.

7*

For brevity let—

$$f = \frac{1}{m} - c \qquad\qquad h = \frac{1}{m} + d$$

The volume of all the mercury being $\frac{r}{m}$, the volume of mercury in the bulb $= \frac{r}{m} - cr = fr$, the weight of mercury in the bulb $= (1 - mc)r$, the volume of glass in the bulb $= g - fr$, the total volume of glass in the whole thermometer $= h - \frac{r}{m} - dx = b - hr$.

Then, since the total weight of all the glass $= a - x$, the weight of glass in the bulb $= \frac{(a - x) \times (g - fx)}{b - hx}$ if the specific gravity of the glass in the bulb be assumed to be the same as that in the stem, and the moment of the whole bulb is $k \left\{ \frac{(a - x) \times g - fx}{b - hx} + 1 - mc)x \right\}$; also the moment of the mercury in the stem $= lmcx$, and the sum of these two values $= a$; whence, by solving the quadratic for x, the weight of all the mercury is determined; and hence, assuming a value for the specific gravity of the glass in the bulb, the separate weights of mercury and glass in the bulb may be obtained. The specific heat of mercury is already known with great accuracy, and that of the glass can be either taken from tables or determined with close approximation from the instrument itself. The product of the weight of the mercury by its specific heat, plus the product of the weight of the glass by *its* specific heat, must be very nearly equal to the effective water equivalent of the bulb. It would be exactly equal to this were it not for a certain amount of heat transferred to or from the bulb by conduction along the stem, which cannot be easily determined. It is small, but not negligible, and without it our results must be below the truth. It is here at first neglected. A numerical example, showing the results actually obtained, is now given. (The measures were taken with a vernier calliper.)

Dimensions of thermometer " Baudin =747."

Mean diameter of spherical bulb	tc .2683
Stem graduated in C. from 20° to + 71 C.; 0 is 13.0 cm. from junction of stem and bulb.	
Length of cylindrical part of stem	25 .591
Average diameter of cylindrical part of stem	0 .140
Length of b	0 .226
Volume of b of bore	0 .0005
Volume of b of stem	0 .0589
Volume of tapering portion of stem at bulb end	0 .0091
Volume of tapering portion of stem at ring end	0 .0041
Volume of ring	0 .0218
Specific gravity of the glass (approximate)	2.85

The geometrical center of figure of the stem (exclusive of bulb) was found by calculation from these measurements to fall at $+10^\circ .1$ on the stem. The thermometer was accordingly balanced on a fulcrum at this point by a weight whose moment a equal to that of the mercury and bulb was found to be 80,151 gramme-centimeters. The values of all the symbols employed in the computation are as follows:

$$a = 18.775 \qquad b = 5.255 \qquad c = \frac{59.6}{5550 \times 13.6} = .000790 \qquad d = \frac{134.0 - 59.6}{5550 \times 13.6} = .000986$$

$$f = \frac{1}{13.6} - c = .072739 \qquad g = 0.4767 \qquad h = \frac{1}{13.6} + d = .074515 \qquad k = 15.884$$

$$l = 8.511 \qquad m = 13.6 \qquad a = 80,151 \text{ (temperature during experiment} = +2^\circ \text{ C).}$$

From the equation

$$k \left\{ \frac{(a - x) \times (g - fx)}{b - hx} + (l - mc)x \right\} + lmcx = a$$

we obtain, by solution of the quadratic for x,

$$x = + \sqrt{\frac{bn - agk}{chkm - hk + fk - chlm} + \frac{A^2}{4}} - \frac{A}{2}$$

where for brevity

$$A = \frac{bclm + hn + hk - (bckm + afk + gk)}{chlm + fk - (chlm + hk)}$$

Substituting numbers for symbols, we obtain the total weight of mercury in the thermometer, $x = 4.70$ grammes.

With this value the approximate water equivalent of the bulb may be found thus:

At 30° C. the weight of mercury in stem $= 0.06$ grammes, leaving weight of mercury in bulb $= 4.64$ grammes.

As the density of mercury at 30 C. is about 13.54, the volume of 4.64 grammes of mercury $= \frac{4.64}{13.54} = 0.3427$ cc. The external volume of bulb (assuming it to be a sphere of 0.4815 c. m. radius) is $\frac{4}{3} \pi \times (.1815)^3 = 0.4688$ cc.

The difference of these volumes gives the volume of glass in bulb $= .4688 - .3427 = 0.1261$ c. c.; whence weight of glass in bulb $= .1261 \times 2.83 = 0.3569$ grammes.

Taking the specific heat of mercury $= .0333$, and the specific heat of glass $= .198$, we have:

Water equivalent of mercury .... $4.64 \times .0333 = 0.1545$
Water equivalent of glass .... $0.3569 \times .198 = 0.0707$

Total water equivalent of bulb 0.2252

By a similar application of method I to thermometer "Green 4571", the following results were obtained:

Grammes.
Weight of mercury in bulb 9.80
Weight of glass in bulb 0.890
Water equivalent of bulb 0.1885

SECOND METHOD.

*Determination of the water equivalent of the bulb of thermometer "Green 4571," by balancing at several points on the stem, both in air and in water.**

In this method we seek to discriminate between the weights of mercury and glass in the thermometer by the difference in the buoyant effect of water on them. The principle of the method may be illustrated by the following example: Suppose that we knew the exact center of gravity of the glass in the thermometer, and made this point the fulcrum; suppose, also, that we balanced the mercury in the bulb and stem by a rider, of the same density as mercury, placed at a suitable distance from the fulcrum upon the part of the stem opposite to the bulb. Under these circumstances the whole apparatus might be immersed in water and the balance would remain undisturbed. Next, suppose the fulcrum situated between the center of gravity of the glass and the bulb of the thermometer. In this position the mercury would be balanced by the combined weight of the rider and a portion of the glass stem at the extremity opposite to the bulb. (This last is only approximately determinate.) If now the apparatus were immersed in water, the counterbalancing portion of the stem, being of *smaller* specific gravity than the rider, would suffer a greater proportionate diminution of weight than the rider or mercury; the distance of the counterpoise would therefore have to be increased in order to restore the equilibrium; and hence if we were to neglect the action of the balancing part of the stem, the calculated weight of mercury would appear to be *greater* from the water experiment than from that in air. On the other hand, if the fulcrum were placed too near the ring end of the thermometer, the apparent weight of mercury would be *less* from the water experiment than from that in air. The mode of procedure suggested by these considerations is to shift the fulcrum and determine experimentally the point at

---

* The writer proposed this method, but all the present details of its application are due to Mr. F. W. Very.

which the mercury is balanced by the same weight in water as in air. The moment of this weight is then equal to the moment of the mercury. In what goes before we have supposed the counterpoise to be of the same density as the mercury ; but this is not necessary, provided its density be known. It is desirable, however, to use a dense metal rider, since the greater the difference between the specific gravity of the counterpoise and the glass, the greater will be the difference between the water and the air values for the same displacement of the fulcrum.

Let W = total weight of mercury.

w_1 = weight of mercury in the bulb.

w_2 = weight of mercury in the stem.

w_3 = weight of rider.

l_1 = lever arm of mercury in the bulb.

l_2 = lever arm of mercury in the stem.

l = lever arm of rider in air.

l_4 lever arm of rider in water.

l_3 = lever arm of empty and uncounterbalanced part of bore.

r = volume of empty and uncounterbalanced part of bore.

c = apparent expansion of mercury for 1 centigrade degree.

a = length of mercury column in centigrade degrees.

S = specific gravity of mercury.

s = specific gravity of rider.

Then if the experiment be performed in air we have

$$(1)\qquad w_1\,l_1 + w_2\,l_2 = w_3\,l_3$$

and

$$(2)\qquad w_2 = ac\,W$$

whence

$$(3)\qquad W = w_1 + w_2 = \frac{w_3\,l}{l_1 + ac\,(l_2 - l_1)}$$

If the weighing be made under water,

$$(4)\qquad W = \frac{w_3\,l_4 \times \left(1 - \frac{1}{s}\right) - r\,l_3}{(l_1 + ac\,l_2 - ac\,l_1) \times \left(1 - \frac{1}{S}\right)}$$

When the fulcrum is properly placed (3) is equal to (4).

In the actual performance of the experiment, the stem of the thermometer, placed horizontally, was firmly attached by fine copper wire to the beam of a balance from which the scale pans had been removed. Any degree of the stem could be brought opposite to the index of the beam, and the equilibrium could be restored by a metal rider sliding along the graduated thermometer stem, whose divisions served to measure its distance from the fulcrum.

As any marked inequality in the *density* of the balance arms, though of very little importance in weighings conducted in air, would have seriously affected those made under water, the beam was tested for such inequality, which was found to be altogether negligible.

Dimensions of thermometer "Green 1571":

The bulb and 13 mm. of the stem of this thermometer are blackened. The diameter of the stem is but slightly diminished where it joins the bulb.

Mean diameter of spherical bulb	$1^{cm}.247$.
Stem graduated in ½° C. from − 10° to + 60° C.	
10° is 7.5^{cm}. from junction of stem and bulb.	
Length of cylindrical part of stem	$35^{cm}.10$.
Average diameter of cylindrical part of stem	$0^{cm}.506$.
Length of 1	$0^{cm}.182$.
Volume of 1 of bore	$0^{cm}.00012$.
Volume of 1 of stem	$0^{cm}.13436$.
Specific gravity of the glass (approximately)	2.7.

The weight of the rider used was $w_3 = 21.03$ grammes, and its density, $s = 9.065$. The density of mercury at 10° C. was taken to be $S = 13.57$. Hence $1 - \frac{1}{s} = 0.88969$, and $1 - \frac{1}{S} = 0.92631$.

The following are the results of experiments with various positions of the fulcrum. W is calculated from formulæ (3) and (4), in which the influence of the glass is neglected. The apparent expansion of mercury is taken to be $c = \frac{1}{5450}$. The correction (rl_s) for buoyancy in water of the empty and unbalanced part of bore is here negligible.

TABLE 60.

Fulcrum at —	h in centimeters	Weighing in air. Temperature = 24.4 C., σ = 30.2.			Weighing in water. Temperature = 19.0 C., σ = 15.8.		
		w, h in gramme centimeters	l in centimeters	W. in grammes	$w, h, l, \left(\frac{1}{s}\right)$ in gramme centimeters	l_s in centimeters	W. in grammes
28.0	16.92	158.39	9.06	9.338	152.68	12.18	9.724
28.5	17.16	167.54	9.30	9.738	159.01	12.72	9.981
29.0	17.40	176.69	9.54	10.128	165.10	12.96	10.213
29.5	17.64	185.84	9.78	10.509	171.76	13.20	10.492
30.0	17.88	194.99	10.02	10.878	178.12	13.44	10.775
30.5	18.12	204.14	10.26	11.287	184.49	13.68	10.971
31.0	18.36	213.29	10.50	11.586	190.85	13.92	11.200
31.5	18.60	222.44	10.74	11.937	197.21	14.16	11.421
32.0	18.84	231.59	10.98	12.259	203.57	14.40	11.613
32.5	19.08	240.74	11.22	12.583	209.94	14.64	11.855

Fig. 7.

Determination of Mercury in Thermometer, Green, No. 4571.

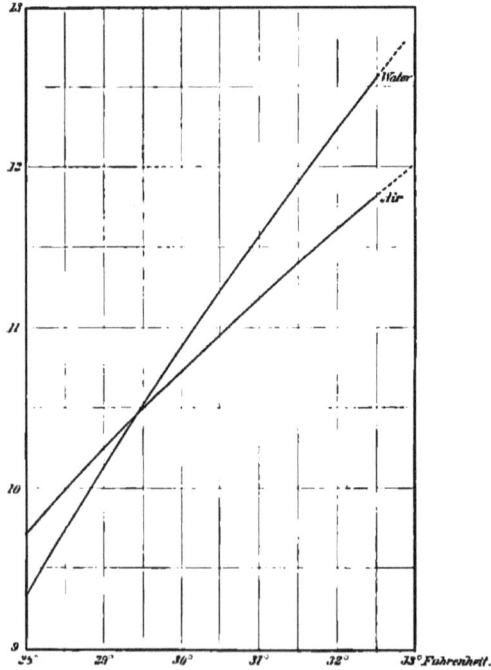

Using the numbers in the fifth and eighth columns for graphical construction, we obtain two slightly curved lines (see Fig. 7), which intersect at the point whose ordinate is 10.45 grammes, falling opposite the point on the stem registering 29°.12, which is the true center of gravity of the glass portion of the thermometer; and 10.45 grammes (which is the weight which would have been obtained either in air or in water if the fulcrum had been placed at 29°.42) is therefore the true weight of the mercury.

From this we calculate the water equivalent of the bulb. At 30° C. the weight of mercury in stem = 0.06 grammes; leaving the weight of mercury in bulb = 10.39 grammes, occupying at 30° C. a volume of 0.7672 cc.

The external volume of the bulb (assuming it to be a sphere of 0.6235 cm. radius) is $\frac{4}{3} \pi \times$ $(.6235)^3 = 1.0184$ cc. The difference of these volumes gives the volume of glass in the bulb $1.0184 - .7672 = 0.2512$ cc., and its weight $= .2512 \times 2.78 = 0.6983$ grammes, whence:

Water equivalent of mercury.. $10.39 \times .0333 = 0.3460$
Water equivalent of glass.. $0.6983 \times .198 = 0.1383$

Total water equivalent of bulb... 0.4443

By a similar application of method II to thermometer " Baudin 8737 " the following results were obtained :

Grammes.
Weight of mercury in bulb.. 4.87
Weight of glass in bulb.. 0.3314
Water equivalent of bulb.. 0.2278

THIRD METHOD.

Determination of the water equivalents of the bulbs of thermometers "Baudin 8737" and "Green 4571" by direct measurements of specific heat.

A cylindrical cup, slightly larger than the bulb of the thermometer, was made of very thin, highly polished steel and attached by a fine stem to an ebonite base, the whole being inclosed in a cylindrical case of reflecting metal. This cup, containing either mercury or water, constituted the calorimeter.

The thermometer was first heated to about 60° C. Its bulb was then held directly over the calorimeter, into which it was plunged at the instant it had cooled to a certain recorded degree. Under these conditions the temperature of the thermometer falls, at first rapidly, becoming nearly stationary in about one half minute. This stationary point was assumed to be the temperature of the mixture.

Calling W = weight of the thermometer bulb and small part of stem.

x = specific heat of thermometer bulb and small part of stem.

ω = weight of steel cup.

r = specific heat of steel cup.

w = weight of liquid in cup.

y = specific heat of liquid in cup.

T = original temperature of thermometer at immersion.

θ = maximum temperature of mixture after immersion.

t = temperature of liquid and cup before immersion.

the thermometer bulb, cooling through $(T-\theta)$, loses $Wx \times (T-\theta)$ calories, and the cup and contents gain $(\omega r + wy) \times (\theta-t)$ calories. Neglecting provisionally all losses or gains by radiation or convection, we may equate these expressions:

$$Wx (T-\theta) = (\omega r + wy)(\theta-t)$$

The water equivalent of the thermometer bulb is then

$$Wx = \frac{(\omega r + wy)(\theta-t)}{T-\theta}$$

We give an actual experiment in full:

Thermometer " Baudin 8737 "; 19.05 grammes of mercury being used in a steel cup weighing 2.51 grammes, the thermometer fell upon immersion from 58°.0 C. to 28°.8, and the cup rose from 20°.85 to 28°.8; whence the water equivalent of bulb=

$$\frac{[(2.51 \times .117) + (19.05 \times .033)] \times 7.95}{29.2} = 0.2511$$

The following results were obtained ; using mercury (mean of 3 experiments), water equivalent= .2546, using water (mean of 2 experiments), water equivalent=.2522. From all experiments with calorimeter Wx=.2536.

The experiments were varied in several ways, beside the substitution of mercury for water. In some of the trials the bulb alone was dipped; in others, a portion of the slender glass stem (about 3 mm.) was also included. Losses or gains by radiation and convection were confined within known limits by insuring that the initial temperature of the calorimeter was in some cases that of the place of experiment; in others, that it was cooler, so that the final temperature should be that of the room. Thus the calorimeter and its contents were in some instances radiating heat to the inclosure, while in others they were receiving heat from the inclosure; or, in other words, part of the results just cited were designedly too small, and others too large. Hence we may be sure that the true result lies between certain narrow limits. What these limits are may be seen from the following table, which shows the variations produced by these changes:

TABLE 61.

Liquid used in calorimeter.	Stem near bulb. (Whether included or not.)	Relative temperature of inclosure.	Water equivalent.
			Cal.
Mercury	Heat received from stem	Cool calorimeter radiating heat	0.241
	Heat not received from stem	Cool calorimeter radiating heat	0.2515
	Heat received from stem	Warm calorimeter receiving heat	0.261
Water	Heat received from stem	Warm calorimeter receiving heat ..	0.2346
	Heat not received from stem ...	Cool calorimeter radiating heat	0.2597

The value 0.2536 cal. (mean of the above 5) has been adopted as the water equivalent of the bulb of "Baudin 8737."

By similar calorimetrical experiments the water equivalent of "Green 4571" has been found to be 0.1971 cal. This value is also adopted.

DETERMINATION OF THE WATER EQUIVALENT OF THERMOMETER "GREEN 4572."

This thermometer, which was one of those used in actinometric measurements on Mount Whitney, was, after the return of the expedition, broken in transit to New Haven, where it was being sent for rating. Three methods, however, remained open for determining its constants.

First. The bulb having fortunately remained, its dimensions were carefully measured and its mercury contents weighed at the Yale Horological Bureau, so that this determination of its weight may be considered at least as accurate as any, though it was obtained at the sacrifice of the instrument.

Second. The equatorial diameter of the bulb of "Green 4571" as measured by a vernier caliper at Allegheny was 1.247 cm., and that of "Green 4572" was 1.207 cm. These thermometers, made at the same time by the same maker as near alike as possible, may be safely assumed to have their water equivalents nearly in the ratio of the volumes of the bulbs, and if we multiply 0.1971 (the adopted water equivalent of "Green 4571") by $\frac{(1.207)^3}{(1.247)^3}$, we get .1808 for the *approximate* water equivalent of "Green 4572." We prefer to treat these two values, however, thus obtained as check values, and to rely for the adopted one on the reduction to the calorimetric standard.

Third. A considerable number of simultaneous actinometer readings having been taken with the thermometer "Green 4572," and with the very similar "Green 4571," it was possible to reduce the measures made with the broken thermometer to the standard of those made with the unbroken one.

The diameter of the bulb of "Green 4572" was 1.207 cm., and the weight of mercury contained in it was 10.60 grammes; whence we calculate:

	cc.
Volume of bulb..	0.9205
Volume of mercury..	0.7794
Volume of glass..	0.1414
	Grammes.
Weight of glass	2.8 × .1414 = 0.3958
Water equivalent of glass	.198 × .3958 = 0.0784
Water equivalent of mercury	.0333 × 10.60 = 0.3530
Total water equivalent of bulb..	0.4314

For the comparison of the thermometers "Green 4571" and "Green 4572," which were used in actinometers No. 2 and No. 3 respectively, we have the following initial rates determined during synchronous readings of the instruments:

TABLE 62.

(Station Lone Pine. Observers, Sergeant, Dobbins and Nancy.)

Initial rate of heating.

Date	Local time.	Actinometer No. 2, thermometer "G. 151.	Actinometer No. 3, thermometer "G. 572.
	$h. \ m.$ $h. \ m$		
1891.			
Aug. 3	11 55 to 12 05	3 611 C.	3 717 C.
3	12 05 to 12 55	3 641	3 751
3	0 55 to 1 50	2 945	3 048
3	1 50 to 3 00	2 901	2 915
4	7 00 to 7 50	3 951	4 127
4	7 50 to 8 00	4 274	4 500
4	11 30 to 12 00	4 361	4 174
4	12 00 to 12 50	3 427	3 431
4	4 00 to 4 50	2 973	3 419
4	4 50 to 5 00	2 911	3 883
5	7 50 to 7 50	3 129	2 946
5	7 50 to 8 00	3 975	3 732
5	11 30 to 12 00	4 573	3 707
5	12 00 to 12 50	7 636	3 599
5	4 00 to 4 50	2 572	2 787
5	4 50 to 5 00	2 475	2 586
	Average initial rate	3 186	3 239

Reduction of "Green 1572" to calorimetric standard.

Calling $(m \ \theta_0)_2$ the initial rate of heating of exposed thermometer in No. 2,
$(m \ \theta_0)_3$ the initial rate of heating of exposed thermometer in No. 3,
M_2 the water equivalent of bulb of exposed thermometer in No. 2,
M_3 the water equivalent of bulb of exposed thermometer in No. 3,
S_2 the area of medium section of bulb in No. 2,
S_3 the area of medium section of bulb in No. 3,
R_2 the solar radiation in calories in No. 2,
R_3 the solar radiation in calories in No. 3,

we have

$$R_2 = (m \ \theta_0)_2 \times \frac{M_2}{S_2}$$

$$R_3 = (m \ \theta_0)_3 \times \frac{M_3}{S_3}$$

in which, for synchronous exposures,

$$R_2 = R_3,$$

whence

$$M_3 = \frac{(m \ \theta_0)_2 \ S_3 \ M_2}{(m \ \theta_0)_3 \ S_2}$$

In the present case, $(m \ \theta_0)_2 = 3.186$,
$(m \ \theta_0)_3 = 3.239$,
$S_2 = 1.221$ sq. cm.,
$S_3 = 1.111$ sq. cm.,
$M_2 = 0.4971$, and
$M_3 = $ water equivalent of bulb of "Green 1572,"
$= 0.4580$, which is the value finally adopted.

It will be seen that the water equivalents of thermometer bulbs, determined by direct calorimetrical observations, are in every case greater than those inferred from weighings. The difference may perhaps be taken as a rough measure of the amount of heat conducted by the stem, which will vary, owing to the differences in the form and blackening of the stem, already alluded to. Our direct calorimetric method has taken a partial account of this conduction in the stem, while our check methods have not taken any, and some small correction should be added to the latter to make them agree with the former. There is some doubt, however, as to how far this correction should be included in the direct value; for, while this is taken in theory to depend only on the initial rate, which is supposed to be the same whether the heat is lost by radiation or conduction, it is hard to admit that in practice it is a matter of entire indifference whether the bulb is insulated or attached to a conducting support.

The amount of heat conducted by the stem was determined by the following experiment, which, however, strictly applies only to the Green thermometers, which have not so small a neck as the Baudin. The stem of thermometer, "Green 4571," was passed transversely through a piece of rubber tubing, so that the bulb and 1.5 cm. of the stem projected beneath the tube and were protected from radiation by multiple card-board screens. A stream of hot water was then made to flow through the tube, heating the included portion of the stem. When the thermometer was inverted at intervals, so as to distribute the heated mercury in the bulb (an essential precaution), it was found that the flow of heat from an excess of 15° C. in the temperature of the stem, along a distance $= 1.5$ cm. of the stem, by its conduction caused the bulb to attain a stationary temperature of 5°.75 in excess of that of the surrounding air at the end of 12 minutes. Now, supposing (what is but approximately true) the rise to have followed the logarithmic law, we have to inquire, in order to find what the radiation was at the end of 12 minutes, what is the initial rate of cooling of this thermometer for an excess of temperature of 5°.75 above its surroundings. This rate is about 1°.25. We conclude, then, that, when the equilibrium was established, the flow was sufficient to heat the bulb 1°.25 per minute in the above experiments.

If, now, we ask what the flow would be the other way—that is, if the bulb were 15° hotter than the stem—it does not seem probable that it would be materially different; for, though the stem is a poorer conductor than the bulb, it is still able to conduct heat so as to produce this result through 1.5 cm. of glass, and it radiates freely. We conclude, then, that the bulb, which loses heat by immediate contact with the glass, will lose its heat by this conduction at the rate of 1°.25 per minute when its temperature of excess is 15°, or that the loss of heat by conduction along the stem is 8⅓ per cent.

FINAL VALUES.

Water equivalent of "Baudin 8737."

By weighing (Method I), 0.2252) Mean	0.2255
(Method II), 0.2274)	
Correction for conduction along stem, 8 per cent	.0181
Check value	0.2446
Adopted value by direct calorimetrical measurement	0.2536

(The check value is 96 per cent of the adopted value.)

Water equivalent of "Green 4571."

By weighing (Method I), 0.4885) Mean	0.4864
(Method II), 0.4843)	
Correction for conduction along stem, 8 per cent	0.0389
Check value	0.5253
Adopted value by direct calorimetrical measurements	0.4971

(The check value is 106 per cent. of the adopted value.)

Water equivalent of "Green 4572."

By direct weighing	0.4314
Correction for conduction along stem, 8 per cent	0.0345
	0.4659
By ratio of bulb volume	0.4508
Mean check value	0.4584
Adopted value by reduction to calorimetrical standard	0.4580

We have adopted, then, these values for what may be called the effective water equivalents, namely: "Baudin 8737," 0.2536; "Green 4571," 0.4971; "Green 4572," 0.4580; with the admission that, owing to difficulties inherent in the method, there can be little confidence in the figures beyond the second decimal place (and some uncertainty, perhaps, even in regard to these), while yet it must, we hope, be clear that no noteworthy error in the solar constant can be due to this cause.

CHAPTER VII.

TABLES OF RESULTS OF ACTINOMETER OBSERVATIONS.

A full example of the actual rise and fall of the thermometer from minute to minute in the heating and cooling curves has already been given for August 25 (Mountain Camp). Every one of the observations which follow, was taken with such minute readings, though the results are only here given for every fifth minute, reduced by M. Violle's method. Wherever the customary mode of reduction has been departed from in any way, the resulting numbers have been distinguished by enclosing them in parentheses. The changes are in all cases trifling, and do not require detailed explanation. We have, using his notation:

$\theta_0 =$ Temperature of final excess.

$m =$ A quantity, which he assumes to be a constant, and the reciprocal of the subtangent of a logarithmic curve.

$m\theta_0 =$ The initial rate of heating.

m, as actually determined here, however (see Chapter VIII), will require subsequent correction.

TABLE 63.

[Lone Pine August 18, 1881. Observer, W. C. Day. Actinometer No 2. Medium aperture.]

Time.	Water thermometer.	Sun thermometer.	Difference.	Exposure.	Time.	Water thermometer.	Sun thermometer.	Difference.	Exposure.
7 00ᵐ A. M.	17 .64	17 .80	0 .16	Shade.	7 30ᵐ A. M.	18 .98	19 .48	0 .50	Shade.
05	17 .80	26 .93	9 .07	Sun.	35	19 .20	28 .89	9 .69	Sun.
10	18 .08	29 .70	11 .62	Sun.	40	19 .42	31 .70	12 .28	Sun.
15	18 .31	70 .80	12 .49	Sun.	45	19 .69	32 .51	12 .82	Sun.
20	18 .53	22 .60	4 .07	Shade.	50	19 .95	24 .15	4 .20	Shade.
25	18 .76	20 .19	1 .43	Shade.	55	20 .20	21 .60	1 .40	Shade.
30	18 .98	19 .48	0 .50	Shade.	8 00	20 .48	20 .90	0 .42	Shade.

θ_0 12 .56, m .223, $m\theta_0$ 2 .89.
Sky, clear, very good. Wind, calm.

θ_0 13 .58, m .227, $m\theta_0$ 3 .08.
Sky, clear. Wind, calm.

Time.	Water thermometer.	Sun thermometer.	Difference.	Exposure.	Time.	Water thermometer.	Sun thermometer.	Difference.	Exposure.
11 30ᵐ A. M.	30 .48	30 .50	0 .02	Shade.	12 00ᵐ M.	30 .77	31 .50	0 .73	Shade.
35	30 .54	41 .20	10 .66	Sun.	12 05 P. M.	30 .80	42 .40	11 .60	Sun.
40	30 .60	44 .29	13 .69	Sun.	10	30 .83	44 .40	13 .57	Sun.
45	30 .66	44 .96	14 .30	Sun.	15	30 .86	45 .12	14 .26	Sun.
50	30 .70	35 .31	4 .61	Shade.	20	30 .88	35 .40	4 .52	Shade.
55	30 .74	32 .48	1 .74	Shade.	25	30 .90	32 .69	1 .79	Shade.
12 00 M.	30 .77	31 .50	0 .73	Shade.	30	30 .92	31 .70	0 .78	Shade.

θ_0 14 .99, m .218, $m\theta_0$ 3 .27.
Sky, very clear. Wind, almost calm.

θ_0 15 .29, m .214, $m\theta_0$ 3 .34.
Sky, clear. Wind, almost calm.

Time.	Water thermometer.	Sun thermometer.	Difference.	Exposure.	Time.	Water thermometer.	Sun thermometer.	Difference.	Exposure.
1 00ᵐ P. M.	31 .62	32 .05	0 .43	Shade.	1 30ᵐ P. M.	31 .35	32 .20	0 .85	Shade.
05	31 .57	41 .20	9 .63	Sun.	35	31 .30	40 .79	9 .40	Sun.
10	31 .53	43 .66	12 .13	Sun.	40	31 .24	42 .90	11 .66	Sun.
15	31 .48	44 .15	12 .67	Sun.	45	31 .28	43 .29	12 .44	Sun.
20	31 .44	35 .65	4 .21	Shade.	50	31 .12	35 .12	4 .00	Shade.
25	31 .39	33 .09	1 .70	Shade.	55	31 .06	32 .70	1 .64	Shade.
30	31 .35	32 .20	0 .85	Shade.	2 00	30 .99	31 .80	0 .81	Shade.

θ_0 15 .58, m .209, $m\theta_0$ 2 .64.
Sky, very clear. Wind light.

θ_0 15 .15, m .202, $m\theta_0$ 2 .66.
Sky, very clear. Wind, light, diminishing.

TABLE 64.

[Lone Pine, August 19, 1881. Observer from 7ʰ to 8ʰ, A. C. Dobbins; remaining series by W. C. Day. Actinometer No. 2. Medium aperture.]

Time.	Water ther- mometer.	Sun ther- mometer.	Difference.	Exposure.	Time.	Water ther- mometer.	Sun ther- mometer.	Difference.	Exposure.
7ʰ00ᵐ A. M.	16°.52	16°.52	0°.00	Shade.	7ʰ30ᵐ A. M.	17°.42	18°.20	0°.58	Shade.
05	16.49	25.48	8.99	Sun.	35	17.87	27.69	9.82	Sun.
10	16.66	28.44	11.78	Sun.	40	18.13	30.59	12.46	Sun.
15	16.87	29.48	12.61	Sun	45	18.40	31.51	13.11	Sun.
20	17.11	21.28	4.17	Shade.	50	18.68	22.90	4.22	Shade.
25	17.36	18.87	1.51	Shade.	55	18.98	20.48	1.50	Shade
30	17.62	18.20	0.58	Shade.	× 00	19.50	19.72	0.42	Shade.

$\theta_0 = 13°.06$; $m = (.217)$; $m\,\theta_0 = 2.83$. Sky, very light haze. Wind, calm to gentle.

$\theta_0 = 13°.81$; $m = (.222)$; $m\,\theta_0 = 3.07$. Sky, very light haze. Wind, gentle to calm.

11ʰ30ᵐ A. M.	31°.04	31°.04	0°.18	Shade.	12ʰ00ᵐ M.	32°.58	33°.41	0°.83	Shade.
35	31.96	43.12	11.16	Sun.	05 P. M.	32.69	43.90	11.21	Sun.
40	32.10	45.94	13.84	Sun.	10	32.80	46.61	13.81	Sun.
45	32.25	46.73	14.48	Sun.	15	32.90	47.40	14.50	Sun.
50	32.37	37.08	4.71	Shade.	20	33.00	37.80	4.80	Shade.
55	32.18	34.29	1.81	Shade.	25	33.09	34.90	1.81	Shade.
12.00 M.	32.58	33.41	0.83	Shade.	30	33.18	34.01	0.83	Shade.

$\theta_0 = 15°.37$; $m = .215$; $m\,\theta_0 = 32.21$. Sky, very clear. Wind, calm.

$\theta_0 = 15°.57$; $m = .215$; $m\,\theta_0 = 32.35$. Sky, clear. Wind, calm.

4ʰ00ᵐ P. M.	33°.82	34°.30	0°.48	Shade.	4ʰ30ᵐ P. M.	33°.26	34°.23	0°.97	Shade.
05	33.73	43.80	10.07	Sun.	35	33.16	42.86	9.70	Sun.
10	33.63	46.20	12.57	Sun.	40	33.01	44.90	11.84	Sun.
15	33.54	46.68	13.14	Sun.	45	32.97	45.30	12.33	Sun.
20	33.44	37.80	4.46	Shade.	50	32.88	36.97	4.09	Shade.
25	33.35	35.20	1.85	Shade.	55	32.78	34.49	1.71	Shade.
30	33.26	34.23	0.97	Shade.	5.00	32.68	33.60	0.92	Shade.

$\theta_0 = 14°.17$; $m = .205$; $m\,\theta_0 = 2.91$. Sky, clear. Wind, very slight breeze.

$\theta_0 = 13°.47$; $m = .208$; $m\,\theta_0 = 2°.80$. Sky, clear. Wind, very slight breeze.

TABLE 65.

[Lone Pine, August 20, 1881. Observer, A. C. D. Actinometer No. 2. Medium aperture.]

Time.	Water ther- mometer.	Sun ther- mometer.	Difference.	Exposure.	Time.	Water ther- mometer.	Sun ther- mometer.	Difference.	Exposure.
7ʰ00ᵐ A. M.	16°.10	16°.16	0°.06	Shade.	7ʰ30ᵐ A. M.	17°.33	17°.80	0°.47	Shade.
05	16.74	25.58	9.34	Sun.	35	17.63	27.65	10.02	Sun.
10	16.39	28.52	12.13	Sun.	40	17.94	30.76	12.82	Sun.
15	16.59	29.50	12.91	Sun.	45	18.27	31.90	13.63	Sun.
20	16.83	20.94	4.11	Shade.	50	18.71	22.91	4.20	Shade.
25	17.09	18.51	1.42	Shade.	55	19.10	20.47	1.37	Shade.
30	17.33	17.80	0.47	Shade.	× 00	19.50	19.70	0.20	Shade.

$\theta_0 = 13°.34$; $m = .227$; $m\,\theta_0 = 3°.03$. Sky, clear, little haze about Inyo mountains. Wind, gentle.

$\theta_0 = 14°.09$; $m = (.232)$; $m\,\theta_0 = 3.26$. Sky, clear. Wind, gentle.

11ʰ30ᵐ A. M.	33°.06	33°.35	0°.29	Shade.	12ʰ00ᵐ M.	33°.53	34°.50	0°.97	Shade.
35	33.14	44.30	11.16	Sun.	05 P. M.	33.61	44.90	11.29	Sun.
40	33.22	47.11	13.89	Sun.	10	33.69	47.53	13.84	Sun.
45	33.30	47.97	14.67	Sun.	15	33.77	48.30	14.53	Sun.
50	33.37	38.05	4.68	Shade.	20	33.84	38.41	4.57	Shade.
55	33.45	35.39	1.94	Shade.	25	33.92	35.80	1.88	Shade.
12.00 M.	33.53	34.50	0.97	Shade.	30	34.00	34.91	0.91	Shade.

$\theta_0 = 15°.57$; $m = .211$; $m\,\theta_0 = 3°.29$. Sky, slight haze. Wind, gentle to fresh.

$\theta_0 = 15°.63$; $m = .216$; $m\,\theta_0 = 3°.38$. Sky, light haze. Wind, gentle to fresh.

4ʰ00ᵐ P. M.	33°.53	34°.03	0°.50	Shade.	4ʰ30ᵐ P. M.	33°.30	34.30	1°.00	Shade.
05	33.51	43.66	10.15	Sun.	35	33.24	42.90	9.66	Sun.
10	33.48	46.00	12.52	Sun.	40	33.17	45.12	11.95	Sun.
15	33.44	46.34	12.90	Sun.	45	33.09	45.35	12.26	Sun.
20	33.40	37.71	4.31	Shade.	50	33.02	37.14	4.12	Shade.
25	33.36	35.19	1.83	Shade.	55	32.94	34.70	1.76	Shade.
30	33.30	34.30	1.00	Shade.	5.00	32.84	32.81	0.07	Shade.

$\theta_0 = 14°.03$; $m = .205$; $m\,\theta_0 = 2°.88$. Sky, light haze. Wind, brisk, variable.

$\theta_0 = 13°.50$; $m = .206$; $m\,\theta_0 = 2°.78$. Sky, light haze. Wind, gentle to fresh.

TABLE 66.

[Lone Pine, August 21, 1881. Observer, A. C. D. Actinometer No. 2. Medium aperture.]

Time	Water ther-mometer.	Sun ther-mometer.	Difference.	Exposure.	Time	Water ther-mometer.	Sun ther-mometer.	Difference	Exposure
7 00ᵐ A. M.	18°.75	18°.75	0°.00	Shade.	7 30ᵐ A. M.	19°.84	20°.20	0°.36	Shade.
05	18.87	28.11	9.24	Sun.	35	20.07	29.68	9.61	Sun.
10	19.00	30.63	11.63	Sun.	40	20.33	32.60	12.27	Sun.
15	19.18	31.71	12.53	Sun.	45	20.64	33.74	13.10	Sun.
20	19.38	23.35	3.97	Shade.	50	21.00	25.09	4.09	Shade.
25	19.62	23.93	4.31	Shade.	55	21.41	22.65	1.24	Shade.
30	19.84	20.20	0.36	Shade.	8 00	21.82	22.00	0.18	Shade.

$\theta_0 = 12°.86$; $m = 0.229$; $m\theta_0 = 2°.93$.
Sky, light smoke or haze. Wind, gentle.

$\theta_0 = 13°.49$, $m = 0.239$; $m\theta_0 = 3°.22$.
Sky, light smoke or haze. Wind, gentle to calm.

11 30ᵐ A. M.	32°.71	32°.90	0°.19	Shade.	12 00ᵐ M.	33°.20	33°.94	0°.74	Shade.
35	32.75	43.50	10.75	Sun.	05 P. M.	33.25	44.29	11.04	Sun.
40	32.80	46.51	13.71	Sun.	10	33.30	47.05	13.75	Sun.
45	32.90	47.20	14.30	Sun.	15	33.42	47.80	14.38	Sun.
50	33.00	37.50	4.50	Shade.	20	33.53	38.42	4.89	Shade.
55	33.10	34.71	1.61	Shade.	25	33.64	35.50	1.86	Shade.
12 00	33.20	33.94	0.74	Shade.	30	33.76	34.59	0.83	Shade.

$\theta_0 = 15°.24$; $m = 0.224$; $m\theta_0 = 3°.41$.
Sky, very light haze or smoke. Wind, fresh, steady.

$\theta_0 = 15°.17$, $m = 0.212$; $m\theta_0 = 3°.28$.
Sky, very light haze or smoke. Wind, fresh, steady.

4 00ᵐ P. M.	34°.07	34°.64	0°.57	Shade.	4 30ᵐ P. M.	33°.67	34°.70	1°.03	Shade.
05	34.01	43.72	9.71	Sun.	35	33.60	42.82	9.22	Sun.
10	33.95	45.82	11.87	Sun.	40	33.52	44.82	11.30	Sun.
15	33.89	46.41	12.52	Sun.	45	33.43	45.22	11.77	Sun.
20	33.82	38.00	4.18	Shade.	50	33.37	37.40	4.03	Shade.
25	33.74	35.51	1.77	Shade.	55	33.29	35.06	1.77	Shade.
30	33.67	34.70	1.03	Shade.	5 00	33.20	34.19	0.99	Shade.

$\theta_0 = 13°.54$, $m = 0.203$, $m\theta_0 = 2°.75$.
Sky, light haze or smoke. Wind, fresh to brisk.

$\theta_0 = 12°.97$, $m = 0.201$, $m\theta_0 = 2°.61$.
Sky, light haze or smoke. Wind, fresh to brisk, decreasing.

TABLE 67.

[Lone Pine, August 22, 1881. Observer, A. C. D. Actinometer No. 2. Medium aperture.]

Time	Water ther-mometer.	Sun ther-mometer.	Difference.	Exposure.	Time	Water ther-mometer.	Sun ther-mometer.	Difference.	Exposure.
7 00ᵐ A. M.	18°.89	18°.89	0°.00	Shade.	7 30ᵐ A. M.	19°.98	20°.32	0°.34	Shade.
05	18.97	27.84	8.87	Sun.	35	20.20	29.71	9.51	Sun.
10	19.14	30.70	11.56	Sun.	40	20.45	32.59	12.14	Sun.
15	19.34	31.58	12.24	Sun.	45	20.76	33.65	12.89	Sun.
20	19.54	23.41	3.87	Shade.	50	21.08	25.13	4.05	Shade.
25	19.76	21.06	1.30	Shade.	55	21.40	22.70	1.30	Shade.
30	19.98	20.32	0.34	Shade.	8 00	21.73	22.09	0.36	Shade.

$\theta_0 = 12°.61$; $m = 0.227$; $m\theta_0 = 2°.86$.
Sky, light smoke. Wind, gentle.

$\theta_0 = 13°.37$; $m = 0.233$; $m\theta_0 = 3°.12$.
Sky, light smoke. Wind, gentle.

11 30ᵐ A. M.	33°.00	33°.04	0°.04	Shade.	12 00ᵐ M.	32°.85	33°.67	0°.82	Shade.
35	32.97	44.00	11.03	Sun.	05 P. M.	32.87	43.96	11.09	Sun.
40	32.90	46.62	13.72	Sun.	10	32.90	46.69	13.79	Sun.
45	32.85	47.45	14.60	Sun.	15	32.95	47.27	14.39	Sun.
50	32.80	37.43	4.63	Shade.	20	32.88	37.48	4.60	Shade.
55	32.83	34.62	1.79	Shade.	25	32.88	34.62	1.74	Shade.
12 00	32.85	33.67	0.82	Shade.	30	32.88	33.72	0.84	Shade.

$\theta_0 = 15°.53$; $m = 0.218$; $m\theta_0 = 3°.39$.
Sky, light haze. Wind, fresh to gentle.

$\theta_0 = 15°.12$, $m = 0.218$; $m\theta_0 = 3°.36$.
Sky, light haze. Wind, fresh.

4 00ᵐ P. M.	33°.92	34°.50	0°.58	Shade.	4 30ᵐ P. M.	33°.18	34°.26	1°.06	Shade.
05	33.77	43.35	9.58	Sun.	35	33.07	42.40	9.33	Sun.
10	33.64	45.65	12.01	Sun.	40	32.97	44.38	11.41	Sun.
15	33.52	46.26	12.74	Sun.	45	32.87	44.80	11.93	Sun.
20	33.40	37.73	4.33	Shade.	50	32.78	36.85	4.07	Shade.
25	33.29	35.18	1.89	Shade.	55	32.89	34.50	1.81	Shade.
30	33.18	34.26	1.08	Shade.	5 00	32.70	33.60	1.01	Shade.

$\theta_0 = 13°.74$, $m = 0.200$, $m\theta_0 = 2°.75$.
Sky, light smoke. Wind, gentle to fresh.

$\theta_0 = 13°.14$; $m = 0.201$, $m\theta_0 = 2°.64$.
Sky, light smoke. Wind, light to gentle.

TABLE 68.

[Lone Pine, August 23, 1881. Observer, A. C. D. Actinometer No. 2. Medium aperture.]

Time.	Water ther-mometer.	Sun ther-mometer.	Difference.	Exposure.	Time.	Water ther-mometer.	Sun ther-mometer.	Difference.	Exposure.
7ʰ00ᵐ A. M.	17.55	17.40	− 0.15	Shade.	7ʰ30ᵐ A. M.	19.27	19.60	0.33	Shade.
05	17.94	26.59	+ 8.75	Sun.	35	19.56	20.23	0.67	Sun.
10	18.12	29.70	11.88	Sun.	40	19.86	31.65	11.79	Sun.
15	18.41	30.80	12.79	Sun.	45	20.18	32.90	12.72	Sun.
20	18.69	22.58	3.89	Shade.	50	20.50	24.50	4.00	Shade.
25	18.98	20.26	1.28	Shade.	55	20.82	22.10	1.28	Shade.
30	19.27	19.60	0.33	Shade.	8 00	21.16	21.45	0.29	Shade.

θ₀=12°62; m =.2290; mθ₀=2°.89.
Sky, light smoke, a few cirrus and cirrocumuli. Wind, light.

θ₀=13.20, m =.2309; mθ₀=2°.98.
Sky, light haze, passing cirri. Wind, light.

Time.	Water ther-mometer.	Sun ther-mometer.	Difference.	Exposure.	Time.	Water ther-mometer.	Sun ther-mometer.	Difference.	Exposure.
11ʰ30ᵐ A. M.	31.30	31.30	0.00	Shade.	12ʰ00ᵐ M.	31.98	32.66	0.64	Shade.
35	31.40	42.32	10.92	Sun.	05	32.00	43.23	11.29	Sun.
40	31.56	45.25	13.75	Sun.	10	32.10	45.98	13.88	Sun.
45	31.65	46.09	14.44	Sun.	15	32.25	46.68	14.43	Sun.
50	31.75	36.19	4.44	Shade.	20	32.40	36.90	4.50	Shade.
55	31.85	33.50	1.65	Shade.	25	32.50	33.15	1.65	Shade.
12 00 M.	31.98	32.66	0.68	Shade.	30	32.60	33.31	0.71	Shade.

θ₀=15.30; m =.226; mθ₀=3.45.
Sky, clear. Wind, gentle to fresh.

θ₀=15.58, m =.235; mθ₀=3.46.
Sky, clear. Wind, gentle to fresh, increasing.

Time.	Water ther-mometer.	Sun ther-mometer.	Difference.	Exposure.	Time.	Water ther-mometer.	Sun ther-mometer.	Difference.	Exposure.
4ʰ00ᵐ P. M.	33.80	34.82	0.76	Shade.	4ʰ30ᵐ P. M.	32.03	34.03	1.00	Shade.
05	33.72	43.95	10.23	Sun.	35	32.88	42.85	9.97	Sun.
10	33.58	46.29	12.71	Sun.	40	32.74	44.80	12.06	Sun.
15	33.44	46.64	14.20	Sun.	45	32.60	45.03	12.43	Sun.
20	33.30	37.72	4.42	Shade.	50	32.40	36.70	4.24	Shade.
25	33.16	35.06	1.90	Shade.	55	32.32	34.12	1.80	Shade.
30	33.03	34.93	1.90	Shade.	5 00	32.14	33.18	1.00	Shade.

θ₀=14.36, m =.205; mθ₀=2°.94.
Sky, clear. Wind, fresh to brisk, decreasing.

θ₀=13.60; m =.204; mθ₀=2.79.
Sky, clear. Wind, fresh to gentle.

TABLE 69.

[Lone Pine, August 24, 1881. Observer, A. C. D. Actinometer No. 2. Medium aperture.]

Time.	Water ther-mometer.	Sun ther-mometer.	Difference.	Exposure.	Time.	Water ther-mometer.	Sun ther-mometer.	Difference.	Exposure.
First set of observations not taken.					7ʰ30ᵐ A. M.	21.46	22.30	0.84	Shade.
					35	21.80	31.72	9.92	Sun.
					40	22.14	34.42	12.28	Sun.
					45	22.50	35.40	12.90	Sun.
					50	22.81	26.90	4.06	Shade.
					55	22.20	24.60	1.40	Shade.
					8 00	24.56	24.90	0.34	Shade.

θ₀=15.66, m =.224; mθ₀=3°.11.
Sky, clear, light smoke on mountains. Wind, light to gentle.

Time.	Water ther-mometer.	Sun ther-mometer.	Difference.	Exposure.	Time.	Water ther-mometer.	Sun ther-mometer.	Difference.	Exposure.
11ʰ30ᵐ A. M.	31.00	31.16	0.16	Shade.	12ʰ00ᵐ M.	31.27	32.09	0.82	Shade.
35	31.00	42.09	11.09	Sun.	05	31.37	42.50	11.13	Sun.
40	31.00	44.79	13.79	Sun.	10	31.43	45.26	13.83	Sun.
45	31.00	45.51	14.51	Sun.	15	31.50	46.00	14.50	Sun.
50	31.10	35.75	4.65	Shade.	20	31.55	36.20	4.65	Shade.
55	31.18	32.98	1.80	Shade.	25	31.60	33.42	1.82	Shade.
12 00 M.	31.27	32.09	0.82	Shade.	30	31.66	32.56	0.90	Shade.

θ₀=15.52; m =.217; mθ₀=3°.37.
Sky, slightly smoky. Wind, gentle to fresh.

θ₀=15.53, m =.215; mθ₀=3°.31.
Sky, slightly smoky. Wind, gentle.

Time.	Water ther-mometer.	Sun ther-mometer.	Difference.	Exposure.	Time.	Water ther-mometer.	Sun ther-mometer.	Difference.	Exposure.
4ʰ00ᵐ P. M.	32.80	33.25	0.45	Shade.	4ʰ30ᵐ P. M.	32.25	33.30	1.05	Shade.
05	32.71	42.81	10.10	Sun.	35	32.15	42.05	9.90	Sun.
10	32.82	45.13	12.51	Sun.	40	32.06	44.15	12.00	Sun.
15	32.92	45.77	13.25	Sun.	45	31.97	44.49	12.52	Sun.
20	32.83	36.88	4.45	Shade.	50	31.88	36.07	4.19	Shade.
25	32.34	34.23	1.89	Shade.	55	31.79	33.40	1.61	Shade.
30	32.25	33.30	1.05	Shade.	5 00	31.70	32.70	1.00	Shade.

θ₀=14.24, m =.203; mθ₀=2°.89.
Sky, light haze or smoke. Wind, brisk, variable.

θ₀=13.77, m =.205; mθ₀=2°.82.
Sky, light haze. Wind, brisk.

TABLE 70.

[Lone Pine August 25, 1881. Observer, A. C. D. Actinometer No. 2. Medium aperture.]

Time	Water ther-mometer	Sun ther-mometer	Difference	Exposure	Time	Water ther-mometer	Sun ther-mometer	Difference	Exposure

Sky clear, slight haze on mountains. Wind calm.

Sky clear. Wind fresh.

Sky clear, haze on mountains. Wind gentle to fresh.

Sky clear, haze on mountains. Wind, gentle to fresh.

TABLE 71.

[Lone Pine August 27, 1881. Observer, A. C. D. Actinometer No. 2. Medium aperture.]

Time	Water ther-mometer	Sun ther-mometer	Difference	Exposure	Time	Water ther-mometer	Sun ther-mometer	Difference	Exposure

Sky clear, little haze on mountains. Wind gentle to very light.

Sky clear, haze on mountains. Wind, gentle to very light.

Sky, light haze. Wind light.

Sky, light haze. Wind light.

Sky, light haze. Wind light.

Sky, light haze. Wind, light.

TABLE 72.

[Lone Pine, August 28, 1881. Observer, A. C. D. Actinometer No. 2. Medium aperture.]

Time	Water ther- mometer.	Sun ther- mometer.	Difference.	Exposure.	Time.	Water ther- mometer.	Sun ther- mometer.	Difference.	Exposure.
7ʰ00ᵐ A. M.	13 .60	13°.68	0°.08	Shade.	7ʰ30ᵐ A. M.	15°.00	15°.49	0 .49	Shade.
05	13 .84	22 .75	8 .91	Sun.	35	15 .26	25 .22	9 .96	Sun.
10	14 .07	25 .88	11 .84	Sun.	40	15 .51	28 .00	12 .49	Sun.
15	14 .31	26 .50	12 .19	Sun.	45	15 .76	29 .10	13 .34	Sun.
20	14 .54	18 .62	4 .08	Shade.	50	16 .01	20 .21	4 .20	Shade.
25	14 .79	16 .18	1 .41	Shade.	55	16 .26	17 .76	1 .50	Shade.
30	15 .00	15 .49	0 .49	Shade.	8 00	16 .51	17 .00	0 .49	Shade.

&ₒ = 12 .94; m = .224; m &ₒ = 2 .90.
Sky, clear, haze on mountains.   Wind, calm to light.

&ₒ = 13°.97; m = (.223); m &ₒ = 3 .12
Sky, clear, haze on mountains.   Wind, calm to light

11ʰ30ᵐ A. M	29°.03	29°.85	0 .18	Shade.	12ʰ00ᵐ M.	29 .08	29 .92	0°.84	Shade.
35	29 .04	30 .86	10 .82	Sun.	05 P. M.	29 .10	40 .08	10 .90	Sun.
40	29 .04	42 .77	13 .73	Sun.	10	29 .12	42 .54	13 .42	Sun.
45	29 .05	43 .50	14 .45	Sun.	15	29 .15	43 .11	13 .96	Sun.
50	29 .06	33 .64	4 .58	Shade.	20	29 .18	33 .54	4 .36	Shade.
55	29 .07	30 .81	1 .74	Shade.	25	29 .21	30 .88	1 .67	Shade.
12 00 M.	29 .08	29 .92	0 .84	Shade.	30	29 .25	29 .82	0 .67	Shade.

&ₒ = 15 .11; m = .216; m &ₒ = 3°.26.
Sky, clear, light haze on horizon.   Wind, gentle to fresh.

&ₒ = 14 .95; m = .224; m &ₒ = 3 .35.
Sky, clear, light haze on horizon.   Wind, light to fresh.

No afternoon observations.

TABLE 73.

[Lone Pine, August 29, 1881. Observer, A. C. D. Actinometer No. 2. Medium aperture.]

Time.	Water ther- mometer.	Sun ther- mometer.	Difference.	Exposure.	Time.	Water ther- mometer.	Sun ther- mometer.	Difference.	Exposure.
7ʰ00ᵐ A. M.	13°.70	13°.70	0°.00	Shade.	7ʰ30ᵐ A. M.	15°.95	16°.30	0°.35	Shade.
05	14 .07	22 .79	8 .72	Sun.	35	16 .34	25 .81	9 .47	Sun.
10	14 .44	25 .85	11 .41	Sun.	40	16 .74	28 .77	12 .03	Sun.
15	14 .82	27 .03	12 .21	Sun.	45	17 .15	30 .01	12 .86	Sun.
20	15 .30	19 .05	3 .85	Shade.	50	17 .55	21 .55	4 .00	Shade.
25	15 .86	16 .90	1 .34	Shade.	55	17 .95	19 .20	1 .25	Shade.
30	15 .95	16 .30	0 .35	Shade.	8 00	18 .35	18 .59	0 .74	Shade.

&ₒ = 12 .52; m = .724; m &ₒ = 2 .80.
Sky, perceptible haze.   Wind, calm to very light.

&ₒ = 13°.27; m = .236); m &ₒ = 3°.12.
Sky, perceptible haze   Wind, calm to very light.

11ʰ30ᵐ A. M.	28°.60	28°.80	0°.20	Shade.	12ʰ00ᵐ M.	28 .54	29°.47	0°.93	Shade.
35	28 .61	39 .63	11 .08	Sun.	05 P. M.	28 .50	39 .45	10 .95	Sun.
40	28 .62	42 .36	13 .74	Sun.	10	28 .50	42 .42	13 .62	Sun.
45	28 .61	43 .00	14 .39	Sun.	15	28 .52	42 .74	14 .22	Sun.
50	28 .60	33 .25	4 .65	Shade.	20	28 .60	33 .21	4 .61	Shade.
55	28 .58	30 .45	1 .87	Shade.	25	28 .70	30 .60	1 .90	Shade.
12 00	28 .54	29 .47	0 .93	Shade.	30	28 .82	29 .75	0 .93	Shade.

&ₒ = 15 .31; m = .212; m &ₒ = 3°.25.
Sky, clear.   Wind, fresh.

&ₒ = 15°.35; m = .212; m &ₒ = 3°.25.
Sky, clear.   Wind, fresh.

4ʰ00ᵐ P. M	29 .70	30 .00	0 .30	Shade.	4ʰ30ᵐ P. M.	29°.57	30 .38	0 .81	Shade.
05	29 .69	37 .81	8 .12	Sun.	35	29 .54	36 .97	7 .41	Sun.
10	29 .68	40 .20	10 .52	Sun.	40	29 .51	38 .86	9 .35	Sun.
15	29 .67	40 .52	10 .89	Sun.	45	29 .50	39 .30	9 .80	Sun.
20	29 .64	31 .31	3 .67	Shade.	50	29 .50	32 .42	3 .72	Shade.
25	29 .60	31 .14	1 .54	Shade.	55	29 .50	30 .90	1 .40	Shade.
30	29 .57	30 .38	0 .81	Shade.	5 00	29 .50	30 .26	0 .76	Shade.

&ₒ = 11 .67; m = .204; m &ₒ = 29.38.
Sky, smoke from fire in cañon intervening.   Wind, fresh to brisk, variable.

&ₒ = 10°.67; m = .204; m &ₒ = 2 .18.
Sky, smoke from fire in cañon intervening.   Wind, fresh to brisk, variable.

TABLE 74.

(Lone Pine, August 30, 1881. Observer, A. C. D. Actinometer No. 2. Medium aperture.)

Time.	Water ther-mometer.	Sun ther-mometer.	Difference.	Exposure.	Time.	Water ther-mometer.	Sun ther-mometer.	Difference.	Exposure.
7ʰ00ᵐ A. M.	16.92	17.00	0.08	Shade.	7ʰ30ᵐ A. M.	18.08	18.50	0.42	Shade.
05	17.09	25.42	8.33	Sun.	35	18.28	27.78	9.50	Sun.
10	17.28	28.40	11.12	Sun.	40	18.49	30.61	12.12	Sun.
15	17.50	29.36	11.86	Sun.	45	18.72	31.72	13.00	Sun.
20	17.68	21.41	3.73	Shade.	50	18.98	23.12	4.14	Shade.
25	17.88	19.10	1.22	Shade.	55	19.24	20.60	1.36	Shade.
30	18.08	18.50	0.42	Shade.	8.00	19.54	19.82	0.28	Shade.

θ. 12 16, m 230, mθ. 2.80.
Sky, light smoke. Wind, fresh.

θ. 13 16, m 229, mθ. 3.08.
Sky, light smoke. Wind, fresh.

10ʰ30ᵐ A. M.	27.09	27.60	0.05	Shade.	12ʰ00ᵐ M.	28.74	29.40	0.66	Shade.
35	27.88	38.18	10.30	Sun.	05	28.73	39.50	10.71	Sun.
40	28.07	41.10	13.03	Sun.	10	28.80	42.28	13.48	Sun.
45	28.25	42.45	13.80	Sun.	15	28.80	11.02	14.22	Sun.
50	28.41	32.54	4.40	Shade.	20	28.80	43.35	4.55	Shade.
55	28.60	30.16	1.56	Shade.	25	28.80	30.56	1.76	Shade.
12.00 M.	28.74	29.40	0.66	Shade.	30	28.80	29.35	0.66	Shade.

θ. 14.39, m 221, mθ. 3.18.
Sky, slightly smoky, a few cirri. Wind, calm to light.

θ. 15.07, m 221, mθ. 3.35.
Sky, slightly smoky, a few cirri. Wind, gentle to fresh.

4ʰ00ᵐ P. M.	28.66	29.02	0.36	Shade.	4ʰ30ᵐ P. M.	28.23	29.14	0.91	Shade.
05	28.59	37.50	8.91	Sun.	35	28.15	36.56	8.41	Sun.
10	28.52	39.80	11.28	Sun.	40	28.08	38.25	10.17	Sun.
15	28.44	40.16	11.72	Sun.	45	28.00	38.53	10.53	Sun.
20	28.07	32.42	4.05	Shade.	50	27.93	31.52	3.59	Shade.
25	28.30	30.09	1.79	Shade.	55	27.86	29.40	1.54	Shade.
30	28.23	29.14	0.91	Shade.	5.00	27.78	28.60	0.82	Shade.

θ. 12.08, m 202, mθ. 2.56.
Sky, smoke from fire in canon less dense than on previous afternoon. Wind, fresh.

θ. 11.63, m 205, mθ. 2.58.
Sky, smoke from forest fire less dense than on previous afternoon. Wind, fresh to gentle.

TABLE 75.

(Lone Pine, August 31, 1881. Observer, A. C. D. Actinometer No. 2. Medium aperture.)

Time.	Water ther-mometer.	Sun ther-mometer.	Difference.	Exposure.	Time.	Water ther-mometer.	Sun ther-mometer.	Difference.	Exposure.
7ʰ00ᵐ A. M.	13.36	13.37	0.01	Shade.	7ʰ30ᵐ A. M.	14.90	15.41	0.51	Shade.
05	13.62	22.40	8.78	Sun.	35	15.19	24.83	9.64	Sun.
10	13.87	25.30	11.43	Sun.	40	15.51	28.00	12.49	Sun.
15	14.12	26.45	12.33	Sun.	45	15.81	29.40	13.26	Sun.
20	14.58	18.50	3.92	Shade.	50	16.16	20.37	4.21	Shade.
25	14.64	16.10	1.46	Shade.	55	16.48	17.90	1.42	Shade.
30	14.90	15.41	0.54	Shade.	8.00	16.82	17.21	0.39	Shade.

θ. 12.69, m 222, mθ. 2.82.
Sky, light smoke. Wind, calm or nearly so.

θ. 13.80, m 227, mθ. 3.43.
Sky, light smoke. Wind, calm or nearly so.

11ʰ30ᵐ A. M.	28.50	0.15	Shade.		12ʰ00ᵐ M.	28.84	29.70	0.86	Shade.
35	28.55	39.65	11.10	Sun.	05	28.80	40.19	11.29	Sun.
40	28.61	42.41	13.80	Sun.	10	28.85	42.65	13.83	Sun.
45	28.67	43.22	14.55	Sun.	15	29.01	43.65	14.64	Sun.
50	28.73	33.42	4.69	Shade.	20	29.07	33.76	4.69	Shade.
55	28.78	30.64	1.86	Shade.	25	29.12	30.96	1.84	Shade.
12.00 M.	28.84	29.70	0.86	Shade.	30	29.18	30.05	0.87	Shade.

θ. 15.39, m 214, mθ. 3.29.
Sky, slightly smoky. Wind, gentle to fresh.

θ. 15.68, m 216, mθ. 3.39.
Sky, slightly smoky. Wind, gentle to fresh.

4ʰ00ᵐ P. M.	29.85	30.11	0.26	Shade.	4ʰ30ᵐ P. M.	29.37	30.71	0.34	Shade.
05	29.76	30.27	9.51	Sun.	35	29.31	38.35	9.02	Sun.
10	29.68	41.52	11.84	Sun.	40	29.30	38.45	11.15	Sun.
15	29.60	41.90	12.30	Sun.	45	29.30	40.76	11.34	Sun.
20	29.52	33.56	4.04	Shade.	50	29.16	32.63	3.47	Shade.
25	29.41	31.16	1.72	Shade.	55	29.11	30.76	1.65	Shade.
30	29.37	30.30	0.93	Shade.	5.00	29.07	29.94	0.84	Shade.

θ. 13.23, m 206, mθ. 2.73.
Sky, slightly smoky. Wind, fresh.

θ. 12.65; m 207, mθ. 2.62.
Sky, slightly smoky. Wind, light to fresh.

TABLE 76.

[Lone Pine, September 5, 1881. Observer, A. C. D. Actinometer No. 2. Medium aperture.]

Time	Water ther- mometer.	Sun ther mometer.	Difference.	Exposure.	Time	Water ther- mometer.	Sun ther- mometer.	Difference.	Exposure.
7ʰ00ᵐ A. M.	20.10	20.10	0.00	Shade.	7ʰ50ᵐ A. M.	21.92	22.26	0.34	Shade.
05	20.58	29.36	8.88	Sun.	55	22.22	31.80	9.58	Sun.
10	20.68	32.27	11.58	Sun.	40	22.50	34.70	12.20	Sun.
15	21.00	33.70	12.50	Sun.	45	22.79	35.70	12.91	Sun.
20	21.30	23.24	3.94	Shade.	50	23.06	27.18	4.12	Shade.
25	21.62	22.90	1.28	Shade.	55	23.32	24.80	1.48	Shade.
30	21.92	22.26	0.34	Shade.	8 00	23.56	24.02	0.46	Shade

θ. 12.88, ω 0.229, αθ. 2.90,
Sky, clear, light smoke on mountains. Wind, gentle to brisk.

θ. 13.50, ω. 0.231, αθ. 2.98.
Sky, clear, light smoke on mountains. Wind, gentle to fresh.

11ʰ30ᵐ A. M.	35.16	35.19	0.03	Shade.	12ʰ00ᵐ M.	34.88	35.89	1.04	Shade.
35	35.12	46.12	11.00	Sun.	05 P. M.	34.82	46.22	11.40	Sun.
40	35.07	49.00	13.93	Sun.	10	34.79	48.90	14.11	Sun.
45	35.01	49.72	14.71	Sun.	15	34.77	49.60	14.83	Sun.
50	34.97	39.50	4.53	Shade.	20	34.77	39.49	4.72	Shade.
55	34.91	36.88	1.97	Shade.	25	34.78	36.63	1.85	Shade.
12 00	34.88	35.90	1.04	Shade.	30	34.78	35.66	0.88	Shade.

θ. 15.75, ω .210, αθ. 3.31.
Sky, clear. Wind, light.

θ. 13.92, ω .217, αθ. 3.46.
Sky, clear. Wind, light.

4ʰ00ᵐ P. M.	34.90	35.50	0.60	Shade.	4ʰ30ᵐ P. M	34.06	35.20	1.14	Shade.
05	34.76	44.72	9.96	Sun.	35	33.92	42.85	8.93	Sun.
10	34.62	46.82	12.20	Sun.	40	33.78	45.10	11.32	Sun.
15	34.48	47.20	12.72	Sun.	45	33.64	45.40	11.76	Sun.
20	34.34	38.72	4.38	Shade.	50	33.50	37.62	4.12	Shade.
25	34.20	36.29	2.00	Shade.	55	33.35	35.20	1.84	Shade.
30	34.06	35.20	1.14	Shade.	5 00	33.22	34.22	1.00	Shade.

θ. 13.93, ω 107, αθ.–2.75.
Sky, clear. Wind, calm.

θ. 12.97, ω .198, αθ. 2.57.
Sky, clear. Wind, calm.

TABLE 77.

Summary of Lone Pine actinometer observations.

[The method pursued in deducing the corrected from the uncorrected observations is explained in the next chapter.]
[Instrument small actinometer, No. 2.]

Date.	Duration of experiment. 7ᵒ00ᵐ to 7ᵒ30ᵐ		Duration of experiment 7ᵒ30ᵐ to 8ᵒ00ᵐ		Duration of experiment. 11ᵒ30ᵐ to 12ᵒ00ᵐ		Duration of experiment. 12ᵒ00ᵐ to 12ᵒ30ᵐ		Duration of experiment. 4ᵒ00ᵐ to 4ᵒ30ᵐ		Duration of experiment. 4ᵒ30ᵐ to 5ᵒ00ᵐ		Summation of daily readings
	Uncorrected observations.	Corrected observations.	Uncorrected observations.	Corrected observations.	Uncorrected observations.	Corrected observations.	Uncorrected observations.	Corrected observations.	Uncorrected observations.	Corrected observations.	Uncorrected observations.	Corrected observations.	
1881.	*Cal.*	*Cal.*	*Cal.*	*Cal.*	*Cal.*	*Cal.*	*Cal.*	*Cal.*	*Cal.*	*Cal.*	*Cal.*	*Cal.*	
Aug. 18	1.337	1.418	1.255	1.551	1.350	1.653	1.349	1.349	1.987	1.155	1.427	1.081	9.105
19	1.453	1.418	1.248	1.541	1.345	1.652	1.362	1.703	1.182	1.450	1.441	1.103	9.296
20	1.765	1.67	1.527	1.654	1.558	1.671	1.371	1.747	1.170	1.445	1.352	1.302	9.700
21	1.203	1.107	1.312	1.619	1.390	1.736	1.786	1.672	1.119	1.582	1.661	1.805	9.183
22	1.363	1.173	1.269	1.567	1.579	1.760	1.712	1.112	1.119	1.582	1.675	1.923	9.241
23	1.177	1.148	1.272	1.575	1.166	1.79	1.109	1.352	1.120	1.530	1.154	1.784	9.589
24		1.449	1.258	1.566	1.371	1.714	1.253	1.700	1.177	1.454	1.149	1.111	9.254
25	1.200	1.177	1.504	1.610	1.192	1.710	1.105	1.758		1.424	1.119	1.377	9.354
27	1.141	1.406	1.273	1.366	1.403	1.744	1.448	1.269	1.122	1.447	1.114	1.369	9.321
28	1.580	1.454	1.569	1.567	1.459	1.462	1.355	1.704		1.420		1.5550	9.162
29	1.141	1.463	1.275	1.575	1.421	1.652	1.524	1.657		1.420		1.5551	9.865
30	1.329	1.401	1.254	1.547	1.291	1.417	1.354	1.697		1.420		1.5551	9.840
31	1.118	1.412	1.275	1.572	1.541	1.671	1.370	1.525	1.110	1.571	1.466	1.312	9.075
Sept. 5	1.552	1.454	1.211	1.493	1.347	1.655	1.447	1.760	1.148	1.481	1.406	1.266	9.066
Mean	1.372	1.441	1.274	1.571	1.356	1.696	1.374	1.718	1.152	1.425	1.402	1.355	

7ᵒ00ᵐ to 7ᵒ30ᵐ | 7ᵒ30ᵐ to 8 00ᵐ | 11ᵒ30ᵐ to 12ᵒ00ᵐ | 12ᵒ00ᵐ to 12ᵒ30ᵐ | 4ᵒ00ᵐ to 4 30ᵐ | 4ᵒ30ᵐ to 5ᵒ00ᵐ

Mean time of exposure to sun	7ʰ15ᵐ a. m.	7ʰ58ᵐ a. m.	11ʰ58ᵐ a. m.	12ʰ08ᵐ p. m.	4ʰ06ᵐ p. m.	4ʰ58ᵐ p. m.
Sun's mean hour angle	45°54′	4°24′	0°24°	16°06′	46°06′	46°30°
Sun's mean zenith distance	70°42′	61°33′	26°56°	26°07′	60°58′	66°50°
Air mass	19.76	15.58	7.42	7.39	13.66	16.84

TABLE 78.

[Mountain Camp, August 21 1881. Observer, J. J. Nauty. Actinometer No 3. Medium aperture.]

Time.	Water ther-mometer.	Sun ther-mometer.	Difference.	Exposure.	Time.	Water ther-mometer.	Sun ther-mometer.	Difference.	Exposure.
7ʰ a. m. Sun hid by high cliffs on the east until the time for second series.					7ʰ 42ᵐ A. M.	15 .90	15 .90	00 .90	Shade.
					47	15 .90	1.96 .76	11 .76	Sun.
					52	15 .90	120 .80	11 .80	Sun.
					57	14 .98	130 .70	15 .70	Sun.
					8 02	14 .96	120 .90	15 .90	Shade.
					07	14 .94	130 .90	12 .90	Shade.
					12	14 .90	15 .90	9 .90	Shade.

r. = 16 .50 m. 215. rob.
Sky clear. Wind, calm.

11ʰ 40ᵐ a. m. Instrument not adjusted. Observations worthless.					12ʰ 48ᵐ P. M.	21 .90	21 .70	0 .42	Shade.
					1ᵐ	21 .98	34 .25	12 .25	Sun.
					5	22 .00	37 .18	15 .18	Sun.
					15	22 .02	38 .16	16 .11	Sun.
					20	22 .01	27 .12	5 .08	Shade.
					25	22 .02	23 .71	1 .69	Shade.
					40	22 .00	122 .63	00 .63	Shade.

m. 16 .96 m. 231 rob. = 3 .92.
Sky clear. Wind, calm.

4ʰ 00ᵐ P. M.	21 .22	21 .26	0 .04	Shade	4 .00 P. M.	20 .31	21 .18	0 .87	Shade.
05	21 .02	21 .53	10 .51	Sun	5	20 .11	30 .13	10 .02	Sun.
10	20 .88	34 .52	12 .51	Sun.	10	19 .88	32 .71	12 .73	Sun.
15	20 .71	34 .81	11 .10	Sun.	15	19 .81	23 .21	13 .13	Sun.
20	20 .56	25 .52	4 .96	Shade.	20	19 .66	24 .22	4 .56	Shade.
25	20 .41	22 .40	1 .90	Shade.	25	19 .53	24 .26	1 .83	Shade.
30	20 .71	21 .18	0 .87	Shade	30	19 .47	20 .41	0 .94	Shade.

θ. = 14 .97 m. 205. rob. 3 .07.
Sky, clear. Wind, fresh.

θ = 14 .45 m = 207. rob. 2 .99.
Sky, clear. Wind, fresh, variable.

TABLE 79.

[Mountain Camp, August 22, 1881. Observer J. J. N. Actinometer No 3. Medium aperture.]

Time.	Water ther-mometer.	Sun ther-mometer.	Difference.	Exposure.	Time.	Water ther-mometer.	Sun ther-mometer.	Difference.	Exposure.
7ʰ a. m. Sun hidden by cliffs.					7 42 a. m. Instrument not adjusted. Observations worthless.				
11ʰ 30ᵐ A. M.	21 .80	23 .98	2 .18	Shade.	12ʰ 00 M.	21 .95	22 .52	0 .57	Shade.
35	21 .90	33 .96	12 .06	Sun	5 P. M.	21 .90	33 .52	11 .52	Sun.
40	21 .95	36 .03	14 .03	Sun	10	21 .97	33 .49	11 .52	Sun.
45	22 .00	37 .31	15 .31	Sun	15	21 .97	37 .12	15 .15	Sun
50	22 .00	26 .78	4 .78	Shade	20	21 .96	26 .66	4 .70	Shade.
55	21 .99	23 .63	1 .64	Shade.	25	21 .94	23 .56	1 .62	Shade.
12 00 M.	21 .95	22 .72	0 .72	Shade.	30	21 .90	22 .51	0 .61	Shade.

θ. = 16 .35 m. 273 rob. 3 .81.
Sky, clear. Wind, fresh variable.

θ = 1 .38 m. 271 rob. 3 .80.
Sky, clear. Wind, fresh, variable.

4ʰ 00ᵐ P. M.	21 .07	21 .73	0 .66	Shade	4 .3 P. M.	20 .13	21 .06	0 .93	Shade.
05	20 .91	31 .52	10 .61	Sun.	5	19 .97	30 .51	10 .54	Sun.
10	20 .78	34 .13	13 .37	Sun	10	19 .82	32 .50	12 .68	Sun
15	20 .59	34 .70	14 .11	Sun	15	19 .66	32 .83	13 .19	Sun
20	20 .41	25 .30	4 .80	Shade.	20	19 .50	24 .16	4 .60	Shade.
25	20 .28	22 .30	1 .92	Shade.	25	19 .34	21 .16	1 .82	Shade.
30	20 .13	21 .06	0 .93	Shade.	5 00	19 .18	20 .14	0 .86	Shade.

θ.=15 .14 m. 207. rob.=3 .13.
Sky, clear. Wind, fresh variable.

θ. 14 .40 m. 208, rob. 3 .90.
Sky, clear. Wind, fresh variable.

TABLE 80.

[Mountain Camp, August 23, 1884.   Observer, J. J. N.   Actinometer No. 3.   Medium aperture.]

Time.	Water ther-mometer.	Sun ther-mometer.	Difference.	Exposure.	Time.	Water ther-mometer.	Sun ther-mometer.	Difference.	Exposure.
7ʰ a. m.	Sun hidden by cliffs.				7ʰ42ᵐ A. M.	18 .77	19 .66	0 .89	Shade.
					47	18 .54	30 .47	11 .93	Sun.
					52	18 .31	33 .30	14 .99	Sun.
					57	18 .49	34 .48	16 .29	Sun.
					8 02	18 .06	23 .55	5 .49	Shade.
					07	18 .15	20 .13	1 .98	Shade.
					12	18 .22	18 .96	0 .74	Shade.

θ₀: 17° 20; m = −.218; mθ₀ = 3.75.
Sky, slightly cloudy.   Wind, light.

11ʰ30ᵐ A. M.	21 .60	21 .78	0 .18	Shade.	12ʰ00ᵐ M.	21 .78	22 .49	0 .71	Shade.
35	21 .70	33 .88	12 .18	Sun.	05 P. M.	21 .71	34 .18	12 .47	Sun.
40	21 .74	37 .47	15 .73	Sun.	10	21 .65	37 .45	15 .89	Sun.
45	21 .75	39 .35	16 .60	Sun	15	21 .62	38 .22	16 .60	Sun.
50	21 .76	27 .05	5 .29	Shade.	20	21 .60	27 .00	5 .40	Shade.
55	21 .77	23 .70	1 .93	Shade.	25	21 .63	23 .53	1 .90	Shade.
12 00 M.	21 .78	22 .49	0 .71	Shade.	30	21 .66	22 .37	0 .71	Shade.

θ₀: (17 .48); m = .224; mθ₀ 3 .91.
Sky, clear.   Wind, fresh, variable.

θ₀ 17 .55; m = .224; mθ₀ = 3 .93.
Sky, clear.   Wind, fresh, variable.

4ʰ00ᵐ P. M.	20 .32	20 .47	0 .15	Shade.	4ʰ30ᵐ P. M.	19 .66	20 .50	0 .84	Shade.
05	20 .22	31 .51	11 .29	Sun.	35	19 .51	30 .52	10 .98	Sun.
10	20 .12	34 .70	14 .58	Sun.	40	19 .42	33 .04	13 .62	Sun.
15	20 .00	35 .52	15 .52	Sun	45	19 .30	33 .90	11 .60	Sun.
20	19 .90	26 .14	5 .24	Shade.	50	19 .17	24 .14	4 .97	Shade.
25	19 .79	21 .78	1 .99	Shade.	55	19 .04	20 .97	1 .93	Shade.
30	19 .66	20 .50	0 .84	Shade.	5 00	18 .90	19 .70	0 .83	Shade.

θ₀ 16 .26; m = .212; mθ₀ 3 .45.
Sky, clear.   Wind, light, variable.

θ₀ 15 .50; m = .212; mθ₀ = 3 .31.
Sky, clear.   Wind, light, variable.

TABLE 81.

[Mountain Camp, August 24, 1884.   Observer, J. J. N.   Actinometer No. 3.   Medium aperture.]

Time.	Water ther-mometer.	Sun ther-mometer.	Difference.	Exposure.	Time.	Water ther-mometer.	Sun ther-mometer.	Difference.	Exposure.
7ʰ A. M.	Sun hidden by cliffs.				7ʰ42ᵐ A. M.	....	(0 .00)	Shade.	
					47	19 .14	30 .65	11 .51	Sun.
					52	19 .02	33 .86	14 .40	Sun
					57	19 .00	34 .77	15 .77	Sun.
					8 02	19 .00	24 .06	5 .06	Shade.
					07	19 .08	20 .86	1 .78	Shade.
					12	19 .12	19 .77	0 .65	Shade.

θ₀ = 16° 29   m = .223   mθ₀ = 3 .64.
Sky, clear.   Wind, calm.

11ʰ30ᵐ A. M.	20 .18	20 .23	0 .05	Shade.	12ʰ00ᵐ M.	20 .54	21 .24	0 .70	Shade.
35	20 .54	32 .55	12 .01	Sun.	05	20 .51	32 .88	12 .37	Sun.
40	20 .51	35 .80	15 .29	Sun.	10	20 .43	35 .87	15 .44	Sun.
45	20 .50	36 .83	16 .33	Sun.	15	20 .33	36 .52	16 .19	Sun.
50	20 .50	25 .72	5 .22	Shade.	20	20 .20	25 .44	5 .24	Shade
55	20 .51	22 .40	1 .89	Shade.	25	20 .16	21 .93	1 .77	Shade.
12 00 M.	20 .54	21 .24	0 .70	Shade.	30	20 .15	20 .80	0 .65	Shade.

θ₀ = 17 .18; m .224; mθ₀ = 3 .85.
Sky, clear.   Wind, light

θ₀ = 17 .14; m = .227; mθ₀ = 3 .89.
Sky, clear.   Wind, light.

4ʰ00ᵐ P. M.	20 .36	.50	0 .20	Shade.	4ʰ30ᵐ P. M.	19 .70	20 .44	0 .74	Shade.
05	20 .18	31 .32	11 .14	Sun.	35	19 .59	30 .34	10 .75	Sun.
10	20 .07	34 .17	14 .10	Sun.	40	19 .43	33 .10	13 .67	Sun.
15	19 .99	34 .89	14 .96	Sun	45	19 .23	33 .88	14 .65	Sun.
20	19 .86	24 .73	4 .87	Shade.	50	19 .12	23 .97	4 .85	Shade.
25	19 .76	21 .55	1 .79	Shade.	55	19 .01	20 .80	1 .79	Shade.
30	19 .71	20 .44	0 .73	Shade.	5 00	18 .90	19 .62	0 .72	Shade.

θ₀ 15 .60; m = .219; mθ₀ 3 .44.
Sky, clear.   Wind, brisk, variable.

θ₀ = 15 .45; m = .217; mθ₀ = 3 .35.
Sky, clear.   Wind, fresh, variable.

TABLE 82.

[Mountain Camp, August 25, 1881. Observer, J. J. N. Actinometer No. 3. Medium aperture.]

Time.	Water ther-mometer.	Sun ther-mometer.	Difference.	Exposure	Time.	Water ther-mometer.	Sun ther-mometer.	Difference.	Exposure
7ʰ A. M. Sun hidden by cliffs.					7:45 A. M.	26 .05	26 .78	0 .73	Shade.
					47	25 .24	37 .00	11 .76	Sun.
					52	24 .50	39 .64	15 .09	Sun.
					57	24 .45	39 .76	15 .31	Sun.
					8 02	21 .22	29 .11	4 .89	Shade.
					07	21 .00	25 .90	1 .90	Shade.
					12	23 .79	24 .80	0 .81	Shade.

$\theta_a = 16 .46, \ m = 219, \ m\theta_a = 3 .60.$
Sky clear. Wind, light.

Time.	Water ther-mometer.	Sun ther-mometer.	Difference.	Exposure	Time.	Water ther-mometer.	Sun ther-mometer.	Difference.	Exposure
11ʰ30ᵐ A. M.	20 .10	20 .15	0 .05	Shade.	12ʰ00ᵐ M.	20 .06	20 .78	0 .72	Shade.
35	20 .10	32 .50	12 .40	Sun.	05 P. M.	20 .05	32 .49	12 .44	Sun.
40	20 .09	33 .72	13 .63	Sun.	10	20 .05	33 .68	13 .63	Sun.
45	20 .08	36 .50	16 .42	Sun.	15	20 .04	36 .49	16 .45	Sun.
50	20 .08	23 .38	5 .30	Shade.	20	20 .03	25 .43	5 .40	Shade.
55	20 .07	22 .00	1 .93	Shade.	25	20 .03	21 .93	1 .90	Shade.
12 00 M.	20 .06	20 .78	0 .72	Shade.	30	20 .02	20 .75	0 .73	Shade.

$\theta_a = 17 .22; \ m = 223, \ m\theta_a = 3 .82.$
Sky, clear. Wind, light.

$\theta_a = 17 .41, \ m = 223; \ m\theta_a \ 3 .89.$
Sky, clear. Wind, light.

Time.	Water ther-mometer.	Sun ther-mometer.	Difference.	Exposure	Time.	Water ther-mometer.	Sun ther-mometer.	Difference.	Exposure
4ʰ00ᵐ P. M.	19 .42	20 .1	0 .7	Shade.	4ʰ30ᵐ P. M.	18 .73	19 .46	0 .73	Shade.
05	19 .22	30 .5	11 .31	Sun.	35	18 .64	29 .66	11 .02	Sun.
10	19 .07	33 .51	14 .41	Sun.	40	18 .57	32 .73	14 .16	Sun.
15	18 .96	34 .55	15 .59	Sun.	45	18 .41	33 .40	14 .99	Sun.
20	18 .87	23 .92	5 .05	Shade.	50	18 .32	23 .35	5 .03	Shade.
25	18 .80	20 .70	1 .90	Shade.	55	18 .18	20 .08	1 .90	Shade.
30	18 .73	19 .46	0 .73	Shade.	5 00	18 .10	18 .80	0 .76	Shade.

$\theta_a = 16 .29, \ m = 219, \ m\theta_a = 3 .77.$
Sky, clear. Wind, fresh, variable.

$\theta_a = 15 .90, \ m = 215, \ m\theta_a = 3 .42.$
Sky, clear. Wind, fresh, variable.

TABLE 83.

[Mountain Camp, August 26, 1881. Observer, J. J. N. Actinometer No. 3. Medium aperture.]

Time.	Water ther-mometer.	Sun ther-mometer.	Difference.	Exposure	Time.	Water ther-mometer.	Sun ther-mometer.	Difference.	Exposure
7ʰ A. M. Sun hidden by cliffs.					7:45 A. M.	19 .38	20 .76	1 .38	Shade.
					50	18 .46	30 .66	12 .20	Sun.
					55	17 .93	32 .45	14 .52	Sun.
					8 00	17 .77	31 .14	14 .36	Sun.
					05	17 .60	22 .80	5 .21	Shade.
					10	17 .64	19 .71	2 .06	Shade.
					15	17 .59	18 .62	1 .03	Shade.

$\theta_a = 16 .79, \ m = 210, \ m\theta_a = 3 .53.$
Sky, clear. Wind, light.

Time.	Water ther-mometer.	Sun ther-mometer.	Difference.	Exposure	Time.	Water ther-mometer.	Sun ther-mometer.	Difference.	Exposure
11ʰ30ᵐ A. M.	15 .97	16 .10	0 .13	Shade.	12ʰ00ᵐ M.	16 .15	17 .04	0 .89	Shade.
35	16 .02	26 .00	11 .98	Sun.	05 P. M.	16 .14	28 .39	12 .25	Sun.
40	16 .68	31 .25	15 .16	Sun.	10	16 .12	31 .60	15 .48	Sun.
45	16 .11	32 .23	16 .12	Sun.	15	16 .09	32 .36	16 .27	Sun.
50	16 .12	21 .56	5 .20	Shade.	20	16 .66	21 .43	5 .13	Shade.
55	16 .16	18 .18	2 .02	Shade.	25	16 .10	18 .13	2 .03	Shade.
12 00 M.	16 .14	17 .04	0 .90	Shade.	30	16 .15	16 .95	0 .80	Shade.

$\theta_a = 16 .92, \ m = 214, \ m\theta_a = 3 .62.$
Sky, clear. Wind, brisk, variable.

$\theta_a = 17 .31, \ m = 218, \ m\theta_a = 3 .76.$
Sky, clear. Wind, fresh, variable.

Time.	Water ther-mometer.	Sun ther-mometer.	Difference.	Exposure	Time.	Water ther-mometer.	Sun ther-mometer.	Difference.	Exposure
4ʰ00ᵐ P. M.	17 .48	17 .76	0 .28	Shade.	4ʰ30ᵐ P. M.	16 .59	17 .55	0 .96	Shade.
05	17 .36	28 .52	11 .16	Sun.	35	16 .37	27 .14	10 .77	Sun.
10	17 .23	31 .33	14 .10	Sun.	40	16 .19	29 .69	13 .50	Sun.
15	17 .06	32 .08	15 .02	Sun.	45	16 .03	30 .28	14 .25	Sun.
20	16 .93	21 .97	5 .04	Shade.	50	15 .88	20 .75	4 .87	Shade.
25	16 .70	18 .80	2 .01	Shade.	55	15 .73	17 .76	2 .03	Shade.
30	16 .55	17 .55	0 .97	Shade.	5 00	15 .56	16 .52	0 .96	Shade.

$\theta_a = 15 .90; \ m = 208, \ m\theta_a = 3 .31.$
Sky, clear. Wind, fresh, variable.

$\theta_a = 15 .40, \ m = 206, \ m\theta_a = 3 .17.$
Sky, clear. Wind, fresh, variable.

TABLE 84.

[Mountain Camp, September 6, 1881. Observer, J. J. N. Actinometer No. 1. Largest aperture.].

Time.	Water ther- mometer.	Sun ther- mometer.	Difference.	Exposure.	Time.	Water ther- mometer.	Sun ther- mometer.	Difference.	Exposure.
7ʰ 42ᵐ. The earlier observations were rendered worthless by an accident to the instrument.					8ʰ15ᵐ A. M.	15°.02	15°.18	0°.16	Shade.
					20	15.20	27.08	11.88	Sun.
					25	15.33	(28.88)	(13.55)	Sun.
					30	15.44	(30.36)	(14.92)	Sun.
					35	15.53	18.78	3.25	Shade.
					40	15.61	16.41	0.80	Shade.
					45	15.64	15.94	0.30	Shade.

θ₀ 15.34 m 295 m θ₀ 4′.42.
Sky, deep blue. Wind, calm.

11ʰ30ᵐ A. M.	19°.34	19°.34	0°.00	Shade.	12ʰ01ᵐ P. M.	19°.77	(19°.88)	(6°.11)	Shade.
35	19.42	32.10	12.68	Sun.	06	19.86	32.84	12.98	Sun.
40	19.48	34.38	14.90	Sun.	11	19.93	35.18	15.25	Sun.
45	19.56	34.94	15.38	Sun.	16	20.00	(35.44)	(15.44)	Sun.
50	19.63	22.90	3.27	Shade.	21	20.05	(23.38)	(3.33)	Shade.
55	19.70	20.46	0.76	Shade.	26	20.12	(20.86)	(0.74)	Shade.
12 00 M.	19.77	19.88	0.11	Shade.	31	20.20	(20.36)	(0.16)	Shade.

θ₀ 15.62; w₀ .302; m θ₀ 4′ 72.
Sky, deep blue. Wind, calm.

θ₀ 15.86; w₀ 308; m θ₀ 4.89.
Sky, deep blue. Wind, calm.

4ʰ00ᵐ P. M.	21°.47	21°.64	0°.17	Shade.
05	21.48	(30.82)	(9.37)	Sun.
10	21.43	(33.86)	(12.43)	Sun.
15	21.40	(33.97)	(12.57)	Sun.
20	21.35	(24.11)	(2.76)	Shade.
25	21.29	(21.95)	(0.66)	Shade.
30	21.22	(31.40)	(0.18)	Shade.

θ₀ 13.36; w₀ .293; m θ₀ 3′.91.
Sky, partly cloudy. Wind, fresh.

NOTE.—The observations of this day made with the large actinometer (No. 1) are not directly comparable with the rest.
The first and last series are not included in the final summary, the instrument having been disturbed. They are, however, not altogether valueless.

TABLE 85.

Summary of actinometer observations at Mountain Camp.

(The method pursued in deducing the corrected from the uncorrected observations is explained in the next chapter.)

[Instrument, small actinometer No. 3.]

Date.	Duration of experiment. 7ʰ 42ᵐ to 8ʰ 12ᵐ.		Duration of experiment. 11ʰ 30ᵐ to 12ʰ 00ᵐ.		Duration of experiment. 12ʰ 00ᵐ to 12ʰ 30ᵐ.		Duration of experiment. 4ʰ 00ᵐ to 4ʰ 30ᵐ.		Duration of experiment. 4ʰ 30ᵐ to 5ʰ 00ᵐ.		Summation of daily readings.
	Uncorrected observations.	Corrected observations.	Uncorrected observations.	Corrected observations.	Uncorrected observations.	Corrected observations.	Uncorrected observations.	Corrected observations.	Uncorrected observations.	Corrected observations.	
1881.	Cal.	Cal.	Cal.	Cal.	Cal.	Cal.	Cal.	Cal.	Cal.	Cal.	
Aug. 21	1.428	1.729	......	(1.882)	1.569	1.931	1.229	1.493	1.198	1.451	8.486
22		(1.752)	1.925	1.878	1.472	1.841	1.253	1.523	1.199	1.452	8.418
23	1.502	1.819	1.568	1.930	1.573	1.905	1.390	1.677	1.323	1.601	8.962
24	1.454	1.760	1.540	1.895	1.558	1.918	1.376	1.666	1.343	1.625	8.864
25	1.443	1.746	1.531	1.884	1.556	1.916	1.429	1.736	1.369	1.658	8.940
26	1.412	1.709	1.450	1.784	1.514	1.862	1.324	1.607	1.271	1.538	8.506
Sept. 6		(1.752)	1.560	1.920	1.615	1.989	......	(1.617)	......	(1.554)	8.892
Mean	1.418	1.732	1.529	1.882	1.561	1.909	1.332	1.617	1.284	1.554	......

	7ʰ 42ᵐ to 8ʰ 12ᵐ.	11ʰ 30ᵐ to 12ʰ 00ᵐ.	12ʰ 00ᵐ to 12ʰ 30ᵐ.	4ʰ 00ᵐ to 4ʰ 30ᵐ.	4ʰ 30ᵐ to 5ʰ 00ᵐ.
Mean time of exposure to sun...............	7ʰ 56ᵐ a. m.	11ʰ 38ᵐ a. m.	12ʰ 08ᵐ p. m.	4ʰ 08ᵐ p. m.	4ʰ 38ᵐ p. m.
Sun's mean hour angle :..................	4ʰ 12ᵐ	0ʰ 34ᵐ	0ʰ 34ᵐ	4ʰ 06ᵐ	4ʰ 36ᵐ
Sun's mean zenith distance	62° 09′	26.36′	26 07′	60°38′	66 56′
Airmass	10.72	5.61	5.58	10.33	12.73

The observations of September 6 made with actinometer No. 1 have here been reduced to the standard of No. 3, in order to render them comparable with the rest.

Acr Yo J K: Whitney Aug 21.   SHOWING ATTENTION OF OBSERVER DISTURBED

Acr No 2, Lone Pine, Aug 9   CLOUDS.

Acr Yo 1 Lone Pine Aug 5   HAZE

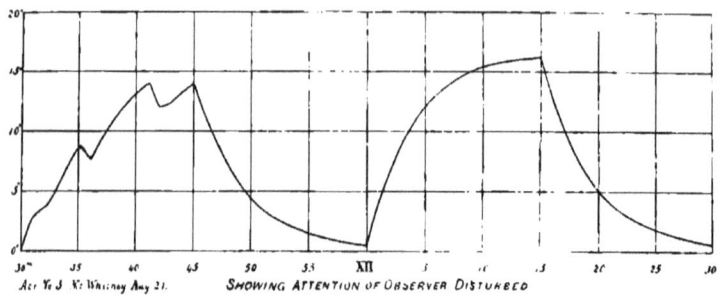

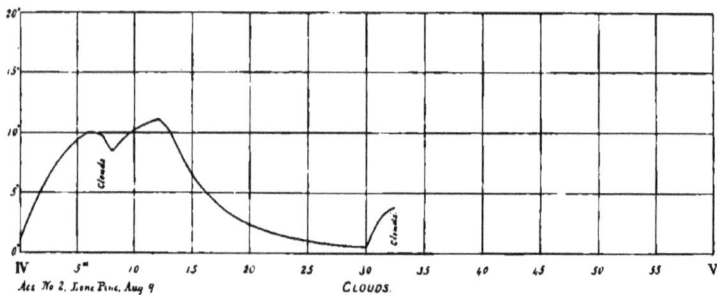

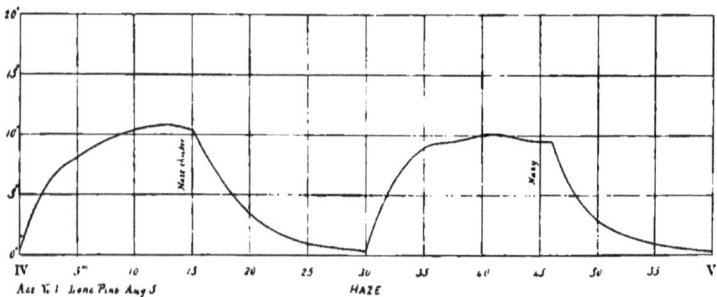

PLATE II
IMPERFECT AND IRREGULAR ACTINOMETER CURVES.

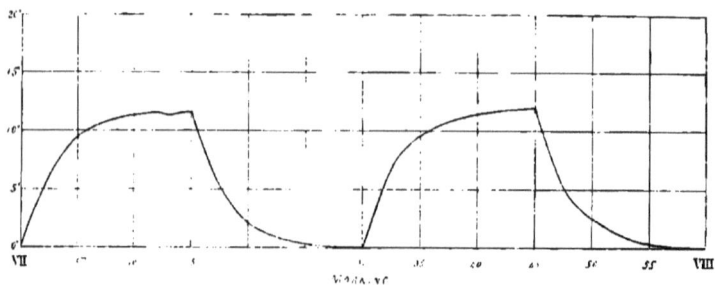

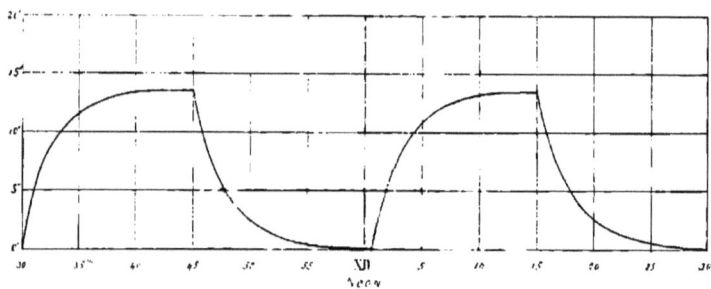

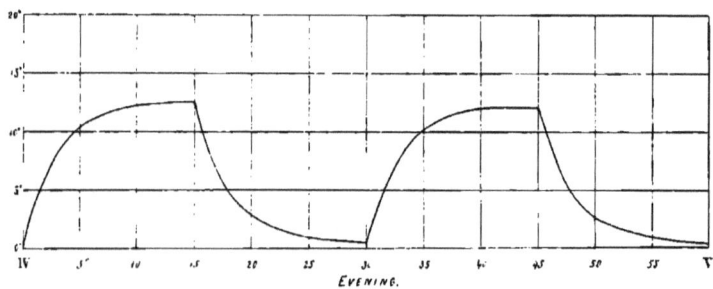

PLATE III
ACTINOMETER CURVES FOR AUGUST 4TH, 1881
LONE PINE—ACTINOMETER No1.

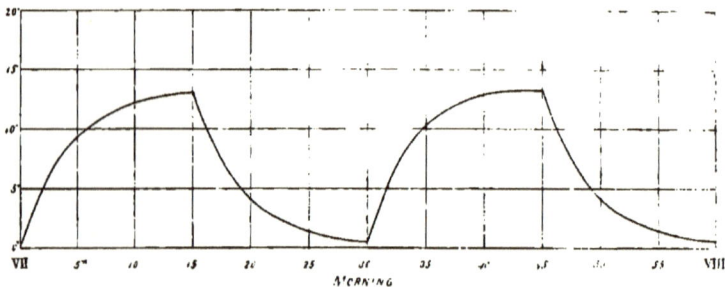

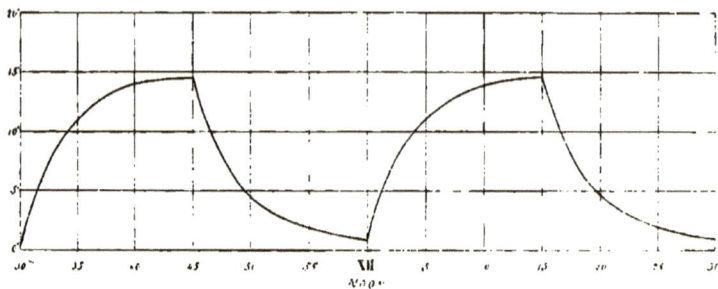

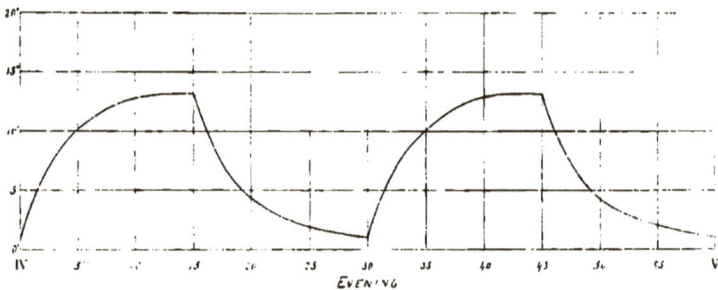

PLATE IV
ACTINOMETER CURVES FOR AUGUST 4TH, 1881.
LONE PINE-ACTINOMETER No2.

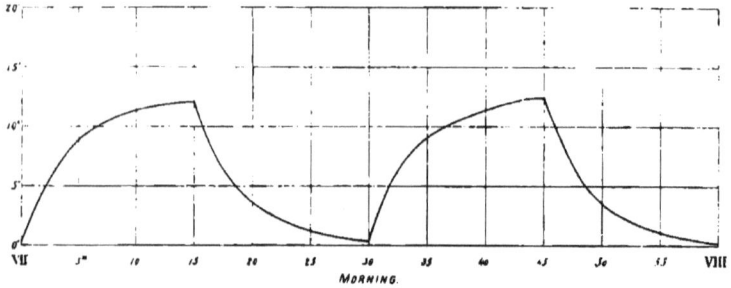

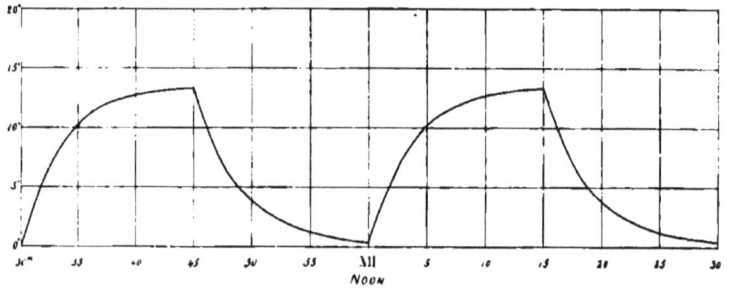

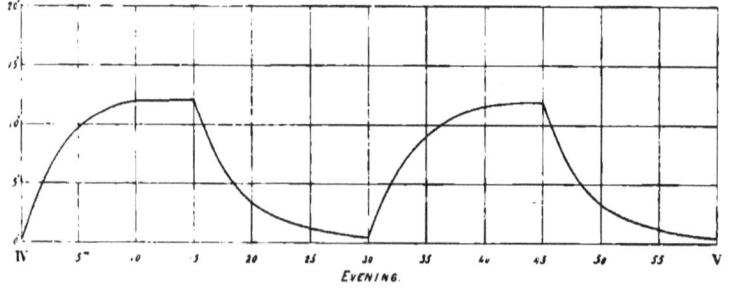

PLATE V.
ACTINOMETER CURVES FOR AUGUST 4TH, 1881.
LONE PINE—ACTINOMETER No.3.

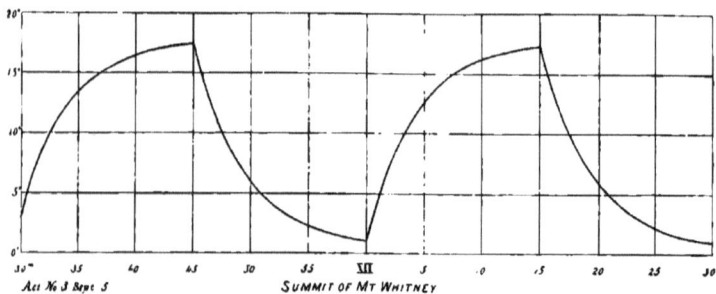

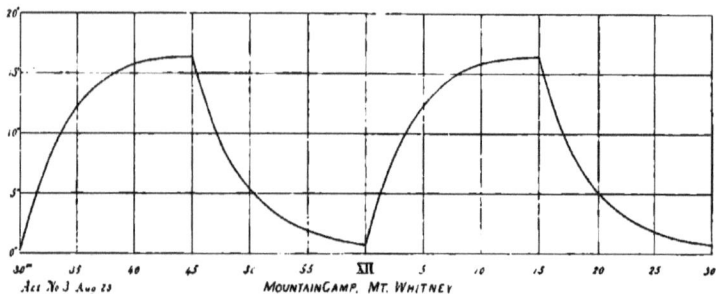

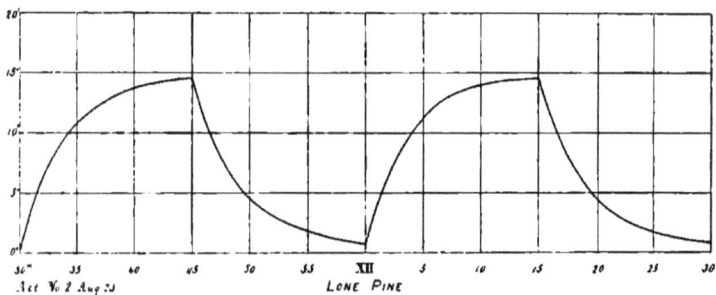

PLATE VI
ACTINOMETER CURVES OF MT. WHITNEY.

Observations with the actinometer on the Peak of Mount Whitney.

Peak of Mount Whitney, September 5 1881.   Observer, J. E. Keeler.   Actinometer No. 3.   Medium aperture.

Time.	Water ther-mometer.	Sun ther-mometer.	Difference.	Exposure.	Time.	Water ther-mometer.	Sun ther-mometer.	Difference.	Exposure
					8ᵒ 00ᵐ A. M	18 . 20	18 . 90	0 . 70	Shade.
					05	17 . 55	29 . 40	11 . 85	Sun.
					10	17 . 60	32 . 00	15 . 00	Sun.
					15	16 . 10	32 . 51	5 . 41	Sun.
					20	15 . 87	21 . 08	5 . 21	Shade.
					25	15 . 58	17 . 20	1 . 82	Shade.
					30	15 . 60	15 . 68	0 . 08	Shade.

θ₀ 16 . 87, m 224, m θ² 3 . 75.
Sky, deep violet.   Wind, very light.

Time.	Water ther-mometer.	Sun ther-mometer.	Difference.	Exposure.	Time.	Water ther-mometer.	Sun ther-mometer.	Difference.	Exposure
11 . 30 A. M	14 . 80	15 . 30	1 . 00	Shade.	12 . 00 M	14 . 29	15 . 30	1 . 01	Shade.
25	14 . 80	27 . 84	1 . 04	Sun.	05 P. M	14 . 22	27 . 03	12 . 81	Sun.
40	14 . 71	31 . 06	16 . 35	Sun.	10	14 . 15	30 . 50	16 . 35	Sun.
45	14 . 61	32 . 06	17 . 45	Sun	15	14 . 09	31 . 32	17 . 23	Sun.
50	14 . 48	20 . 45	5 . 97	Shade.	20	14 . 54	12 . 94	5 . 80	Shade.
55	14 . 48	16 . 70	2 . 42	Shade.	25	14 . 99	16 . 12	2 . 13	Shade.
12 . 00 M.	14 . 29	15 . 30	1 . 01	Shade.	30	13 . 94	14 . 90	0 . 86	Shade.

θ₀ 14 . 61, m 210, m θ² 3 . 01.
Sky, deep violet.   Wind, very light breeze.

θ₀ 14 . 30, m 216, m θ² 2 . 97.
Sky, deep violet.   Wind, very light breeze.

From the above we have for September 5:

At—		8ʰ 08ᵐ A. M	11ʰ 37ᵐ A. M.	12ʰ 07ᵐ P. M.
		Cal.	Cal.	Cal.
Uncorrected observations		1 . 533	1 . 567	1 . 591
Corrected observations		1 . 941	1 . 926	1 . 951

The preceding observations have all been examined, and represented by means of graphical constructions, accompanying the original reductions and serving as a check upon them.

Where the observations were interrupted or sensibly affected by haze, clouds, or other cause, the curve exhibits the defect in kind and degree in a striking manner. We have omitted a great number thus defective, but give one or two examples of defective curves in illustration. (See Plate No. 11, "Imperfect and irregular actinometer curves.")

We give, as examples fairly typical of a great number, three plates, showing the readings of actinometers Nos. 1, 2, and 3 on August 4, at Lone Pine (Plates III, IV, and V), and also a plate (No. VI) showing curves of observations at the Peak of Whitney September 5, Mountain Camp on August 23, and Lone Pine on August 25.

It is to be observed that the smoothness and uniformity of these curves in general is due not only to the exquisitely clear sky and absence of all ordinary disturbing causes, but to the skill of the observers, who were very thoroughly drilled and practiced before these were taken.

Nothing has been done to smooth the curves, which as now engraved faithfully represent the accuracy of the original observer.

CHAPTER VIII.

ACTINOMETER CORRECTIONS.

CORRECTION TO THE RESULTS OF ACTINOMETRIC REDUCTIONS.

A great deal of labor had been already spent in reducing the actinometric observations by the method proposed by M. Violle, when it became clear that the actual initial rise of the thermometer was in every case greater than this method made it. It was then necessary to apply a correction to the results thus obtained. It is here called "correction A." The necessity of correction A, it will be seen, would not arise with direct observation by the method which was finally adopted. It is, therefore, special to the observations made and reduced by M. Violle's method.

A second correction arises from the imperfect conductivity of mercury. It is called here "correction B."

A third correction must be made for the imperfect absorption of heat by the thermometer bulb. It is here called "correction C."

A fourth correction ("correction D") is due to the fact that Violle's method demands in theory an unlimitedly long exposure, and that in practice, when we limit this exposure to 15 minutes, the results are too small. This correction, then, is special to the observations made by this method.

All the above corrections are instrumental ones, and all are additive.

A fifth correction ("correction E") is due to the fact that the portion of the thermometer's heat lost by convection and conduction varies as the air is rarer or denser. This correction is instrumental and is negative.

The necessity of the sixth correction ("correction F") is indicated by M. Violle. It arises from the fact that the actinometer registers radiations from the portion of sky immediately around the sun with those from the sun itself. It is insignificant in amount as compared with the others, and is subtractive. It is the only one of the preceding list which M. Violle applies. • • • Though most of the above corrections are here applied for the first time, they affect the value of the solar constant most materially. Their method of determination as well as their value is therefore given in detail.

Determination of the first correction (A), whose application shall reduce the initial rates, inferred by M. Violle's method, to the true initial rates which would be given by direct observation.

Since the losses of temperature by radiation, convection, and conduction are more considerable as the difference between the temperature of the thermometer and of its inclosure increases, the errors produced by neglect or erroneous estimation of these losses will be least if we make our experiments of short duration and allow the temperature of the exposed bulb to vary only slightly from that of the inclosure.

Two methods of procedure suggest themselves.

100

DETERMINATION OF CORRECTION A, BY FIRST METHOD.

Take the readings of the first three minutes of an exposure. If the sky remains clear we may compare the initial rate of heating, calculated from these three minutes, with that of the complete series.

Let n = the greatest attainable excess of temperature, and θ_1, θ_2, θ_3, the excesses at the end of 1, 2, and 3 minutes. Since, under the conditions of the whole experiment, radiation is approximately proportional to the excess of temperature; for a brief time we may treat it as exactly proportional, without sensible error; that is, we may graphically represent these excesses by the ordinates of a logarithmic curve; and (since the lengths of three equidistant ordinates must be in geometrical progression) to determine the axes of such a curve to be passed through these points, in case θ_1, θ_2, θ_3, are not already in geometric progression, they must be made so by the addition of a constant, n. The common ratio is

$$a = \frac{n - \theta_1}{n - \theta_2} = \frac{n - \theta_2}{n - \theta_3}$$

whence

$$n = \frac{\theta^2_2 - \theta_1 \theta_3}{2\theta_2 - (\theta_1 + \theta_3)}$$

and since $a = \varepsilon^m$, where ε is the Napierian base,

$$m = \frac{\log a}{\log \varepsilon} = \frac{\log a}{0.4343}$$

and the initial rate of heating per minute is represented by the product $m \times n$, if the sun thermometer starts exactly at the temperature of its environment at the instant of exposure, or by $n \times (n - \theta)$, if there is an excess of temperature, θ. For example, take the following observation made on Mount Whitney, from 11ʰ 30ᵐ to 12ʰ 00ᵐ, August 23, 1881:

$$t_0 = 0^{m}, \quad \theta = 0.48 = \text{excess of temperature at the instant of exposure.}$$

$$t_1 = 1^{m}, \quad \theta_1 = 3.60 \ \big\} \ n - \theta_1 = 12.33, \log (n - \theta_1) = 1.0909$$
$$t_2 = 2^{m}, \quad \theta_2 = 6.60 \ \big\} \ n = 15.93 \ \big\{ \ n - \theta_2 = 9.33, \log (n - \theta_2) = 0.9699$$
$$t_3 = 3^{m}, \quad \theta_3 = 8.87 \ \big\} \ n - \theta_3 = 7.06, \log (n - \theta_3) = 0.8488$$

From (1) and (2):

$$\log a = 0.1210$$

From (2) and (3):

$$\log a = 0.1211$$

and

$$m = \frac{0.1211}{0.4343} = 0.2788 \qquad m \times (n - \theta) = 0.2788 \times 15.75 = 17.391$$

The entire series, including 15 minutes' exposure to the solar radiation and an equal time for cooling, gave, when reduced by Violle's method, an initial rate of 37.914 per minute.

The following table gives the results of both methods of computation for a considerable number of observations.

TABLE 86.

Station.	Date.	Hour.	Initial rate from first, second, and third minutes.	Initial rate by Violle's method.
	1881.			
Mountain Camp (Whitney)	Aug. 21	12 10 -12 40	4 208	3 908
	Aug. 22	11 40 -12 00	4 594	3 914
	Aug. 23	12 00 -12 30	4 895	3 920
	Aug. 24	11 30 -12 00	4 745	3 848
	Aug. 24	12 00 -12 30	4 710	3 880
	Aug. 25	11 30 -12 00	4 620	4 573
		P. M.		
Mountain Camp (Whitney)	Aug. 21	4 30 - 5 00	3 603	2 994
	Aug. 22	4 30 - 5 00	4 758	3 905
	Aug. 24	4 30 - 5 00	3 506	3 305
	Aug. 24	4 30 - 5 00	3 205	3 53
	Aug. 25	4 50 - 5 00	3 854	3 448
	Aug. 26	4 30 - 5 00	4 747	3 173
Lone Pine	Aug. 21	12 00 - 12 30	3 120	3 279
	Aug. 23	11 30 - 12 00	3 079	3 434
	Aug. 24	12 00 -12 30	2 687	3 160
	Aug. 24	11 30 -12 00	4 782	3 368
	Aug. 24	12 00 -12 30	3 870	3 338
		P. M.		
Lone Pine.....................	Aug. 21	11 30 -12 00	3 903	3 420
	Aug. 21	1 30 - 5 00	3 228	2 607
	Aug. 22	1 30 - 5 00	3 136	2 611
	Aug. 23	1 30 - 5 00	3 524	2 719
	Aug. 24	1 30 - 5 00	3 056	2 823
	Aug. 25	1 30 - 5 00	3 658	2 554
	Aug. 27	1 30 - 5 00	3 401	2 705
Mountain Camp	Sept. 5	11 30 12 00	3 511	5 157
	Sept. 5	12 00 -12 30	4 587	4 196
	Sept. 6	11 30 -12 00	3 720	4 718
	Sept. 6	12 01 -12 31	5 424	4 887
	1882.			
Allegheny	Mar. 1	11 50½-12 4 0½	4 574	4 664
	Mar. 1	12 00½ -12 50½	4 5½2	4 164
	Mar. 1	12 52½ - 1 42½	5 605	4 200
Mean	....		4 083	3 566

The average initial rate from 34 observations is by the first method 4°.083, and by Violle's method 3°.566, and the ratio of these numbers is 1.145, whence a correction of 11.5 per cent. ought to be added to a result deduced by Violle's method of computation, according to this comparison.

DETERMINATION OF CORRECTION A, BY SECOND METHOD.

By the second plan we obtain direct observations of the initial rate of heating for short periods of exposure (15 or 30 seconds), exactly determined by an audible signal from a standard clock, or better still, automatically regulated by an electro-magnetic mechanism, controlled by the clock.

This method requires preferably two actinometers and two observers—one to carry on the ordinary routine, alternately exposing and shading his instrument for 15 minutes—the other to make simultaneous direct observations of the initial rate of the second instrument for short exposures. The mode of procedure in the second instance is as follows: The sun-thermometer is taken out of the case and cooled as much below the temperature of the water as it is expected to rise above it in the course of the experiment. We thus insure that, during the first part of the exposure, the bulb of the sun-thermometer shall be receiving heat from the inclosure as well as from the sun, while, during the second part, it receives heat from the sun but radiates it to the inclosure. The amounts of heat received from or radiated to the surrounding water-jacket by the thermometer are nearly equal, though not exactly so, because the middle temperature is attained in less than half the time of exposure, whence the cooling agencies are slightly more effective than the heating, and the initial rate thus measured *will therefore still be below the truth*, although, as will be seen, it is in all cases larger than that inferred by the usual process. Having cooled the thermometer, the observer transfers it to the case, centers it, reads its temperature, as well as that of the well-mixed water in the surrounding jacket, and exposes it at the beat of a loud sounding relay, which repeats the ticking of the clock. Then, having counted the seconds from 0 to 30, he closes the shutter at the instant of the 30th beat and proceeds to read the temperature attained, which may be done in a comparatively leisurely manner, since, though the reading rises rapidly, it scarcely falls perceptibly for several seconds. This observation is therefore far more accurate than that obtained by a hasty glance while the mercury column is still moving rapidly up, and thus, although the total change is small, the degree of accuracy is comparable with that obtained in the ordinary way. Moreover, a considerable number of observations can be made in a short time. The operations, however, need to be performed with care, and the frequent handling of the thermometer is, of course, attended with increased danger of breakage.

The following special observations have been made by this second method for the purpose of comparing the method of reduction used by M. Violle with the results of direct observation:

TABLE 87.*

[Allegheny, March 29, 1882. Observer, F. W. Very. Measurements made at noon. Sky, milky blue.]

Actinometer No. 1 (large)				Actinometer No. 2 (small)			
Interval of time, t.	Excess of temperature.			Interval of time, t.	Excess of temperature.		
	Heating, θ.	Cooling, θ'.	Sun, $\theta + \theta'$.		Heating, θ.	Cooling, θ'.	Sun, $\theta + \theta'$.
Min.				*Min.*			
0	− 0.18	12.15	11.97	0	0.19	13.80	13.99
5	8.75	2.50	11.25	5	9.25	4.60	13.85
10	11.25	0.55	11.80	10	12.50	2.00	14.50
15	12.15	0.05	12.20	15	13.80	1.15	14.95

From the equation $m\,t\,\log e = \log\theta_0 - \log\theta'$.
For $t = 5$, $m = 210$; for $t = 10$, $m = .307$; for $t = 15$, $m = .366$.
Average $m = .294$, $\theta_0 = \theta = \theta' = 11.84$.
$m\,\theta_0 = 3.47$.

From the equation $m\,t\,\log e = \log\theta_0 - \log\theta'$.
For $t = 5$, $m = .227$; for $t = 10$, $m = .197$; for $t = 15$, $m = .168$.
Average $m = .197$, $\theta_0 = \theta = \theta' = 14.32$.
$m\,\theta_0 = 2.82$.

Initial rate obtained from direct measurements.
3.85.

Ratio = $\dfrac{\text{calculated rate}}{\text{observed rate}} = \dfrac{3.85}{3.47}$.

Initial rate obtained from direct measurements.
3.35.

Ratio = $\dfrac{\text{calculated rate}}{\text{observed rate}} = \dfrac{3.35}{2.82}$ 1.188

The comparisons of March 29 having been made by only one observer, were necessarily not absolutely synchronous. The sky however remained uniform, and the results are considered trustworthy.

TABLE 88.

[Allegheny, October 20, 1882. Observer, J. E. Keeler. Sky, milky blue with occasional thin smoke.]

Actinometer No. 2 (small), measurements made from 11h 30m to 12h 00m.				Actinometer No. 2 (small), measurements made from 12h 00m to 12h 30m.			
Interval of time, t.	Excess of temperature.			Interval of time, t.	Excess of temperature.		
	Heating, θ.	Cooling, θ'.	Sun, $\theta + \theta'$.		Heating, θ.	Cooling, θ'.	Sun, $\theta + \theta'$.
Min.				*Min.*			
0	0.70	13.35	14.00	0	1.22	13.40	14.62
5	10.05	4.67	14.50	5	10.50	4.00	14.99
10	12.50	2.08	14.58	10	13.00	2.00	15.00
15	13.30	1.22	14.52	15	13.40	1.12	14.52

From the equation $m\,t\,\log e = \log\theta_0 - \log\theta'$.
For $t = 5$, $m = .226$; for $t = 10$, $m = .194$; for $t = 15$, $m = .165$.
Average $m = .195$, $\theta_0 = \theta = \theta' = 14.45$.
$m\,\theta_0 = 2.82$.

$m\,\theta_0 = \dfrac{M}{N} = 2.82$. $.070 = 1.148$ cal.

From the equation $m\,t\,\log e = \log\theta_0 - \log\theta'$.
For $t = 5$, $m = .234$; for $t = 10$, $m = .200$; for $t = 15$, $m = .172$.
Average $m = .202$, $\theta_0 = \theta = \theta' = 14.75$.
$m\,\theta_0 = 2.99$.

$m\,\theta_0 = \dfrac{M}{N} = 2.99$. $.070 = 1.217$ cal.

TABLE 89.

[Actinometer No. 1 (large), direct observation of initial rate synchronously with above observation of October 20.]

15 seconds exposure.	30 seconds exposure.	60 seconds exposure.	Remarks
1.00	1.95	4.05	
1.05	1.95	3.95	
0.95 } mean 1.02	2.00 } mean 1.99	3.95 } mean 3.98	Time from 11h 15m to
1.05	2.05	4.05	12h 45m. Observer,
1.05	2.00	3.90	F. W. Very.

Mean ÷ 4 = 4.08.   Mean ÷ 2 = 3.98.   Mean ÷ 1 = 3.98.

Mean of all = 4°.01 (15 observations), 4.01 × .3484 = 1.397 calories.

TABLE 90.

(Allegheny, October 20, 1882. Observer J. E. Keeler. Light breeze. Sky, good blue. A little smoke toward sun.)

Actinometer No. 1 (larger), measurements made from 12ʰ 45ᵐ to 1ʰ 15ᵐ.				Actinometer No. 1 (larger), measurements made from 1ʰ 15ᵐ to 1ʰ 45ᵐ.			
Interval of time, t.	Excess of temperature.			Interval of time, t.	Excess of temperature.		
	Heating, θ.	Cooling, θ'.	Sum, θ + θ'.		Heating, θ.	Cooling, θ'.	Sum, θ + θ'.
Min.				Min.			
0	0 .18	12 .10	12 .28	0	0 .21	12 .20	12 .41
5	10 .53	2 .67	13 .20	5	9 .79	2 .62	12 .41
10	11 .97	0 .69	12 .66	10	11 .71	0 .69	12 .40
15	12 .10	0 .22	12 .32	15	12 .20	0 .20	12 .40

From the equation of t (log θ : log θ = log θ + θ'. For t=5, m = .510; for t=10, m = .291; for t=15, m = .270. Average m = .290, θ + θ' 12 .61. mθ, 3 .66. mθ = M/N .466; and 1.276 calories.

From the equation of t (log θ : log θ = log θ + θ'. For t=5, m = .511; for t=10, m = .289; for t=15, m = .273. Average m = .282, θ + θ' 12 .41. mθ, 3 .62. nθ = M/N .362; and 1.261 calories.

TABLE 91.

Actinometer No. 2 (small), direct observation of initial rate synchronously with above.)

15 seconds exposure.	30 seconds exposure.	60 seconds exposure.	Remarks.
3 .89 0 .85 3 .71 } mean 0 .82	3 .79 3 .75 3 .53 } mean 3 .68	3 .43 3 .55 3 .29 } mean 3 .32	Time from 1ʰ 55ᵐ to 1ʰ 51ᵐ. Observer, F. W. Very.

Mean 4 3 .58   Mean 2 3 .36   Mean 1 3 .33

Mean of all 3 .32 (9 observations), 3.32 .4070 1.352 calories.

The observations of October 20, 1882, may be thus summarized.

With actinometer No. 1 :

(1).—Initial rate by Violle's method .... { 3°.66 / 3°.62 } Mean .... 3°.64

(2).—Initial rate by direct method .... 4 .01 (Mean of 15 observations).

With actinometer No. 2 :

(3).—Initial rate by Violle's method........ .... { 2°.82 / 2°.99 } Mean .... 2°.91

(4).—Initial rate by direct method .... 3°.32 (Mean of 9 observations).

(1) is synchronous with (4), and (2) with (3).

In order to compare the radiations measured synchronously, but by different instruments, the measurements made with actinometer No. 2 have been reduced to the standard of No. 1 by multiplying the results, expressed in calories, by the factor 1.054, whose determination is described further on. We then have—

(1).—Actinometer No. 1, Calories by Violle's method ... ... 1.276 and 1.261
(2).—Actinometer No. 2, Calories by direct observation .... 1.424 and 1.424

And their ratios are .... 1.117 and 1.129
(3).—Actinometer No. 2, Calories by Violle's method .... 1.210 and 1.283
(2).—Actinometer No. 1, Calories by direct observation... .... 1.397 and 1.397

And their ratios are .... 1.154 and 1.088

From the observations of March 29 and October 20, 1882.

The conclusion from the mean of six comparisons by the second method is that the Violle method gives a result which is to be increased by at least 13.2 per cent. to equal the result of direct observation (which is itself too small). Although the number of comparisons by the second method is very much smaller than by the first, they are so much more reliable that equal weights will be given to the mean of each set, and the finally adopted value of correction A is + 13.8 per cent.

Determination of the second correction (correction B) for imperfect conductivity of the mercury in the bulb of the thermometer used for measuring the intensity of solar radiation.

The communication of heat to the mercury within the bulb of a thermometer takes place partly by conduction and partly by convection currents in the liquid. If the heat is applied from below, the convection currents attain their maximum energy; but if the source of heat is above the thermometer, the communication of heat must be largely due to conduction, which, on account of the imperfect conductivity of the mercury, is slow. If the heat is received from the side, convection currents will be free to act, but in their upward course they meet a surface already heated, and must be far less efficient than when they rise from a lower heated hemisphere into a cool upper one.

It follows that the altitude of the sun affects the accuracy of the indications of the solar thermometer, as has been well pointed out by Mr. Ericsson ("Contributions to the Centennial Exhibition." Chap. XVII), and that the most reliable use of the thermometer as a measurer of radiation requires that its lower surface should be exposed to the source of heat. This being impracticable in ordinary actinometric measurements, a correction must be applied to all observations to reduce them to what they would have been with a *nadir sun*. This correction has been determined as follows by Mr. F. W. Very: A beam of sunlight, being kept fixed in a horizontal direction by a heliostat, was received upon a second mirror, which reflected it either upwards, downwards, or horizontally, the actinometer being correspondingly and successively placed above, below, and at the side. (See Fig. 8.) The rise of the

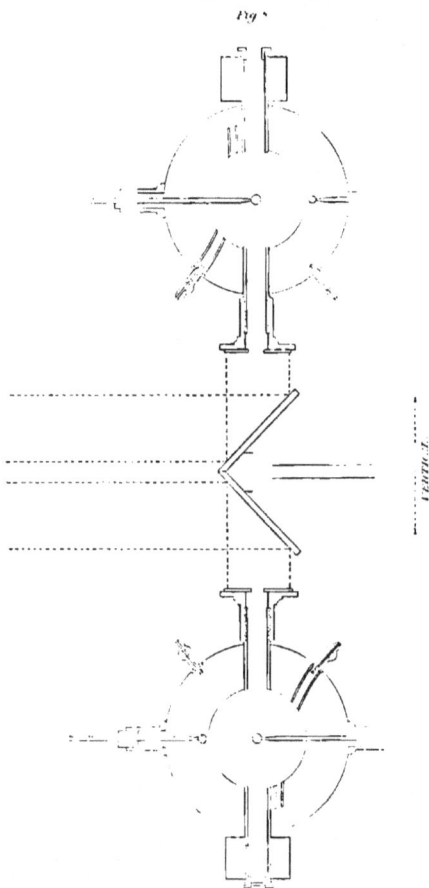

Fig. 8.

Arrangement of apparatus as used in the determination of the correction for nadir sun.

sun thermometer in one minute was noted in each case, and the observations were repeated often enough to eliminate error from atmospheric changes.

An example is here given in full. The components of a pair were taken in as rapid succession as possible. The exposures could be timed with great precision by listening to the beats of a loud-sounding relay, which repeated the ticks of the observatory clock by the observer's side.

TABLE 92.

[Station, Allegheny. Observer, F. W. Very.]

Duration of exposure.	Mean time	Water ther mometer.	Sun ther mometer.	Gain.	" gain in 1 minute.	Ratio of Rad: from zenith divided by Rad: from nadir	Direction from which radiation comes
Sec.			9.56 C.				
30 ...	Nov. 14, 1882	9.50 C.	10.62	1.12	2.24		Nadir.
30 ...	12° 24' P. M.	9.48	9.70 10.55	1.05	2.10	.94	Zenith.
			9.20				
30	12 42	9.22	10.58	1.36	2.76		Nadir.
30		9.30	9.30 10.65	1.35	2.70	.95	Zenith.
			9.25				
30	12 55	9.25	10.60	1.35	2.70		Nadir.
30		9.18	9.20 10.35	1.15	2.50	.93	Zenith.
			9.25				
30	1 02	9 23	10.55	1.30	2.60		Nadir.
30		9.25	9.25 10.45	1.20	2.40	.92	Zenith.
			9.20				
60	1 10	9.25	11.15	1.95	1.95		Zenith.
60		9.25	9.25 11.35	2.10	2.10	.93	Nadir.
			9.20				
60	1 18	9.25	11.60	2.40	2.40		Nadir.
60		9.25	9.25 11.50	2.25	2.25	.94	Zenith.
					Mean ..		

The following is a summary of all the results obtained in two days of experiment, each result here given being usually the mean of 5 determinations like those just cited. Those of the Baudin thermometer (obtained on a favorable day) are entitled to special weight.

TABLE 93.

[Station, Allegheny. Observer, F. W. Very.]

Thermometer.	Direction from which radiation comes.	" gain in 1 minute.		Ratio
Baudin 8737...	Zenith	2.25	92	Radiation from zenith.
	Nadir	2.47		Radiation from nadir.
Do	Horizon	2.43	95	Radiation from horizon.
	Nadir	2.57		Radiation from nadir.
Green 4571....	Zenith	1.29	93	Radiation from zenith.
	Nadir	1.39		Radiation from nadir.
Do.........	Horizon	1.85	93	Radiation from horizon.
	Nadir	1.98		Radiation from nadir.

An inspection of these figures shows that the correction for imperfect conductivity of mercury is not only an appreciable one, but is of considerable importance, where the radiation is received from points above the horizon, which is the condition occurring in ordinary actinometric work. For a constant amount of heat received, then, the reading of a thermometer progressively increases as the sun approaches the horizon. To determine the exact law of increase still more experiments are required, but the present ones show that for the thermometers actually employed this increment is approximately proportional to $1 - \cos \frac{1}{2} z$ (z being the zenith distance). We have, in fact, upon expressing the above observations in terms of the radiation from nadir sun (after giving the Baudin thermometer results double weight)—

TABLE 94.

		Observed correction.	Calculated correction
		Per cent.	Per cent.
Mean observed radiation from—			
Zenith		92.3	4.3	4.3
Horizon		94.3	6.6	5.9
Nadir		100.0	0	0

We shall use, then, in the following reductions the empirical formula—

$$T = t + b \cos \tfrac{1}{2} z$$

where

$b =$ the correction (always additive) to the reading of either thermometer receiving heat from a zenith sun, to reduce it to what the thermometer would record if receiving the *same* heat from a nadir sun; $T =$ the corrected reading; $t =$ the observed reading.

It may be observed that, if this correction be neglected, not only will the direct observation be too small, but as a noon observation in this case is small relatively to an afternoon observation, the resulting heat outside the atmosphere (as determined from the two) will be smaller in an enhanced degree.

Per cent.

Taking the high sun's mean noon zenith distance at Lone Pine, 26° 22′ ...... $b \times \cos \tfrac{1}{2} z =$ 8.08
Taking the low sun's mean zenith distance at Lone Pine, 65° 45′ ... ... $b \times \cos \tfrac{1}{2} z =$ 6.97
Taking the high sun's mean zenith distance at Mountain Camp, 26° 22′ ...... $b \times \cos \tfrac{1}{2} z =$ 8.08
Taking the low sun's mean zenith distance at Mountain Camp, 63° 21′ ...... $b \times \cos \tfrac{1}{2} z =$ 7.06

Third actinometer correction (correction C)—Determination of the amount of heat lost through its imperfect absorption by thermometer bulb.

This imperfect absorption is due to various causes, and firstly to the fact that lamp-black, though the best heat absorber known, is yet partial in its action, selecting the short wave-lengths more than the long, so that an ordinary blackened thermometer bulb is probably less sensitive to a given amount of heat of great wave-length, (extreme invisible or dark heat rays,) than to the same amount of visible heat, as Tyndall has pointed out.

We know little about the matter, physicists being accustomed, save in exceptional instances, to treat the lamp black with which their thermo-piles or thermometers are covered, as a perfect absorbent, or at least as an indifferent one. Some experiments of our own, however, indicate that its absorption of certain radiations is selective in a high degree, but as these are not yet complete we can only conclude that the correction will be additive and that were it applied, the general result would be to increase in some small but perceptible degree the value of the solar constant.

Beside the effect of selective absorption we have that of reflection (also to some extent selective), as the spherical form of the bulb causes the rays which fall nearly tangentially to the sphere to be more reflected than those which strike its surface normally. To determine the amount of this latter effect a special thermometer with a hemispherical lamp-blacked bulb, 0.842 cm. in diameter ("Green 534 1") graduated to one tenth degree Centigrade, was designed for use in the large Violle actinometer. It was at first thought that by reversing the bulb, so as to expose alternately the flat and rounded surfaces, when the maximum excess of temperature was attained, a difference of reading might be detected, owing to the diminished loss by reflection with a flat surface. Upon trial it was found that the highest temperature attained with full exposure to the sun (averaging 11½° C. above that of the inclosure) the gain by the use of the flat side of the bulb over the hemispherical surface was only 0.02. This is probably owing to the flat side of the bulb being necessarily (from the mode of construction) thicker than the other, whence a greater amount of heat is retained by the glass of the flat side and slowly yielded to the mercury by conduction and convection when the flat side is turned down, thus compensating in a great measure for the diminished absorption of the hemispherical surface.

There remained one other method, namely, the direct measurements of initial rates with alternate surfaces.

On June 20, 1882, measurements of this character were made at noon, the sky being a good blue, with occasional cumulus clouds.

[Initial rate per minute from direct measurement. Thermometer "Green 5314" in large actinometer case. Station, Allegheny. Observer, F. W. Very:]

Flat side up.   Round side up.

7 .4	7 .2
7 .8	7 .15
7 .0	7 .2
7 .65	7 .3
7 .2	7 .1
7 .85	7 .8
7 48	7 29

From six pairs of measurements, the exposures being separated by an interval of only a few minutes and the sky apparently continuing uniform, it was found that the initial rate, being $7^\circ.29$ with the round side up, a gain of $0^\circ.19$, or 2.6 per cent., was obtained by using the flat side. This value $+0.026$ is therefore adopted as the factor for the ascertained part of "correction C."

Determination of the actinometer correction for unfinished exposure (correction D).

The actinometer thermometer, when used in the present method, should be one with a bulb sufficiently small to rapidly attain its temperature of equilibrium. The Baudin thermometer, used in the large actinometer No. 1, nearly fulfills this requirement when the exposure is prolonged to fifteen minutes, as has usually been done; but the Green thermometers, used in No. 2 and No. 3, are so large that fifteen minutes are not enough to wholly establish equilibrium. Both the heating and cooling curves are therefore incomplete; and, as will be seen by an inspection of any good set of curves, the value of c_i, obtained by taking the sum of a pair of incomplete heating and cooling curves, will be slightly smaller than that from the same curves completed.

Hence an additive correction must be made to the results of all observations with actinometers Nos. 2 and 3, whenever the time of exposure does not exceed fifteen minutes. The value of this correction has been determined from simultaneous observations carried on at the same station with actinometers Nos. 1 and 2.

TABLE 95.

Synchronous comparisons of actinometers.

Station.	Date	Hour.	Actinometer No. 1.	Actinometer No. 2
	1881	h. m. h. m.	Cal.	Cal.
Lone Pine ...	August 3	7 30 to 8 00	1.353	1.32
Lone Pine ...	August 4	7.00 to 7.30	1.303	1.250
Lone Pine ...	August 4	7 30 to 8.00	1.423	1.332
Lone Pine ...	August 4	4 00 to 4.30	1.24*	1.210
Lone Pine ...	August 4	4.30 to 5.00	1.231	1.19*
Lone Pine ...	August 5	11 30 to 12 00	1.555	1.454
Lone Pine ...	August 5	12 00 to 12 30	1 494	1.488
Lone Pine ...	August 5	1.00 to 1 30	0.993	1.047
Lone Pine ...	August 5 ...	4 30 to 5 02	0.995	1.009
	1882			
Allegheny ...	March 17 ...	1. 10 to 1.42	0.662	0.563
Allegheny ...	March 21 ...	10.25 to 10 55	1 165	1.054
Allegheny ...	March 21 ...	11.21 to 11.53	1.881	0.916
Allegheny ...	March 29 ...	11. 48 to 12.20	1 207	1 807
Means				1 115

From the above comparisons, it is found that to reduce observations from incomplete series with actinometer No. 2 or No. 3 to the effect which would have been observed with the fullest exposure, a correction of $+5.4$ per cent. must be added. The actual value of "correction D," hereafter used, is but 3 per cent.; accordingly, so far as this is concerned, our resulting values of the solar constant will be too small.

It will be observed that all our corrections have thus far been additive, that is, they have in every case increased the final result, and not diminished it. A reason for this predominance of additive corrections is to be found in the universal tendency to the dissipation of thermal energy.

We take such precautions as we can to prevent the loss of heat, and nevertheless at every step of the process, heat *is* lost, for the purpose of our measurement, without any compensating gain. Usually in an investigation, the neglected minute instrumental errors tend to compensate each other. Here and for the above reason, all, or nearly all, the instrumental errors have the same sign. We have considered others too minute or too difficult to determine quantitatively with the same result.

There is another class of errors, however, to which this remark does not apply, and which we now investigate.

Determination of the correction for variation of atmospheric pressure (correction E).

A fifth correction (correction E) arises from the fact that beside the cooling of the thermometer from radiation, it loses heat by contact with the air, and more rapidly as the air is denser. Accordingly, other things being equal, the rate of loss of heat for a given excess of temperature will be less on the mountain (where the barometric pressure is less) than at the sea-level.

To determine this correction, the bulb of the thermometer was sealed within a small copper globe two inches in diameter, blackened within, and from which the air could be exhausted by a Sprengel's pump.

First, the globe being filled with air at a barometric pressure of 731 mm. (that prevailing at the time of the experiment at the station, Allegheny), was cooled during 15 minutes from a temperature of excess of 17° to an excess of little more than 0°, and then the best vacuum attainable by the use of the Sprengel's pump having been made in the copper globe, the same experiment was repeated. The experiments were so conducted that the rate of cooling for each degree of excess could be determined with accuracy. A more particular account of them will be found under another head. (See Appendix.)

The curves by which they were originally represented are not here given. By measurements taken on the larger original sheets, we find $-\frac{dy}{dx}$ for various temperatures of excess (*i. e.* the rate of the thermometer's cooling according to its excess over the temperature of the globe) in vacuo and in air, at a pressure of 731 mm. From a comparison of these results, we obtain the loss by convection in air at this pressure, for various temperatures of excess. Thus with an excess of 10°, the convection amounts to 27 per cent. of the total loss; for 5°, 19 per cent.; for 2½° (which may be taken as the average excess of the sun thermometer during its first minute of heating), 13 per cent., which is very nearly the value of "correction A." "Correction A," then, represents nearly that part of the loss, due to convection. We shall assume it to do so exactly. It is not certain that the diminution of convection is directly proportional to the pressure, but experiments rather indicate that for moderate pressures the diminution is less than for very small ones. If we treat, then, the diminution as proportional to the pressure within the range of these experiments, we may conclude that we have rather over than under estimated the amount of the correction itself.

The mean reading of the barometer at noon at the Mountain Camp was 502, which is 233 mm. below 735, the average pressure of the air at Allegheny. We have, then, the proportion

$$735 : 233 :: A : E$$

whence E = 4.4 per cent.

This value of E is probably too large, hence since its sign is negative, we rather under than over estimate the resulting value of the solar constant.

This correction has been taken account of by M. Soret, but, like all the preceding, has been omitted by M. Violle. It is the only considerable correction whose sign is negative.

Taking into account that the effect of this is to diminish correction A, we find that, if we express it here as an independent correction (E), we have

$$\text{For Lone Pine,} \quad E = -.044$$
$$\text{For Mountain Camp,} \quad = -.044$$

Determination of the actinometer correction for sky radiation (correction F).

This correction must be subtractive, since the actinometer has included a part of the radiation from the sky about the sun with that from the sun itself.

In connection with the actinometric observations, photometric measures of the intensity of the light reflected from the portions of the sky in the immediate vicinity of the sun were made both at Lone Pine and Mount Whitney, in order to determine the correction to be applied to the actinometer readings on account of the reflected radiation.

These measurements were effected by means of an apparatus designed for the purpose, and which we will here name "the comparator." It consists of a wooden box (Fig. 9) 100 cm. long, 10 cm. wide, and 10 cm. deep, blackened on the inside, and provided at the ends with small astronomical telescopes, $S L$, $S' L'$, whose common optical axis is parallel to the central line of the box. M and M' are plane glass mirrors, silvered on the front face, each capable of rotation in two directions about axes, one axis being perpendicular to, and the other coinciding with, the longitudinal axis of the box. M' is provided with a tangent screw, with graduated head, so that it may be moved through any desired angle, in order to bring into view the part of the sky which is to be compared with the sun. B is the screen of a Bunsen photometer, both sides of which, by means of two mirrors inclined at a suitable angle and placed below the screen, may be viewed by an eye at E. This screen is attached to a sliding piece, so that it, together with the viewing aperture E, may be placed in any position between L' and L, its place being read by an index and centimeter scale on the outside of the box.

By means of this arrangement, the intensities of light from two luminous objects, reflected by means of the mirrors M and M' into the box, may be directly compared. The lenses L and L' should be focused so that an image of the object is formed on either side of the screen when placed midway between them. The screen is next to be moved until the two sides appear equally bright. Then, assuming that the lights from L' and L are equal when the telescopes are directed on the same object by means of the mirrors, light from L' : light from L :: $L l^2$: $L' l'^2$.

When the light from the sky is compared with direct sunlight, the latter must be greatly diminished in order to make an observation possible. The original apparatus was provided with a system of unsilvered reflectors, to be placed between S and L, for this purpose; but on trial this was found to effect too great a diminution of light, and a cap, pierced with a small circular aperture to cover the objective S, was substituted for it.

Let m = the ratio of the intensity of light from the solar lens to that from the sky-lens, when both are directed on the sun, and a = the ratio of the amount of light from the solar lens with diaphragm to the amount with full aperture. If, then, the intensity of solar light from the sky-lens be taken as unity, that from the solar lens will be $m a$. Let

L = the intensity of light from any part of the sky, relatively to that from the sun;

P = distance from focus of sky-telescope to point of equal illumination;

Q = distance from focus of solar-telescope to point of equal illumination.

Then, since the lights which have come through the lenses are proportional to the squares of these distances,

$$L = a\, m\, \frac{P^2}{Q^2}$$

In order to determine the value of a, the sky-telescope and mirror were fitted on the end of a piece attached to the box, so that sunlight from the sky-lens, with full aperture, became sufficiently enfeebled by distance to be directly compared with the light from the sun-lens with the

Solar Comparator.

small aperture. In this way it was found that $a = 0.0050$. It was also found by direct observation that $m = 1$. Therefore, for the comparator used,

$$L = .005 \frac{P^2}{Q}$$

Example of the reduction of comparator observations.

[Station, Lone Pine. Observer, Mr. G. F. Davidson. Date, August 5 inst. On this day the reading of the solar lens was at 1 cm. : sky lens 2.5 cm., P : Q 51.9 cm.]

	Distances from sun+limb				Time
	½ diam.	1 diam.	2 diam.	2½ diam.	
Reading	19.4	16.2	12.0	11.6	
P	16.9	13.7	9.5	9.1	
Q	35.9	38.2	42.1	42.5	7.40 to 8.10 a.m.
P²	285.6	187.7	90.25	82.81	
Q²	125	145	179	189	
L	.001162	.000642	.000250	.000225	
Reading	14.5	17.2	12.4	11.2	
P	17.0	14.7	9.9	8.7	
Q	31.9	37.2	42.0	43.2	11.50 a.m. to 12.10 p.m.
P²	289.0	216.1	98.01	75.69	
Q²	1248	1384	1764	1866	
L	.001164	.000779	.000277	.000203	

The afternoon observations of this date (August 5) are very imperfect, but show a great increase in the sky illumination, caused by the presence of haze around the sun.

The above example will indicate the mode of observation and reduction for each day. The results are given in the following table. They are not corrected for diffused light (i. e., non specular reflection) in the mirror.

The silver of the mirrors, however highly polished, does not possess absolute specular reflection, but its surface, with the help of the nearly invisible dust particles, which are incessantly deposited everywhere, diffuses a certain amount of light, which is added to the sky light. This is negligible until we approach within a solar diameter of the solar limb. We have, from this point up to the edge, an increasing amount of light entering the instrument, along with that from the sky, but which has been directly received from the sun. The consequence is that the results here given for sky radiation within half a degree of the sun are too large; but, owing to the difficulty of determining quantitatively the amount of this excess, no correction for it has yet been applied. It may be observed, however, that it is probably owing to the absence of such a correction that the observations appear to show much the same sky radiation immediately around the sun on the mountain as in the valley. The absolute difference is very small, but the relative one is large, and becomes most manifest very close to the sun, for the diffusion has added a constant quantity to the sky heat close to the sun on mountain or in valley, and has thus made the ratio of these values approach unity.

From the following photometric observations, there appears to be in the clear air of our stations no very great difference in the intensity of sky illumination at noon from that at the time of morning and afternoon observation. In determining the correction to be applied to the actinometer readings for the heat reflected from the regions immediately surrounding the sun, therefore, the mean of all the photometric measures will be taken.

TABLE 96.

LONG PINE.—*Summary of comparator observations.*

Time.	Distance from sun's limb.				
	½ diam.	1 diam.	2 diam.	2½ diam.	3 diam.
August 5, noon	.00160	.000164	.000042	....	....
5, p. m	.00120	.000003	.000154	....	....
4, a. m	.00260	.00720	....	....	.000199
4, a. m	.00111	.000277	.000180	....	....
4, m	.00076	.000206	.000046	....	....
4, p. m	.00028	.000013	.000028	....	....
5, a. m	.00162	.000013	.000250	.00225	....
5, m	.00184	.000779	.000277	.000263	....
Mean values of L	.00173	.000081	.000074	.000275	.000199

TABLE 97.

MOUNT WHITNEY.—*Summary of comparator observations.*

Time	Distance from sun's limb.		
	0 diam.	1 diam.	2 diam.
August 20, m	.00218	.00043	.000250
21, a. m	.00123	.000081	.00465
21, p. m	.00120	.000036	.000291
22, a. m	.00162	.000081	....
Mean values of L	.00184	.000050	.000050

From these mean values curves have been plotted, as given on Plate VII, where the unit of length on the abscissæ is a solar diameter, and the ordinates denote intensities of sky illumination expressed in units, each of which is $\frac{1}{1000}$ of the mean solar luminosity. The central column, if prolonged to a height of 1,000 units, would represent the direct radiation from the solar disk, and the curves show the diminution of sky radiation at various distances from the sun's limb.

Determination of the actinometer correction from the above curves for Lone Pine and Mount Whitney observations.

If the curves are rotated about the axis of Y, the amount of light emitted by the sun will be represented by the volume of the cylinder described by the two lines representing the boundaries of the sun's disk, whose equations are $y = \frac{1}{2}$ and $y = -\frac{1}{2}$, or calling the height 1000, $\frac{1}{4}\pi \times 1000$; while the volume of the solid, included between the plane of the axis of X and the surface described by the curve, will represent the amount of light received from adjacent portions of the sky.

The curve of observation (uncorrected for diffused light) coincides quite closely with an equilateral hyperbola whose asymptotes are the lines $x = -\frac{1}{4}$, $y = -\frac{1}{4}$. The equation of the equilateral hyperbola referred to its asymptotes is

$$xy = \frac{a^2}{9}$$

By measurement $a = 1.9$, whence the equation referred to the lines $x = -\frac{1}{4}$, $y = -\frac{1}{4}$ (the dotted lines of the figure), is $xy = 1.805$.

Transferring to the origin O, whose co-ordinates are $x = +\frac{1}{4}$, $y = +\frac{1}{4}$,

$$(x + \tfrac{1}{4})(y + \tfrac{1}{4}) = 1.805$$

or

$$y = \frac{1.805}{x + \tfrac{1}{4}} - \tfrac{1}{4}$$

LONE PINE

MT WHITNEY

ALLEGHENY

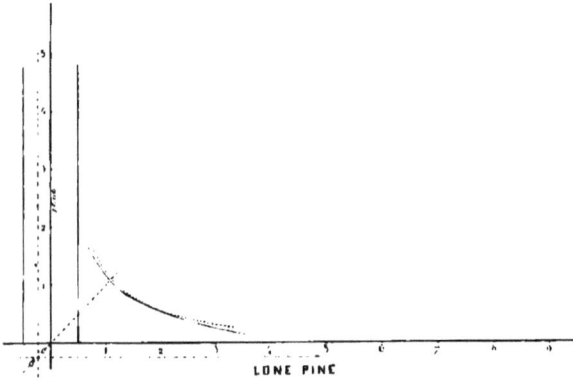

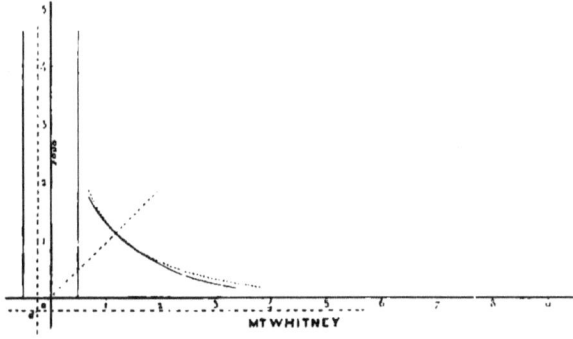

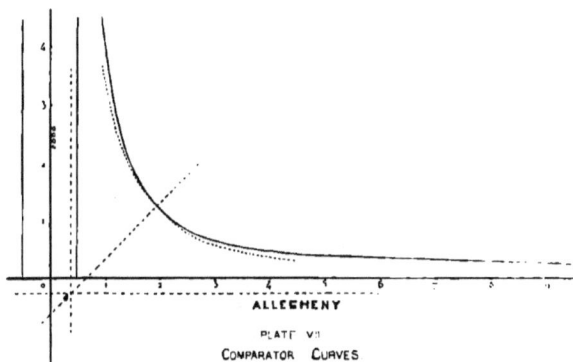

PLATE VII
COMPARATOR CURVES

The dotted line shows the locus of this equation. The volume generated by this curve may be considered to be made up of elementary cylindrical rings whose circumference is $2 \pi x$, whose thickness is dx, and whose height is y. Therefore

$$V = 2 \pi \int x\, y\, dx.$$

Taking the limits of the equation, $X = \frac{1}{2}$, and $X = 1$, the volume becomes

$$V = 3.61 \int_{\frac{1}{2}}^{1} \frac{x\, dx}{x \times \frac{1}{2}} = \frac{\pi}{2} \int_{\frac{1}{2}}^{1} x\, dx = \left[3.61 \div |x - \frac{1}{2} \log_e (x + \frac{1}{2})| - \frac{\pi x^2}{4} \right]_{\frac{1}{2}}^{1}$$

$x = 1, \quad V_1 = 10.11 \pi - .9025 \pi \log_e 1.25$

$x = \frac{1}{2}, \quad V_1 = 1.71 \pi - .9025 \pi \log_e 0.75$

$V = V_1 - V_1 = 8.70 \pi - .9025 \pi (\log_e 1.25 - \log_e .75)$

$V = 8.70 \pi - .9025 \pi \log_e \dfrac{1.25}{0.75} = 8.70 \pi - .9025 \pi \log_e 5.667 = 22.11.$

The volume of the cylinder is $250 \pi = 785.2$; hence sunlight : skylight = 785.2 : 22.11 = 1 : .0285.

The curve of sky illumination on Mount Whitney does not differ greatly from that at Lone Pine. Referred to axes whose equations are $x = -\frac{1}{2}$, $y = -\frac{1}{2}$, it is nearly represented by the equation $xy = 1.805$, the same that has been used to represent the Lone Pine curve. The only noticeable difference is that it approaches closer to the axis of X at a distance from the origin. The volume of the curve generated by its revolution is therefore somewhat less, and consequently the derived actinometer correction, expressed as a quantity to be subtracted, will also be smaller. If the total light received from sun and so much of the adjacent sky as radiates to the thermometer bulb at Lone Pine is represented by 10,285, that from the sun is 10,000; hence $\dfrac{10,000}{10,285} = .9725$ is the fraction of this total amount due to the sun alone. This is the actinometric corrective factor F as determined from the Lone Pine and Mount Whitney observations. We next give the results of similar determinations at Allegheny, a hazy sky being chosen to determine an extreme value for the correction.

Comparator observations of October 3, 1882, made at Allegheny.

J. E. Keeler, observer. Brilliant haze around the sun, sky very hazy and smoky. Simultaneous observations made by Mr. F. W. Very with the large Vodle actinometer, gave the following results; the instrumental actinometer corrections have been applied.]

	Cal.
Exposure to sun from 11h.36m a. m. to 12h.02m p. m., radiation	−0.545
Exposure to sun from 12h.33m p. m. to 12h.23m p. m., radiation	0.736
Exposure to sun from 12h.41m p. m. to 12h.57m p. m., radiation	0.791

The observations with the comparator, made at different times, seem to indicate no similar progressive change in the sky illumination, and therefore the means of observations made at the same distance will be taken.

Distance from the sun's limb.

	½ diam.	1 diam.	2 diam.	5 diam.	13 diam.
Time 11h.36 to 11h.19 a. m.	27.5	21.1	16.5	11.5	10.2
Time 12.06 to 12.11 p. m.	26.1	22.4	16.9		
Time 1.36 to 4.16 p. m.	26.1	21.7	16.9		
Means	26.6	21.7	16.5	11.5	10.2

Reduction of observations of October 3, 1882.

P	25.0	20.1	11.9	9.9	8.6
Q	27.2	12.1	57.3	12.4	1.36
R	.625	16.0	222.0	98.01	73.96
C	759.8	1630	1391	1759	1901
P	.00421	.00924	.00793	.00723	.00791

Observations with comparator at Allegheny October 4, 1882.

[Sky dense uniform haze, mainly produced by smoke. Observer, J. E. Keeler.]

Full apertures on both lens. Comparison of solar images. 10ʰ a. m.		Small diaphragm on sun lens. 11ʰ 05ᵐ a. m.			
28.5	27.9	Distances from sun's limb.			
27.9	28.5				
28.5	28.1	¼ diam.	½ diam.	1 diam.	2 diam.
28.4	28.6				
27.8	27.9	32.0	30.0	24.5	18.5
		31.0	31.5	23.8	18.8
28.1	28.1	32.0	29.0	22.0	18.4
Mean		32.0	30.2	23.4	18.6

Reduction of observations of October 4, 1882.

P...	30.4	28.6	24.8	17.0
Q...	21.8	25.6	30.4	25.2
R...	924.2	818.0	475.2	289.0
Q'...	175.2	557.0	924.2	1250
L...	.00970	.00732	.00256	.00116

Actinometer correction for Allegheny observations.

In determining the actinometer correction for Allegheny, the curve of sky illumination from the observations of October 3, 1882, is used. It is sufficiently nearly represented by a hyperbola whose asymptotes are $x = +0.4$, $y = -0.2$, and whose transverse axis is 2.12. Its equation, referred to these asymptotes (dotted lines in figure 3, Plate VII), is $x\,y = 2.217$.

If in this equation $x = 0.1$, $y = 22.17$. The hyperbola therefore cuts the line, $x = \frac{1}{2}$, at a considerably higher point than the curve of observation; but on the other hand it lies nearer to the line than the latter. Referred to the axes drawn in the figure, the equation becomes $y = \dfrac{2.217}{x - .4} - .2$, the locus of which is represented by the dotted curve. If revolved about the axis of Y, it describes a solid whose volume between x_1 and x_2 is

$$V = 2\pi \int_{x_1}^{x_2} x\, y\, d\,x$$

If the hyperbola $x\,y = \dfrac{a^2}{2}$ be transferred to a new origin $x = p$, $y = q$, its equation becomes

$$(x + p)(y + q) = \frac{a^2}{2}$$

whence

$$y = \frac{a^2}{2\,(x + p)} - q$$

and if rotated around the axis of Y, the portion included between the axis of X and the curve and the two ordinates at x_1, x_2 will describe the volume,—

$$V = 2\pi \int_{x_1}^{x_2} x\, y\, d\,x = \pi\, a^2 \int_{x_1}^{x_2} \frac{x\, d\,x}{x + p} - 2\pi q \int_{x_1}^{x_2} x\, d\,x = \left[\pi\, a^2 \Big(x - p \log_e (x + p) \Big) - \pi\, q\, x^2 \right]_{x_1}^{x_2}$$

$$= \left[1.494\, \pi \Big(x + .4 \log_e (x - .4) \Big) - .2\, \pi\, x^2 \right]_{0.5}^{5.0}$$

where $a = 2.12$, $p = -0.4$, and $q = 0.2$. Taking the limits $x_1 = 0.5$ and $x_2 = 5.0$

$$V_2 = 17.47\, \pi + 1.798\, \pi \log_e 4.6 \qquad V_1 = 2.20\, \pi + 1.798\, \pi \log_e 0.1$$

$$V_2 - V_1 = V = 15.27\, \pi + 1.798\, \pi \log_e 4.6$$

whence

$$V = 69.55$$

The volume of the cylinder being as before, 785.2

$$\text{sunlight} : \text{skylight} \quad 785.2 : 69.55 = 1 : 0.0886$$

whence the actinometer corrective factor $= \dfrac{1}{1.0886} = .9185$ for Allegheny observations.

The angular apertures of both large and medium diaphragms, used on the small actinometer, are so great as to include all of the portion of the sky within the limits of the integration—that is, the correction is the same for both apertures. Owing to diffused light from the sky-lens, mirror, and telescope-tube, these factors, namely, .9725 from the Lone Pine observations and .9185 from those at Allegheny, are probably somewhat too small; that is, they should be nearer unity.

The following subtractive corrections for sky radiation were finally adopted after comparisons of all available observations, both from those made with actinometer and comparator at Lone Pine and Mt. Whitney, and also from those obtained with the same instruments at Allegheny, after applying suitable corrections for the Allegheny sky.

CORRECTION F.

For Lone Pine at noon $= -1$ per cent., with low sun $= -2$ per cent.; for Mount Whitney at noon $= 0$, with low sun $= -1$ per cent.

Summarizing the preceding statements, we have the following adopted values, expressed as multiplying factors, where c is the effect of the solar heat, as directly determined by the globe actinometer through the method proposed by M. Violle. As some of the factors change with the altitude of the sun, the final correction will differ slightly according as the observation is taken at high or low sun, and at Lone Pine or Mountain Camp.

TABLE 98.

	Lone Pine observations.		Mountain Camp observations.	
	High sun.	Low sun.	High sun.	Low sun.
Corr. A	+ .125 c	.125 c	+ .125 c	+ .125 c
Corr. B	.051	+ .070	+ .051	.051
Corr. C	0.6	.026	.026	.026
Corr. D	.050	.050	.050	.050
Corr. E	.011	− .011	.011	.011
Corr. F	.010	− .020	− .000	.010
	+.25 + .021	.254 − .031	+ .275 − .011	.265 .051
	.234 c	+ .240 c	.234 c	+ .211 c

We have then a mean additive correction of about 25 per cent. as the result of all the preceding investigations, as the least we can assign. In passing, however, from the "uncorrected" to the "corrected" observations, we use the exact values above. The final values in the "summaries of actinometer observations" are obtained by applying the above corrections to the values in calories determined from the initial rate ($m\theta_0$) of each day, and the water equivalent of the thermometer used.

CHAPTER IX.

SUMMARY OF RESULTS.

In this summary we have selected the two clearest and best days of synchronous observation. viz, August 23 and August 25, for separate reduction. The meaning of the symbols is $M_{1}z_1$, the air-mass traversed at noon; $M_{1'}z_2$, the air-mass traversed at morning or evening, obtained from the formula $M = \sec z$, for zenith distances less than 65°, and from $M = \dfrac{0.171 \times \text{tabular refraction}}{\text{Cos. Appt. Alt.}}$ for those greater than 65°; β the barometer in decimeters; where n, for brevity, is put equal to $\dfrac{7.6}{M_{n}z_1 - M_{1}z_2}$; C, and C_{1n} the values in calories at high and low sun respectively; a the coefficient of transmission for an entire atmosphere of 7.6^{dm}; E the solar constant expressed in calories. Besides these two days all the noon observations at Lone Pine have been united for comparison with all the morning and evening observations. In like manner the observations at Mountain Camp are reduced, and the noon observations compared with the mean of morning and evening.

TABLE 99.

Reduction of Lone Pine actinometer observations.

COMPUTATION OF a.

Data	Morning and noon.			Evening and noon.		
	From observations of August 25.	From observations of August 23.	From mean of all observations.	From observations of August 23.	From observations of August 25.	From mean of all observations.
M 9	15 76	16 58	15 57	15 99	16 67	15 25
M β	7 37	7 39	7 10	7 37	7 30	7 40
M 9 M 9	8 39	8 59	10 17	8 62	9 24	7 85
n	0 906	0 845	0 745	0 882	0 822	0 968
C	1 496	1 443	1 408	1 437	1 410	1 397
C_1	1 760	1 749	1 707	1 760	1 749	1 707
Log C	0 379	0 184	0 174	0 1575	0 102	0 145
Log C_1	0 2455	0 428	0 222	0 2455	0 2428	0 2522
Log C_1 - log C	- 0 0596	- 0 644	- 0 0058	- 0 0880	- 0 0926	0 0820
Log a	- 0 0610	0 0490	- 0 0401	- 0 0776	- 0 0765	- 0 0812
Tab log a	9 9458	9 9539	9 9599	9 9221	9 9231	9 9158
a	0 8625	0 8998	0 9120	0 8364	0 8378	0 8208

TABLE 100.

COMPUTATION OF E.

Data	August 25.		August 23		Mean of all observations	
	From mean of morning and evening observations.	From noon observations.	From mean of morning and evening observations.	From noon observations.	From mean of morning and evening observations.	From noon observations.
$M4$	15 87	7 37	16 58	7 39	16 41	7 40
$M0$	2 09	97	2 17	97	2 16	98
i n						
Log C	0 1602	0 2455	0 1688	0 2455	0 1618	0 2322
Log a	0 0708	0 0708	- 0 0611	0 0614	- 0 0621	- 0 0621
Log a $\dfrac{M8}{7.6}$	0 1180	- 0 0687	0 1322	0 0586	0 1261	- 0 0669
Log E	0 3142	0 3142	0 3110	0 3024	0 2959	0 2293
E	2 061	2 061	2 046	2 000	1 976	1 964

Computer, A. B. S.

TABLE 101.

Results of Lone Pine actinometer reductions

	a	E	
Dates	Morning and noon	Evening and noon	From means of morning and evening observations	From noon observations
August 24	0.8636	0.8364	2.064	2.063
August 25	0.8503	0.8178	2.046	2.046
Mean of all observations	0.5120	0.8278	1.976	1.964
Mean of results....	0.8503	0.8327	2.058	2.040

Final means, *a* = 0.8630, *E* = 2.049.

TABLE 102.

Reduction of Mountain Camp actinometer observations.

COMPUTATION OF *a*

	Morning and noon			Evening and noon		
Data	From observations of August 23	From observations of August 25	From mean of all observations	From observations of August 23	From observations of August 25	From mean of all observations
$M_r a$	9.82	10.25	10.72	12.34	13.10	11.52
$M_n a$	5.56	5.59	5.59	5.56	5.59	5.59
$M_r a - M_n a$ ..	4.26	4.66	5.13	6.78	7.51	5.93
a	1.79	1.63	1.19	1.32	0.87	1.29
C	1.889	1.746	1.732	1.699	1.697	1.585
C	1.822	1.806	1.805	1.902	1.906	1.895
Log C_r	0.2708	0.2420	0.2385	0.2246	0.2297	0.2000
Log C_n	0.2860	0.2561	0.2776	0.2796	0.2801	0.2776
Log C_r, log C_n	0.0262	-0.0184	0.0311	0.0711	0.0804	0.0776
Log a	0.0168	-0.0021	0.0388	0.0799	0.0880	0.1001
Tab log a ...	9.8621	9.9279	9.8192	9.9581	9.9431	9.8991
a	0.8077	0.8567	0.8803	0.8128	0.8936	0.7942

TABLE 103.

COMPUTATION OF *E*

	August 23		August 25		Mean of all observations	
Data.	From mean of morning and evening observations	From noon observations	From mean of morning and evening observations	From noon observations	From mean of morning and evening observations	From noon observations
Ud ...	11.08	5.56	11.82	5.59	11.12	5.59
$U4$ ...	1.16	0.73	1.55	0.71	1.17	0.71
7.6						
Log C	0.2572	0.2600	0.2356	0.2907	0.2297	0.2776
Log a	-0.0611	-0.0621	-0.0655	0.0553	-0.0651	-0.0651
Log a $\frac{Ud}{7.6}$	0.6993	0.3163	0.0860	0.6611	-0.0108	0.0637
Log E .	0.3298	0.3357	0.3218	0.3213	0.3325	0.3272
E	2.137	2.149	2.098	2.095	2.151	2.155

Computer, A. B. S.

TABLE 104.

Results of Mountain Camp actinometer reductions.

Dates	a.		F.	
	Morning and noon	Evening and noon	From mean of morning and evening observations.	From noon observations.
August 23	0.8677	0.8504	2.157	2.119
August 25	0.8667	0.8801	2.005	2.095
Mean of all observations	0.8807	0.7902	2.151	2.154
Means of results	0.8847	8.899	2.128	2.123

Final means a = 0.8823; F = 2.131.

The simultaneous observations of August 23 and August 25, and the means of all observations at Lone Pine and Mountain Camp, were reduced by the same method as the high and low sun observations at each station, noon being compared with noon and evening with evening. Since the process has been illustrated by the tables already given, we omit the lengthy computations, and simply give the results, as follows:

TABLE 105.

Results from Lone Pine and Mountain Camp comparative actinometer reductions, on the assumption that atmospheric transmission is the same above the mountain as it is between Lone Pine and Mountain Camp.

Computer A. B. S.

Dates	a.		F.	
	Noon.	Evening.	Noon.	Evening.
August 23	0.6730	0.7278	2.57	2.56
August 25	0.6951	0.6769	2.19	2.66
Mean of all observations	0.6445	0.7731	2.62	2.34
Means of results	0.6719	0.7265	2.56	2.53

Final means a = 0.6992; F = 2.705.

In deducing a value of the solar constant by a comparison of Mountain Camp and Lone Pine observations through Pouillet's formula, just employed, we have taken account only of the air-masses, and have tacitly assumed that for the same air-mass we shall always have the same absorption. If it were true that we had one absorption for the air-mass between the top and bottom of the mountain and another and different absorption for an identical mass taken from air above it, our formula would not hold good, had it no other defect than this alone. Let us, then, compare observations taken in the valley and on the mountain, when the mass of air was the same in each. At some time—late or early in the day—on the mountain, in spite of the greater altitude, the mass of air traversed must have been the same as at noon in the valley; and though, as observations were not incessant through the day on the mountain, none may have actually been made at this instant, we can, nevertheless, by interpolation between neighboring values, obtain nearly as trustworthy a result. Thus, if we represent the observations of August 23d graphically, we find that at Lone Pine, noon, the air-mass was 7.37dm and the observed calories 1.760c. On the same day, on the mountain, our interpolation between contiguous values shows that equal air-mass would have been observed through at 9h 30m in the forenoon and at 2h 30m in the afternoon, with a value of 1.881 c. in the first case and 1.878 c. in the second. It appears, then, from this day's observations that like masses of air on the mountain transmit more heat than those in the valley.

In other words, without regard to the greater rarity of the air on the mountain, but comparing a given weight of it with an equal one taken in the valley, the former is, in a sensible degree, more diathermanous. If we repeat the experiment with the observations of August 25, we have the following values:

d. m.   c.

At Lone Pine, noon .... air-mass = 7.39; calories, 1.719
At Mountain Camp, 9.30 a. m .... air-mass = 7.39; calories, 1.836
At Mountain Camp, 2.30 p. m.................................... air-mass = 7.39; calories, 1.830

Finally, if we repeat it for the means of all observations, we find the values—

Lone Pine, noon .... air-mass = 7.40; calories, 1.797
Mountain Camp, 9.30 a. m .... air-mass 7.40; calories, 1.812
Mountain Camp, 2.30 p. m .... air-mass = 7.40; calories, 1.784

So that the evidence from the actinometer alone appears to be conclusive as to the fact that some ingredient present in the lower air, is comparatively absent in the upper,* an inference already drawn from our pyrheliometer results.

Combining the observations just cited, and giving to the mean of all double weight, we have the following, taking the mean of morning and afternoon values (for the reader will not have failed to remark that these values on the mountain systematically differ for an equal altitude of the sun, the same mass of air being found always more diathermanous in the morning than in the afternoon).

For same air-mass (7.39ᵈᵐ) at Lone Pine and Mountain Camp, we have at Lone Pine, 1.731c; at Mountain Camp, 1.833c. The results for various air-masses are represented graphically in Fig. 10.

It appears that in the valley as well as on the mountain, with the same air-mass, we have greater diathermancy in the morning than in the afternoon. (See Fig. 11.)

We have just found from the mean of all our comparisons between observations on the mountain and at Lone Pine, noon being compared with noon and evening with evening, the actinometer value, 2.705c, for the solar constant; but this was on the usual assumption as to the uniformity of the absorption of equal air-masses, an assumption whose fallacy we have just exposed. Reserving a fuller discussion of this point for the chapter on the spectro-bolometer, we may point out here a method of approximately correcting for the effect of this different constitution of like air-masses, with the least possible dependence on hypothesis.

Knowing that the air-mass above Mountain Camp was of the same *quality* as that portion of it included with the part between the stations in a Lone Pine observation, and knowing the comparative transmissibility of two equal air-masses at Lone Pine and Mountain Camp, since the barometer gives us the air-mass above and below, we have sufficient data for introducing two coefficients of transmission.

(In view of the large error which Pouillet's formula involves, irrespective of the one we are now discussing, it does not seem expedient to do more than approximately correct the last value of E, found by comparing observations at Mountain Camp with synchronous ones at Lone Pine.

Let a_1 be the coefficient of transmission for the mass of air above Mountain Camp, a_2 that for the mass of air between Lone Pine and Mountain Camp.

From the observed values of C given above we find the actual ratio of transmissibility at Mountain Camp to that at Lone Pine for the same air-mass.

$$\frac{1.833}{1.731} = 1.06 \text{ (nearly)}.$$

By the comparison of the two stations, the coefficient of transmission for the mass of air between them was found to be $a_2 = 0.70$; but a_1 is obviously larger than this in at least the above ratio.

In order, then, to fix a value of the coefficient for the mass of air above Mountain Camp, a_2 may be multiplied by the ratio of transparency, 1.06, giving $a_1 = 0.70 \times 1.06 = 0.74$. With this and

---

* See Journey to Mount Whitney, page 12.

Fig. 10

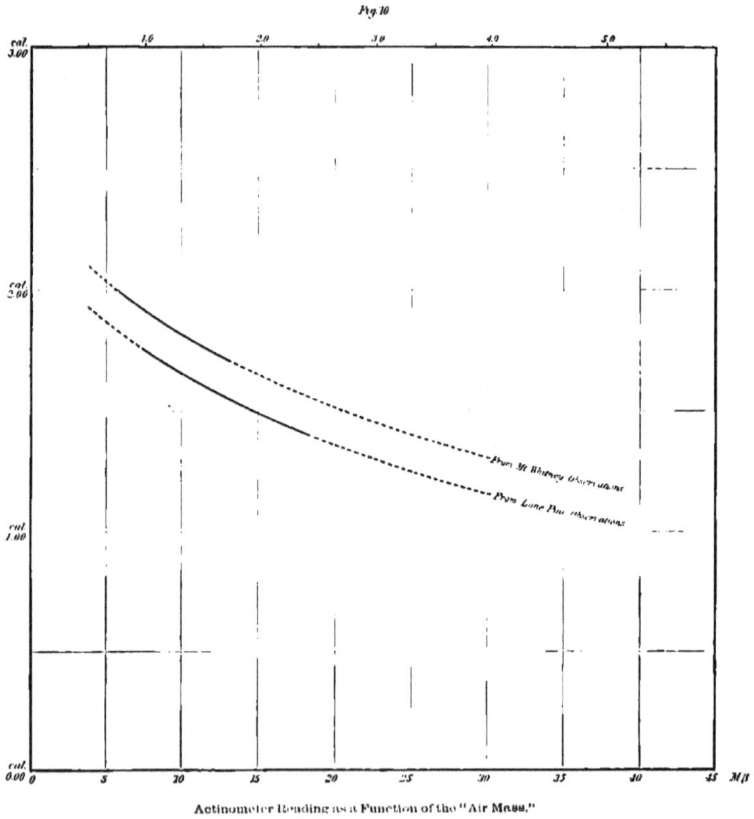

Actinometer Reading as a Function of the "Air Mass."

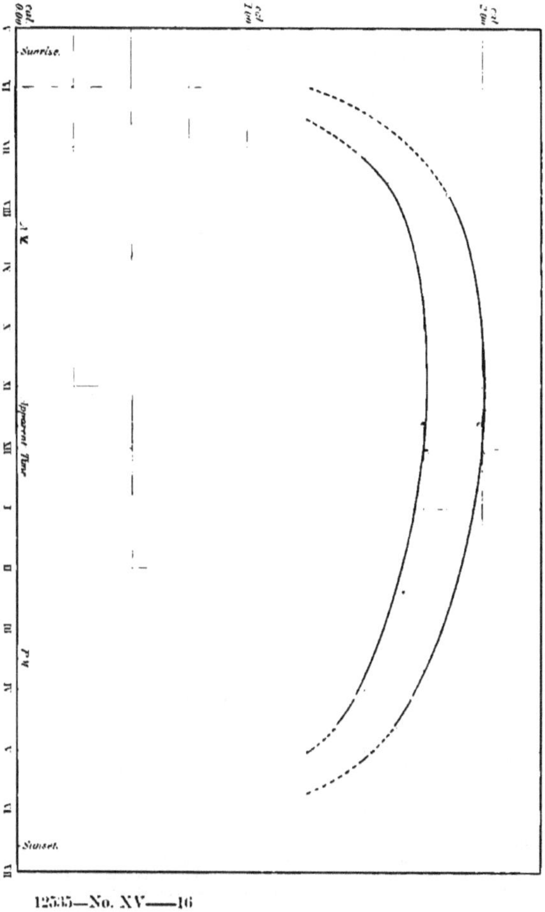

Actinometer Reading as a Function of the Hour Angle.

Fig. 11

the mean actinometer reading at Mountain Camp in calories, $C=1.9$, as arguments the value of E may now be determined.

By actual computation we find $E=2.382$, a value which is perhaps as near the truth as we can reach by these methods, but necessarily much inferior to that which would be attained could we consider the effect of selective absorption.

A very important piece of evidence which these observations at elevated and low contiguous stations have furnished is that the solar constant, as determined by observations of high and low sun at one station, is too small. We say *evidence*, for, however we may have felt assured that this must be the case from indirect observation and inference, we could never, with a single station, have tested this conclusion as we can now; for it will be observed that, with the values of the observed heat at Lone Pine, and the transmissibility determined there, we can calculate the heat received at a certain considerable altitude—that of Mountain Camp—and that by direct experiment we find it too small.

Beyond this (with an exception to be immediately noted) the chief use of the elaborate determination we have just made will, so far as the solar constant is concerned, be found in the ensuing chapters in connection with the work on the spectro-bolometer.

One most important conclusion remains, however, to be drawn, which must rest directly on the evidence of this globe actinometer. We have pointed out at the commencement of this chapter that, owing to the enormous difference between the temperature of the sun and that which is familiar to us at the surface of this planet, the amount to which a body exposed to the direct solar rays will rise above the temperature of its surroundings, is, though rigorously speaking, dependent on the temperature of those surroundings, yet sensibly independent of them within the range of our experiments. This very important remark appears to have been first made by Waterston, and has been confirmed by most careful experiment at the hands of others. According to Mr. Ericsson's experiments, a difference of temperature of nearly a thousand degrees Centigrade made *no* sensible difference whatever in the excess, while M. Violle (who argues for an extremely low temperature of the sun) admits that but a minute difference is observable within a range of 100° C. We conclude, then, that if the temperature of our actinometer globe was that of the absolute zero, or −273° Centigrade, the thermometer in it would either give sensibly the same excess that it does now, or one but slightly greater.

If our thermometer bulb were replaced by the globe of the earth itself, and if the walls of its chamber were represented by empty inter-planetary space, returning no radiation, (and in this respect corresponding to the actinometer walls at a temperature of −273°), the temperature to which the sunward surface of the earth would rise, would be sensibly the same as that to which our thermometer would rise in vacuo, unless we suppose some source of heat for the earth's surface other than that contemplated in what has just preceded. Such other sources as we can suggest, namely, the internal heat of the earth, the friction of the tides, the dynamical effect of the fall of meteorites, the radiation from stars or dark bodies in space, &c., are absolutely insignificant in comparison with the solar heat, and the old idea of a "temperature of space" is founded, as we have endeavored to show in the case of Pouillet's celebrated value, on a supposed necessity which no longer exists.[*] It may be stated confidently that we have no reason to believe, from any ex-

---

[*] I have made experiments where possible, and calculations founded on authentic data, which satisfy me of the truth of this statement. I feel confident that the united heat of all the stars and planets cannot be represented by the ten thousandth part of our small calorie, or anything near as great. The dynamic effect of meteorites may perhaps be admitted to be the most important of those above cited, but this is demonstrably negligible in the present connection.

I cannot here enter upon the results of the measurements of the heat of the heavenly bodies other than the sun; but as Pouillet concludes that the heating effect of the stars alone is considerably over one small calorie per minute per square centimeter, it may be well to point out some considerations which, to the reader who accepts the modern doctrine that light and heat are but different manifestations of a common energy, will prove perhaps as conclusive as any statement I could offer.

The most recent and trustworthy comparisons of the light of Sirius with the light of the sun give the latter approximately 1×10^{10}, that of Sirius being unity, while that of the whole heavens visible to the naked eye is less than 100

perimental evidence, that the heat derived from all sources besides the sun is other than entirely negligible at this stage of the inquiry. There is one important circumstance which causes the thermometer within the actinometer to attain a lower excess than the earth would in space, for our thermometer is losing heat by the conduction and convection of air about it, and space, outside the earth's atmosphere, we must here consider as a void. I have made careful experiments on the heating and cooling of the thermometers used on Mount Whitney, by inclosing them in a vacuum chamber and determining the rates of heating and cooling corresponding to a given excess. These experiments will be found detailed in the appendix. The result, so far as it affects our present purpose, is that the rate of heating or cooling in vacuo is approximately proportional to the excess (as we might anticipate that it would be from the approximate truth of Newton's law of radiation), and that the ratio of this rate to the excess is, though not strictly constant, yet approximately so; so that if, for instance, the excess is , the rate of its radiation at that instant in vacuo is 0.180 per minute; whence it follows that if the initial rate be the highest admissible initial rate observed, as is shown, 5 .7, the final temperature of excess on Mount Whitney in vacuo would be 31 .7. A not dissimilar conclusion was reached by M. Violle, who found that the final temperature of excess of his thermometer on Mount Blanc would have been, if in vacuo, 29 .8. If we adopt, as we shall see reason to do later, a value for the solar constant nearly one-half greater than the highest observed heat on Mount Whitney, we shall conclude that the temperature of final excess would be increased in like proportion, and be not far from 48 C.

We have seen, in the chapter on the ascension of Mount Whitney, that as the air grew rarer the temperature fell, though the sun's direct radiation increased. We might infer, then, from this primitive and common experience, that if the air grew rarer still, the temperature would fall still more, and that *when air was altogether absent, the temperature of the earth under direct sunshine would be excessively low*. We now draw further conclusions from the experiments with the globe actinometer, which we have just detailed, and which show that a small sphere in full sunshine would, in the absence of any atmosphere whatever, attain a final excess of 48 above its surroundings; that the *surface* of the earth, were there no enveloping and heat-storing atmosphere, would (cut off, as it is, from heat within, while still retaining that atmosphere) reach only a corresponding excess above the temperature of its surroundings—in this case —273 . In other words, I believe that if the atmosphere were wholly removed, the temperature of the earth, under the direct solar rays, would not be greatly more than —225 C., and that the same result would follow if the earth, while still retaining that atmosphere, were deprived of the power of selective absorption which it now possesses. In my view, then, these experiments show that, after making every allowance for other sources than solar heat, the temperature of this planet, and the existence of our own and all organized life upon it, is maintained in but slight degree by the direct solar rays, which of themselves are far too feeble to render fluid a planet of frozen mercury, but that the life of the globe is rendered possible by to this little regarded property of selective absorption in our atmosphere.

I expect to take an opportunity elsewhere of enlarging upon the present remarks. I will at present only repeat that I consider that *the temperature of the earth under direct sunshine, even though our atmosphere were present as now, would probably fall to —200 C. if that atmosphere did not possess the quality of selective absorption.*

times that of Sirius. Accordingly, a light greater than that of all the stars down to the seventh magnitude is less than $\frac{1}{4\times10}$ that of the sun. We have no valid reason whatever, in the light of our modern knowledge, to suppose that the ratio of star light to sun light differs materially from that of solar heat to sun heat. If we admit that the heat of the sun is but three calories, then the united heat of the stars will be represented by $\frac{3}{4\times10}$ cal. = 0.00000075, or much less than one erg. It will not materially help the case of those who contend for a sensible heat of the stars, if any such persons there be, to assert that it comes from those beyond the range of the eye of telescopic vision, unless they are prepared to assert that there is not only a sensible light, but a large degree of it, more than comparable to that of all naked-eye stars, from this source, for I have found that the sensitiveness of the eye is at least 10,000,000 times that of our most sensitive heat instrument, whether thermopile or bolometer. Against the existence of hypothetical dark bodies in space radiating heat sufficient in quantity to be considered here, it would seem to be quite superfluous to argue.

CHAPTER X.

THE DETERMINATION OF THE SOLAR CONSTANT BY THE STUDY OF HOMO-GENEOUS RAYS.

I have repeatedly called attention to the fact that the exponential formula of Bouguer and Pouillet still in use ($I = Ap$) is only applicable to an absolutely homogeneous ray, such as we cannot physically isolate, and I have added that if we apply this formula to actual observations with the thermometer on highly composite radiations, we get in every case, so far as the formula is concerned, too small a value. If we admit that the solar radiation is of different wave-lengths, passing from one to another by steps which we cannot consider individually (or at least can only consider as infinitely minute gradations), these statements are susceptible of rigorous demonstration. To show this let us take up the demonstration I have given elsewhere [*] for a particular case, and let us first suppose, in accordance with what has just preceded, that the original intensity of the sun or a star before absorption is represented by A, and that p represents the fraction of the energy transmitted from the sun or star in the zenith, the mass of air above the sea-level being here taken as the unit of mass, and it being supposed that this air is everywhere of the same constitution, according to the ordinary theory. When the energy is transmitted through one such stratum, what was A has become Ap; when it is transmitted through two such strata, Ap^2; through n such strata, Ap^n, and so on—a formula whose fundamental error lies in assuming this coefficient of transmission p to be a constant, for the energy of the sun or star. A is in fact not homogeneous, but composed of an infinite number of radiations, each of which has its own coefficient of transmission. To commence with an ideally simple case, let us suppose the energy to be really composed before absorption of two portions, A and B. Let A have a special coefficient of transmission, a; and B another special to itself, b. Then if we assume (still for considerations of convenience only) that each of these portions is, separately considered, homogeneous, we may write down the results in the form of two geometrical progressions, thus:

TABLE 106.

Original light	Coefficient of transmission	Light after absorption		
		By one stratum	By two strata	By three strata
A	a	Aa	Aa^2	Aa^3
B	b	Bb	Bb^2	Bb^3
Sum $A+B$		$Aa+Bb$	Aa^2+Bb^2	Aa^3+Bb^3
		$\dfrac{Aa+Bb}{A+B}$	$\dfrac{Aa^2+Bb^2}{Aa+Bb}$	$\dfrac{Aa^3+Bb^3}{Aa^2+Bb^2}$

The fractions here are the coefficients of transmission, as deduced from observations at different zenith distances. They evidently differ, and (as will be shown) each is larger than the preceding.

In the above table $Aa + Bb$ is the sum of the two kinds of energy as observed, after absorp-

See Comptes Rendus, tome 92, p. 701 (Mars, 1881).

tion by one unit stratum (see $\frac{z}{z}=1$) by the thermometer or photometer; $Aa^2 + Bb^2$ is the sum of the energies observed after absorption by two strata (see $\frac{z}{z}=2$), &c.; and we are here supposed to know the really dual constitution of the energy, which the thermometer or photometer does not discern. According to the usual hypothesis, the coefficient of transmission, which is the quotient obtained by dividing the value after n absorptions by that after $n-1$ absorptions, or, more generally, that from the expression

$$\left(\frac{\text{Value after } n \text{ absorptions}}{\text{Value after } m \text{ absorptions}} \right)^{\frac{1}{\cdot}}$$

is a constant. It is, in fact, not a constant, as we shall prove later; but we shall first show that if we proceed upon the ordinary assumption that it is such, the value obtained for the original energy of the sun before absorption will always be too small. For if we observe by a method which discriminates between the two radiations, we shall have, if we separately deduce the original energies from our observations of what remains after one and two absorptions,

$$A = \frac{(Aa)^2}{Aa^2} ; \text{ and } B = \frac{(Bb)^2}{Bb^2}$$

whence the true sum

$$A + B = \frac{(Aa)^2}{Aa^2} + \frac{(Bb)^2}{Bb^2}$$

while if we observe by the ordinary method, which makes no discrimination, we shall have the erroneous equation

$$A + B = \frac{(Aa + Bb)^2}{Aa^2 + Bb^2}$$

which is algebraically less than the first or correct value

$$A + B = \frac{(Aa)^2}{Aa^2} + \frac{(Bb)^2}{Bb^2}$$

For the expression

$$\frac{(Aa)^2}{Aa^2} + \frac{(Bb)^2}{Bb^2} > \frac{(Aa + Bb)^2}{Aa^2 + Bb^2}$$

readily reduces to the known form

$$a^2 + b^2 > 2 \, ab$$

Moreover since $a^2 + b^2 - 2ab = (a - b)^2$, *the error increases with the difference between the coefficients.*

Now, in the general case, if we suppose the original radiation L to be composed before absorption, of any number of parts $A_1, A_2, A_3, + \ldots$ having respectively the coefficients of absorption $a_1, a_2, a_3, + \ldots$ the true value of L is given by a series of fractions which may be written in the form

$$L = \Sigma \frac{(Aa)^2}{Aa^2} = \Sigma A$$

whereas the value of the original energy by the customary formula would be

$$L_1 = \frac{\Sigma (Aa)^2}{\Sigma Aa^2}$$

so that all the quantities being positive, by a known theorem $L \cdot L_1$, and for the same values of $A_1, A_2, A \ldots$ this inequality is greater, the greater the difference in the values of the coefficients $a_1, a_2, a_3 \ldots$ But this is stating, in other words, that the true value found by observing separate coefficients of transmission are *always greater* than those found when we do not distinguish between the radiations of which the energy of the sun or star is composed.

We have stated above that the usual hypothesis makes the coefficient of transmission a constant. It will be seen from the above table, however, that it varies from one stratum to the next; that it is least when obtained by observations made near the zenith, and *that it increases progressively as we approach the horizon.*

For since a and b are less than unity, each of the sums $A + B$, $Aa + Bb$, &c., in the above table, is less than the preceding.

It is also evident that their rate of diminution *decreases* as we approach the horizon, since

$$Aa^2 - Aa^3 > Aa^3 - Aa^4 \qquad\qquad Bb^2 - Bb^3 > Bb^3 - Bb^4$$

Hence

$$(Aa^2 + Bb^2) - (Aa^3 + Bb^3) > (Aa^3 + Bb^3) - (Aa^4 + Bb^4)$$

consequently the difference between the numerators of two successive ratios, such as

$$\frac{Aa^3 + Bb^3}{Aa^2 + Bb^2} < \frac{Aa^4 + Bb^4}{Aa^3 + Bb^3}$$

is less than that of their denominators. In other words, although both numerator and denominator decrease in successive ratios, *the ratios themselves increase progressively.*

Further, a simple inspection of the form of the expression

$$Aa^2 - Aa^3 > Aa^3 - Aa^4 \qquad\qquad Bb^2 - Bb^3 > Bb^3 - Bb^4$$

shows that what is there demonstrated for two numbers, and two coefficients A, a, and B, b, is true for any number, even infinite; which is the case we deal with in actual observation.

In other words, it is universally true that when the numbers are positive, and a, b, c, d, &c., proper fractions,

$$\frac{Aa^{n+1} + Bb^{n+1} + Cc^{n+1} + Dd^{n+1} + \ldots\ldots}{Aa^{n} + Bb^{n} + Cc^{n} + Dd^{n} + \ldots\ldots} < \frac{Aa^{n+2} + Bb^{n+2} + Cc^{n+2} + Dd^{n+2} + \ldots\ldots}{Aa^{n+1} + Bb^{n+1} + Cc^{n+1} + Dd^{n+1} + \ldots\ldots}$$

even if the number of corresponding terms be infinite; and hence universally true, that when the separate coefficients of transmission are positive and less than unity, as is the case in nature, the general coefficient of transmission in the customary exponential formula, is (1) not a constant, (2) always too large under any circumstances, (3) always larger and larger as we approach the horizon, and (4) that the *original energy of the sun or star before absorption, as found by the thermometric and photometric processes and formulæ in universal use, is always too small,* a conclusion which we have just reached here by another method.

It seems to be incumbent, then, on those who still use Pouillet's formula, to at least show that though it may give too small a value, the error is a negligible one in practice, but this has not been attempted so far as I know, and the result of actual observation detailed already, shows conclusively that the error is not practicably negligible, but induces a wide departure from truth, and that it is a principal cause that the values of the solar constant already found are too small. I have already observed that dust and the grosser particles in our atmosphere, by scattering a part of the light and heat, exercise a nearly non-selective absorption on that part to which separately the formula of Pouillet would apply with little error, but that there is every degree of fineness in these particles, from the grosser dust to to the water vesicle, and that on the whole the absorption grows more and more selective, even down to the purely selective absorption caused by the molecule itself. But though there may be such a continuous gradation in nature, we cannot follow it in our present formulæ, which would become unmanageably complex at the outset. I accordingly present a numerical illustration in the following table of the way in which we may make a first approximation to a consideration of the actual complexity of the atmosphere's action. I have here supposed the original heat of the sun before absorption (corresponding to our symbol, Pouillet's A) to be divided into 10 parts, and that these are of such a nature as to be acted on with as many degrees of selection, each having a separate coefficient of transmission, so that ten degrees of absorption are represented, ranging from extinction to nearly total transmission. And thus we may grossly typify the far more complex action which actually goes on in nature.

The first column represents the amount and composition of the original energy. The second column represents the intensity in each case under our suppositions when the ray has reached a height of about one-third of the homogeneous atmosphere. The third column represents the intensity attained in each case when the radiation has reached the sea level after a vertical transmission. The fourth column that which would be observed at any late morning or early afternoon observations when sec. $z=2$. The fifth column that for an observation when the sun is still lower, or when sec. $z=3$.

TABLE 107.

$$\frac{Aa + Bb + Cc + \text{etc.}}{Aa + Bb + Cc + \text{etc.}} = p = \frac{1.5}{5.159} = 0.821 \qquad \begin{cases} Ep = 5.159 & p = 0.560 \\ Ep = 1.5 & E = 8.01 \end{cases}$$

$$\frac{Aa^2 + Bb^2 + Cc^2 + \text{etc.}}{Aa + Bb + Cc + \text{etc.}} = p = \frac{2.85}{1.5} = 0.633 \qquad \begin{cases} Ep = 1.5 & p = 0.633 \\ Ep^2 = 2.85 & E = 7.11 \end{cases}$$

$$\frac{Aa^3 + Bb^3 + Cc^3 + \text{etc.}}{Aa + Bb + Cc + \text{etc.}} = p^2 = \frac{2.025}{1.5} = 0.450 \qquad \begin{cases} Ep = 1.5 & p = 0.671 \\ Ep^3 = 2.025 & E = 6.71 \end{cases}$$

We have used the symbols of our preceding formulæ, though here $A = B = C$, etc.

Under the above hypothetical conditions an actual observer on the summit of Mount Whitney provided with an actinometer or pyrheliometer (whose thermometer bulb cannot discriminate between radiations) would find an amount of heat represented by 5.16, and comparing this by Pouillet's formula with the heat at sea level (4.5) he would find the coefficient of transmission 0.56 and for the heat outside the atmosphere $A = 8$, so that under these ideally favorable circumstances his value of the solar constant would be but four fifths of what it should be. We say "ideally favorable" because the construction of our table tacitly assumes that all strata of like density have like transmissibilities, an assumption which in no way represents the complexity of the actual state of things. If our observer is at the sea level, where he finds a heat represented by 4.5 at noon under a vertical sun, and a heat of 2.85 when the sun's zenith distance is such that the mass of air traversed is doubled, he will, by a combination of these two through Pouillet's formula, find the values $a = 0.63$; $A = 7$ nearly. Now, since we are supposed to know here that the actual heat outside the atmosphere was 10, we see that the real coefficient of transmission is represented by $\frac{1.5}{10}$ or that a is actually 15 per cent. instead of 0.63, so that this determination gives a much larger value than the truth for the transmissibility, and a much smaller value than the truth for the solar constant. Finally, if we suppose the observer to compare his noon value with that after the ray had suffered three times its noon absorption, giving an observed heat only a little over 2, we find 0.67 for the transmissibility and 6.7 for the solar constant, so that the error in each case is increased still further.

It may, perhaps, be asked how it is that if the transmissibility, as determined by the thermometer or by ordinary photometric measurements, is really as erroneous as we assert, that we obtain values for it on the whole so harmonious when we observe through very different depths of atmosphere. Our present table shows quite clearly how there *may* be such a coincidence of results by the ordinary method, for we know that the low sun or star observation can seldom be taken with advantage much nearer the zenith than we have here represented it, where sec. $\zeta = 2$, or seldom much farther from it with advantage than where sec. $\zeta = 3$. Now the entire range of values for the transmissibility to be obtained under these conditions is, as we have seen, only from 0.63 to 0.67. No one who knows anything of the difficulty of such observations will affirm, I think, that the difference from the mean of 0.65 could be determined with any certainty even by years of observation. A lifetime of observations at a single station, under ordinary conditions, would probably only confirm the observer in the belief that the true coefficient of transmission in our hypothetical case was about 0.65 per cent., whereas it is, as we see in our imagined instance, but 45 per cent. The same considerations will help us to observe how it is that the method of Forbes and M. Crova, of deducing the solar constant from an empirical curve which strictly represents the facts of observation, though sound in theory and the best the *actinometer* observer at one station possesses, must fail to give a proper result, owing to the insufficiency of the data at the command of the most skillful and assiduous observer by the old method. The results thus obtained at a single station must always be too small then, even under the ideally favorable conditions we have here imagined; but there is every reason to believe that apart from the difficulties we have just mentioned, nature presents many others, and that among these is the fact that there is a systematic difference between the condition of the air observed through, at high and low sun, even on the clearest day. It results from all we have stated, that a great step toward accuracy will be made by measuring on pencils as homogeneous as possible, which we proceed to do with the spectro-bolometer, but that also the same considerations which prevent us from regarding the use of the thermometer as trustworthy, apply, though in a greatly lessened degree, to those with the spectro-bolometer, since the pencils on which this operates cannot be absolutely homogeneous, and their coefficient of transmission can not be absolutely a constant, so that even *its* observation will in theory tend to give somewhat too small a result.

To prevent misapprehension, it may be remarked that our theoretical conclusions here rest on the algebraic demonstration, and that the numerical table is only presented in illustration.

CHAPTER XI.

THE SPECTRO-BOLOMETER.

Before describing the use of the spectro-bolometer it will be convenient to recall the nature of the analogous processes already employed at Allegheny and the first results.

Thus we found, as in Table 10, that the heat in a certain narrow pencil of rays near $\lambda = 0.^s60$ was 621 (on some arbitrary scale), and that the mean coefficient of transmission for this pencil (Table 6) was 0.636; so that $\frac{621}{.636} = 976$ represents the energy which would have been observed in this ray, could we have ascended from *the sea level* to the upper limit of the atmosphere. We have here been obliged to make the provisional assumption that this narrow pencil of rays is homogeneous, and that its rate of transmission is a constant. When we examine our pencil, however, with the spectroscope visually, we find that, irrespective of the solar lines, there are in it a large number of alternations from complete transmission to absolute absorption, familiar to us as the telluric lines, and due wholly to absorption in our atmosphere. If we ascend a lofty mountain with our spectroscope, we find that some of these lines are sensibly as black (and as cold) there as at the sea-level.

To fix our ideas, let us consider the familiar *D* lines. The space between these is (on the scale of ordinary observation) far narrower than the narrowest linear thermopile, or even bolometer strip, yet within these narrow limits we have at least four conspicuous lines, which we know to be cold spaces where telluric absorption has already done its complete work. These particular lines are visible at great altitudes, and hence at the highest point we can observe in our atmosphere we find that some of the rays are already totally extinguished by it, and never reach us. In other words, their coefficient of transmission is so small that for our purposes it may be treated as zero. But the intermediate spaces between the *D* lines we know to be crowded with other lines at sunrise and sunset, and each must be there in fainter degree even at our high sun observation; contributing to diminish the aggregate brightness and warmth, although it may not be separately perceptible. In such a narrow space, then, which we have treated as homogeneous from necessity, we have no real homogeneity. Our bolometer strip or our linear thermopile thus used in differentiating different portions of the spectrum is doubtless a great advance upon the thermometer, which does not discriminate at all, but we should need an infinite minuteness of discrimination to bring all our conclusions to the test of direct experiment.

Still considering the telluric lines between the *D*'s as an example, let us suppose them divided into two typical groups, the first of which is absolutely extinguished before it comes in any way under observation. It is the second alone of these, including all that is relatively bright, then, which has furnished us the coefficient .636 which we have just found, and if we suppose (merely for illustration) that one-third the energy in this narrow group was extinguished before we could observe at all, it would follow that the heat outside the atmosphere in this pencil was certainly over one-half more than our formula gave it, *even when we apply that formula to observations which discriminate between different kinds of heat*. It is not, it is true, probable that in this particular region which we now consider, so much as one third of the heat has disappeared before its descent to the highest mountain top, but the crowds of telluric lines which, as we have just remarked, spring out at sunrise and sunset, blotting out the light between the sodium lines, make us believe that, if we could take the separate action of all into account, there would be little exaggeration in our estimate. I have elsewhere pointed out that the grosser dust particles with which our atmosphere is filled, must, in scattering all rays indiscriminately, exercise a kind of non-selective absorption, and that between this and that of the molecules whose vibration gives us the telluric lines, every kind and degree of absorption must go on in the narrowest pencil that we can ever hope to physically isolate.

It is most plain, however, that we must operate on pencils as narrow as we can, and our apparatus for doing this (the spectro-bolometer*) is shown in elevation in Plate VIII, and in plan in Plate IX.

Two long arms, A A', turn independently about the vertical axis, the angle between them being measured by a graduated circle with two verniers reading to $10''$. One of these arms is directed toward the slit, and the other toward the spectrum formed by the light on leaving the prism which we here suppose to be used. (When the grating is employed the arms are brought as nearly as possible together in the position shown in Plate X.) This latter arm carries at its extremity a concave mirror M, of 98 centimeters focus, and bears on either side of the prism an accurately planed track directed toward the center of the mirror. On either of these tracks slides a carriage with y's. Into these y's, at B, drops either of two like ebonite cylinders, one containing the bolometer, and the other the ordinary reticule and eyepiece. The bolometer used in the measurements of the cold bands on the charts (see Plates XI and XII) exposes to the spectrum a single vertical strip of platinum, ½ mm. wide, covered with lampblack, and placed accurately in the axis of the ebonite cylinder by reversal under a compound microscope‡. The eyepiece also has its cross-wires centered in the axis of the second cylinder, and serves to examine optically the place which will be occupied by the bolometer strip when the bolometer cylinder is in the y's. A simple interchange of the cylinders places the bolometer strip with precision in the part of the spectrum optically pointed on, a moment before. The optical axis of the mirror M exactly bisects the angle between the direction of the arm A' and the central line of the track, so that a ray falling on the center of the mirror from the center of the instrument at P, after reflection falls upon the bolometer strips. C, C' are counterpoises to offset the weight of the arms A, A'.

To adjust the apparatus for observation, the screws at P are loosened, the prism removed, and the arm A' brought around in line with the long tube. The eyepiece being placed in the y's at B, the image of the distant slit is brought upon the central wire, when the reading of the divided circle should be 0° $00'$ $00''$, indicating a deviation of zero. The arm is then moved to one side as in Plate IX, until the mirror intercepts the rays from the prism, which has first been replaced upon its table and adjusted by the screws below. The prism is now carefully set to minimum deviation (usually for the D_1 line), and is then automatically kept in minimum deviation for all other rays by the tail-piece and attachment at P. When the cross wires of the eyepiece are set upon the D_1 line the circle should indicate a deviation of 17° $11'$ $15''$, for the particular prism in question. A bright and pure image of the spectrum about 6 mm. wide and 640 mm. long between the A and H lines is now formed in the principal focus of M near the prism, and the bolometer case being substituted for the eyepiece, the carriage is slid along the track until the central strip, placed vertically and parallel to the Fraunhofer lines, comes exactly into focus. The heat of the solar rays in any part of the spectrum may now be measured by the bolometer (the galvanometer giving a marked deflection as it passes over the leading Fraunhofer lines), and the deviation for that part is indicated by the divided circle to $10''$.

"DIRECTIONS FOR ADJUSTING THE SPECTROMETER FOR COMPARATIVE MERIDIAN AND AFTERNOON MEASUREMENTS WITH THE BOLOMETER, OR FOR MAPPING THE SPECTRUM OF THE PRISM.§

"Arrangement of apparatus.—In the present (June, 1883) arrangement the light comes from the north, and in this case the bolometer carriage slides on the west track, the spectrum is thrown

---

* Made from the writer's designs by W. Grunow, of New York, and first used on this expedition.

† When studying the extended grating infra-red spectrum a wider bolometer is used.

‡ The prism used was one of a glass specially diathermanous to infra-red rays. It was made by A. Hilger, of London, and its optical properties are excellent. Its principal constants are: Size of two polished faces, 53 mm. × 49 mm.; specific gravity, 3.25; refracting angle, 62° 31' 43'; index of refraction for D_1 line, 1.5798; index of refraction for H line, 1.6070.

A rock-salt prism of nearly equal size and great purity, by A. Clark & Sons, has also been used. It is capable of dividing the D line when freshly polished. Prisms of quartz and spar have also been used, and the selective absorption of the glass for different parts of the spectrum determined by comparison.

§ Taken from instructions in Observatory Record Book for the use of the observer.

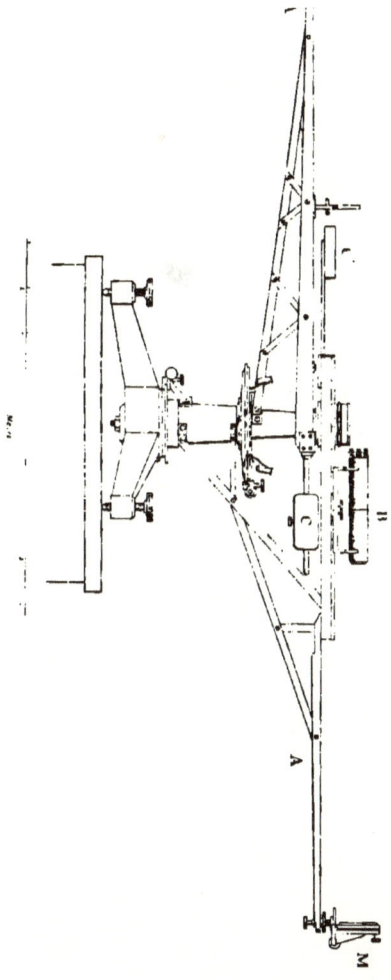

PLATE VIII

SPECTRO-BOLOMETER

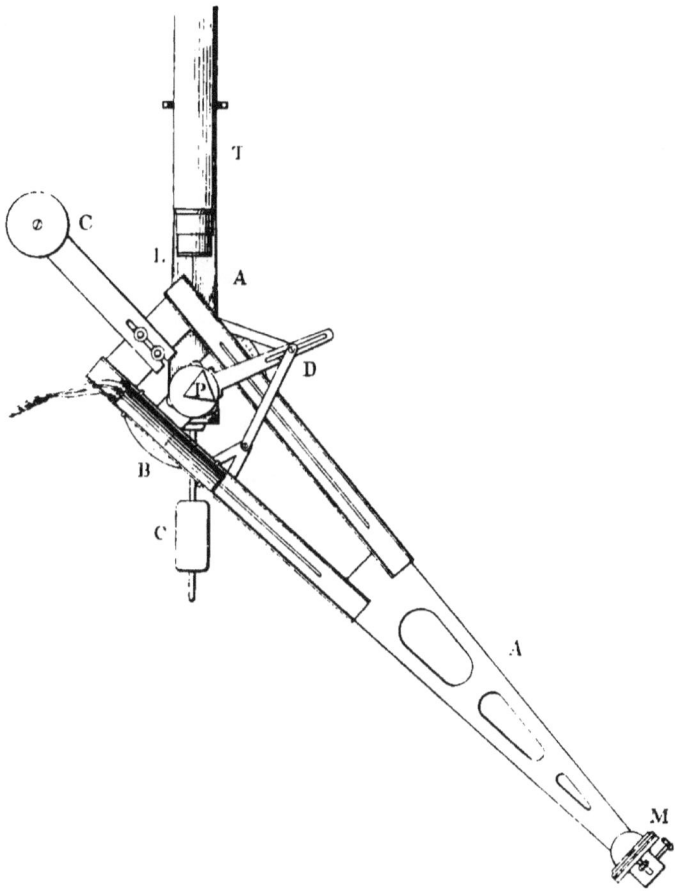

PLATE IX

SPECTRO--BOLOMETER
AS USED FOR MAPPING PRISMATIC SPECTRUM

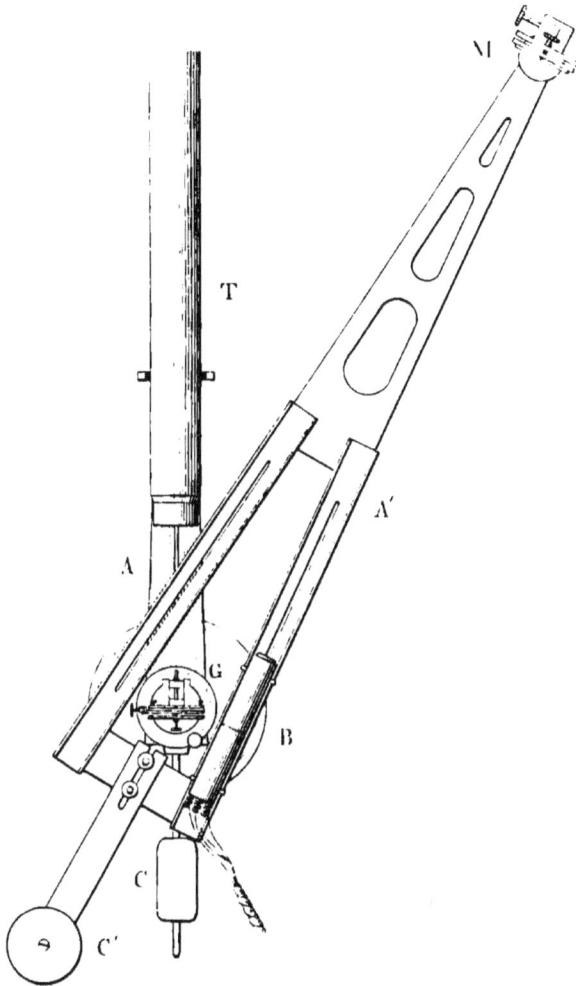

PLATE X

SPECTRO—BOLOMETER (PLAN)

AS USED FOR MAPPING NORMAL SPECTRUM.

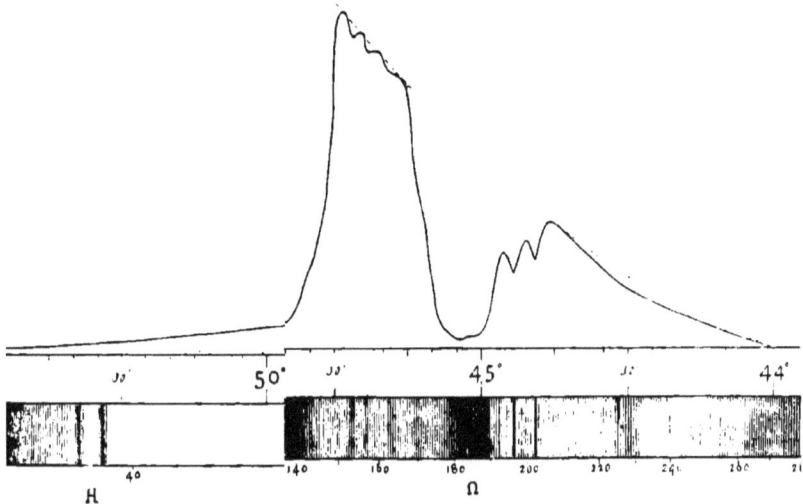

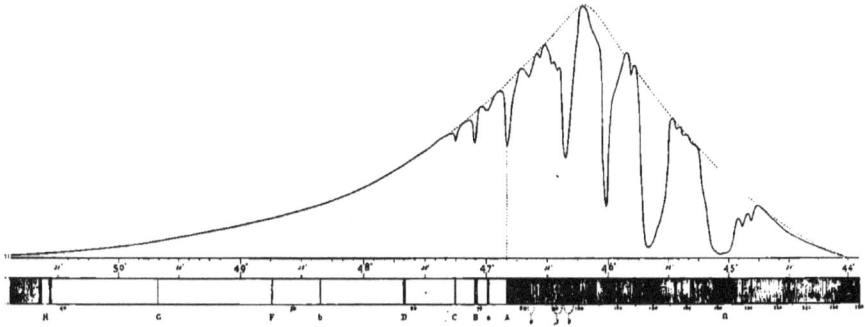

PLATE XI.
PRISMATIC SPECTRUM.
(ENERGY CURVE)

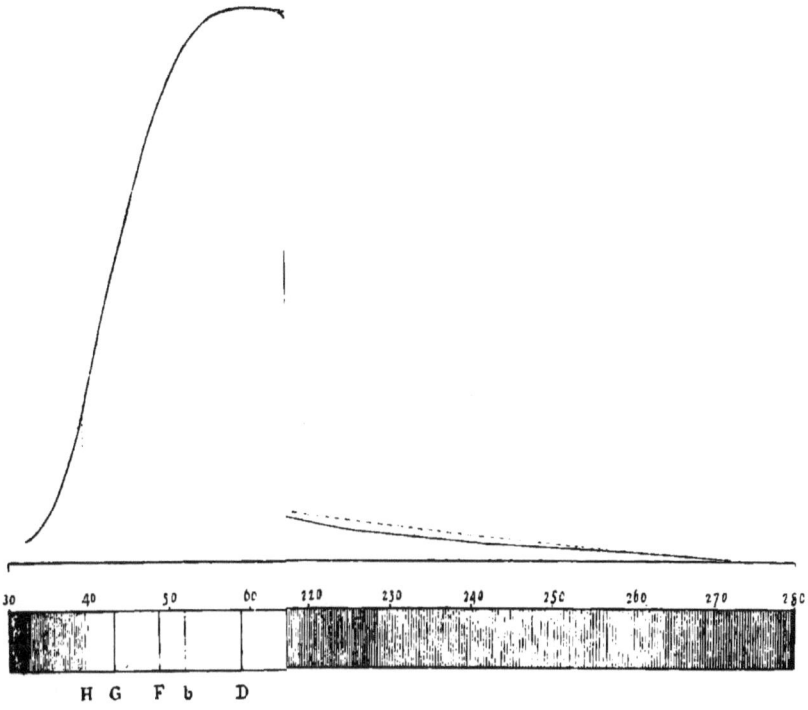

30   40   50   60   110   230   240   250   260   270   280

H G   F b   D

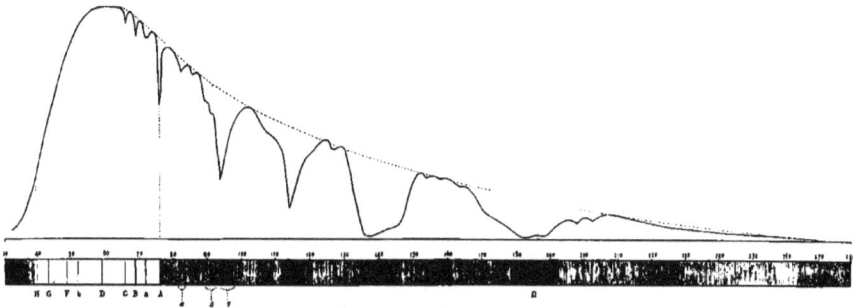

PLATE XII

NORMAL SPECTRUM.
(ENERGY CURVE)

toward the east, and the reading of the circle by the E. vernier, when the image of the slit is thrown on the bolometer, is 0°. The figures on the circle run in such a direction that under these conditions the reading of the circle for any position of the arm on the east side of the instrument is equal to the deviation.

"*Adjustment.*—The spectrometer must first be adjusted by the three foot screws until the plane containing the centers of the prism, mirror, and bolometer passes through the slit in the end of the long tube. It is preferable, though not necessary, in making this adjustment, that the axis of the spectrometer should be vertical, and the slit and central plane 1 meter above the floor, which is supposed to be level.

"The collimating lens having been placed in the tube at its own focal length from the slit, the end of the tube should next be so placed, by means of the adjustable stand, that the circle of light projected from the lens falls fair upon the face of the prism placed over the center of the instrument, normal to the incident rays. The prism being then removed, the circle of light will also fall centrally upon the mirror when the two arms of the instrument are in line. Remove the prism and screw the small sight into the central hole of the prism table, and move the short arm which carries the mirror, preferably by the hand, until the shadow of the sight falls as nearly as possible upon the center of the mirror. The short arm should then be clamped firmly by the lower clamping-screw.

"The reading on the slit should now be made 0°. As it is exceedingly difficult, in this position of the arm, to get the eye down to the observing eyepiece in order to do this, the following method has been hitherto used and found to be equally accurate and much more convenient. The shadow of the sight falling centrally upon the mirror, the light from the mirror should be directed by the two slow motion screws until it is reflected down into the adjusting eyepiece and the image of the slit formed upon the cross-wires, when, upon properly focussing the eyepiece, a sharp image of both slit and cross wires will be projected upon the north wall of the dark room on the west side of the slit itself and at the same height above the floor. The wires should be made to bisect the slit by means of the two slow motion screws of the mirror mount. By this adjustment the optical axis of the mirror has been made to bisect the angle between the slit and the eyepiece.

"The long arm of the spectrometer should now be moved by hand until the reading of the east vernier is 0°, which is therefore the reading on the slit. The tripod having been placed nearly in its true position at the beginning, the change in position will be small. The long collimator being, for the present purpose, disconnected from the long arm, the position of the slit will not have been changed by this manipulation. Now unclasp the short arm and move it to the east, the long arm remaining undisturbed throughout all the subsequent operations, attach the arms of the minimum deviation apparatus to the tail-piece of the prism table, and set the circle by a positive rotation of the tangent screw to a deviation of 47° 41' 15''. Place the prism on its table with its refracting angle over one of the small leveling screws, its back surface at right angles to the tail piece, and at a distance from the center of the table such that the rays leaving the prism appear to proceed from the center of the instrument.

"Turn the prism table to adjust for minimum deviation. The spectrum, after leaving the prism, should strike the center of the mirror and be reflected back upon the cross-wires. If the spectrum (viewed in the manner described below) is not bisected by the horizontal wire, it must be made to do so by means of the leveling screw under the refracting angle of the prism—the mirror must not be disturbed. Note if, on rotating the prism table in both directions past the position for minimum deviation, the spectrum returns on the same path. If not, the screws under the other two angles on the prism required adjustment. The spectrum must then be brought on the horizontal cross-wire again by the first screw, and so on until the spectrum remains bisected by the horizontal wire while the prism table is rotated, when the refracting angle of the prism will be vertical. The slit should now be closed up and an image of the spectrum, together with the cross-wires, projected from the eyepiece upon a piece of white paper held about a foot distant. By proper focussing both the wires and the D lines may be sharply seen. The prism should now be very carefully set to minimum deviation by rotating its table, when the image of D_1 should fall exactly upon the vertical cross-wire. If it does not, but comes within a distance of only about twice that separating the *D* lines (say), the difference is probably due to springing of the long arm in moving by the hand, and possibly to change of temperature, and D_2 may be brought upon the cross-wire by slightly moving the horizontal tangent screw of the small mirror.

"Note the distance of the cross-wires from the mirror on the graduated scale of the bolometer track, take out the eyepiece, and put in the bolometer case holding the required bolometer, and set mark which indicates the position of its strips to the distance just noted. Set the bolometer slit or strip vertical and at right angles to the spectrum, which may be done by means of a small reflector held in front of the aperture of the case. Secure in position by rubber bands, open the slit to the required width (usually two millimeters), and the apparatus, so far as the optical arrangements are concerned, is ready for work. The settings should all be made by a positive rotation of the tangent screw. It is, of course, assumed that the cross-wires of the eyepiece and the bolometer aperture have been previously truly centered."

Plate XIII shows the arrangements adopted in the observations made on the expedition; the solar rays were directed into the instrument by a heliostat, H (made by W. Grunow), having two mirrors both 6 inches in diameter, and silvered by Foucault's process on an optically plane surface. One of these mirrors revolved by clock-work on a polar axis, the other was fixed, and the heliostat as thus used was about 5 meters to the south of the spectro-bolometer, B, which was placed upon a solid pier and covered by the "hospital" tent described in the general account of the expedition. The galvanometer used, G, was an extremely sensitive reflecting one of Sir William Thomson's pattern, and of the most recent construction, by Elliott. A current of 0,0000000052 ampère caused a deflection of one division upon its scales, placed at a distance of one meter.

It will be remembered that most of the observations upon the mountain were made directly in the diffraction spectrum and only carried as far as $\lambda = 1^\mu.2$. This wave-length ($1^\mu.2$) to which we measured with the grating is beyond the limits of the invisible spectrum as those limits were, until very recently known,[*] but below this lies a great region whose extent was first discovered in the course of the very observations now in question.

Towards the close of our stay upon Mount Whitney, when using the prism, a hitherto unknown extension of the infra-red region was observed, and once found, we have been able to continue its observation at Allegheny, and have thus obtained data in continuation of those obtained upon the mountain. Those on the mountain extend, as we have said, only to $1^\mu.2$. Those at Allegheny have been continued in the following year, 1882, to more than double the length on the normal scale, of the previously known spectrum, or to wave-length of rather more than $2^\mu.7$, where the solar spectrum sensibly terminates (at the sea-level). Since these latter observations were taken both for a high and low altitude of the sun daily, we have obtained, with their help, a knowledge not only of the heat in each ray, but of its transmissibility, a knowledge, that is, of the coefficient of transmission throughout the whole of the invisible spectrum which reaches us; and we are able to supplement our determinations on Mount Whitney by them. Immediately after the bolometric observations on Mount Whitney, in 1881, with the grating, we give accordingly those taken at Allegheny, in 1882, with the same apparatus, except that in lieu of the grating the special prism of glass already described as transparent to the lowest infra-red rays, was employed.

All the observations quoted in the following tables are made without reference to local absorptions like the Fraunhofer lines, and give only the general distribution of the solar energy on the normal scale, the grating being employed on the expedition for this purpose. But on the 9th of September, 1881, at Mountain Camp, a hasty review of the spectrum was made by a prism of rock salt, and also with a glass one; and with this the great cold band which we have marked on our charts as Ω was discovered. It was, at first, supposed that this would not be discernible near the sea-level, but on the return to Allegheny it was found to be entirely within the reach of the linear bolometer. Two minute cold lines of even greater wave length were indeed observed and an investigation of the whole spectrum below $\lambda = 0^\mu.7$ at Allegheny, extended our knowledge to the limit $2^\mu.7$; between two and three times that previously known.

_____

Dr. J. W. Draper, so late as 1881 (Proceedings American Academy), regards this wave length as probably exceeding the limits of the extremest infra-red.

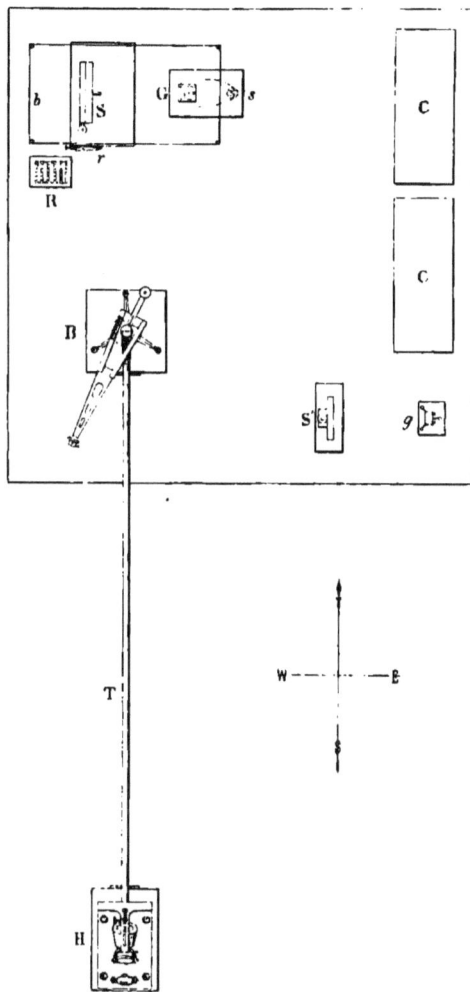

PLATE XIII.

PLAN OF TENT
AND
ARRANGEMENT OF APPARATUS.

There is no more important feature of these bolometer observations than their contradiction of the common, indeed the universal, belief, prevalent at the present time among scientific men, that our atmosphere is least transmissible to the sun's dark heat. It is constantly asserted that the "heat" ray has found easy admission to our atmosphere as "light," that it has fallen upon the surface of the earth and been returned toward space as dark *heat*, which because of its non-transmissibility cannot readily escape again; and the meteorologist has seen in this heat-storing action a preservative of the surface temperature of our planet. In a certain sense this is true, but the physicist, biased, perhaps, by the meteorological evidence, has asserted far too hastily that all the dark solar heat is comparatively non-transmissible by the atmosphere. *The contrary is the case.* It is difficult to say how far physics has here leaned in the past on meteorology, and meteorology on physics, but if the present observations are correct, it seems clear that the present view of both must be modified.

Let me state, to prevent misapprehension, that I do not at all deny that dark heat *radiated from the soil* is arrested by our atmosphere, or that the surface temperature of the earth may be elevated by such a heat-storing action as has been pointed out as possible by Tyndall and by others, but I assert that out to the utmost limit known, at 2ᵘ.7 of the invisible infra-red solar heat spectrum, the rays grow progressively more transmissible as their wave-length increases, (always excepting of course the regions of special absorption known as telluric lines or cold bands).

At the present time (1883) it is almost universally understood, on the contrary, that the light rays are most transmissible, and these dark heat rays least transmissible, so that the transmissibility declines on either side whether toward the infra-red or the violet.

I began these bolometric investigations three years ago, and when I had first perfected apparatus which enabled me to measure with a novel degree of precision in the invisible heat region, nothing surprised, and perhaps I should say disconcerted, me more than to find that the lower invisible heat rays were *more* freely transmitted by our atmosphere than any other; for this was directly the opposite of what I had been taught to believe. The apparent consequence which the reader may perhaps remark I also saw, and asked myself how, if this were so, the atmosphere could play toward the earth the part of the glass cover to the hot-bed, to which it is so usually compared. I need hardly say that it seemed to me unlikely, also, that previous observations by others could be so wrong as to a matter of fact, where that fact was so important; and with all these doubts in mind I resumed my work, and tested in every way my first observations. All observations that I was able to make in this respect agreed, however, and all checks I could apply verified this first result. Within the extremest range of the then known invisible spectrum, though the heat itself diminished, the proportion of that heat transmitted grew greater and greater as the wave-length increased, instead of less and less. When no further pains could be applied at Allegheny, it was largely with this point in view that I resolved to repeat the observations at the base and summit of a lofty mountain, and to actually demonstrate there, either that the heat waves were less transmissible than the light waves, as is usually thought, or were more so, as I had cause to believe. Direct observation, as will be seen by these tables, comparing the absorption between the mountain and valley, demonstrated that the heat was more transmissible.

On Mount Whitney was discovered, as I have said, a great extension of the infra-red spectrum. On my return to Allegheny in 1882, this extension of the spectrum was studied with reference to the same question.

I have now given over two years of most assiduous observation to this point, and if the truth has been missed it is not for want of careful experiment; but I believe that I have not been misled in this respect, and I desire the reader's attention to the nature and amount of the evidence that, within the entire region open to our observation, the solar heat uniformly (with the exception of the telluric lines and cold bands) becomes *more and more transmissible as its wave-length becomes greater.*

"How, then," it may be asked, "can the heat-storing action of the atmosphere be maintained, if the 'dark' heat escapes more easily than it enters as 'light' heat?" I might answer that it was sufficient for me to have demonstrated the fact of this increasing transmissibility, and that the fact was not to be disproved by the contingency that inferences from it might conflict with preconceived opinion. I desire, however, to re-iterate here that I do not say that *all* dark heat escapes more easily than it enters as "light" heat, but only that throughout the solar spectrum, *as far as we know it*, even in the extremest infra-red heat, these heat rays are more absorbed than the light ones. I believe that "dark" heat of the wave-length of that which must be re-radiated from the surface of our planet could only be found in the unabsorbed solar beam before it enters our atmosphere, and that such heat as this has never reached us in the actual sunbeam, and has never been analyzed by the prism. I do not at all question the experiments of Tyndall and others on the non-transmissibility of such heat as that from the Leslie cube, and I do not myself find any conflict between them and my own, but as far as I have been able to learn myself by direct experiment, the wave-length of the heat absorbed and re-radiated by the soil, or that from a source much below the boiling point of water, is such as to preclude the forming of any spectrum of it which we can yet analyze with the grating or the prism.

At present, the great cold band discovered on Mount Whitney is far larger than any previously known in the upper part of the spectrum. Experiments made shortly after the return of the expedition were indecisive as to the nature of these cold bands, but our later ones render it probable that all of them are telluric. While within the whole known infra-red spectral region, then, I find the heat growing more and more transmissible, excepting in the cold bands, I find these cold bands (*i. e.*, the regions of non-transmissibility) growing wider and wider until they obliterate the spectrum beyond $2^\mu.8$, for we may look, I think, on what remains as an unlimitedly extended cold band, possibly varied by one or two traces of spectral energy. It is below this *ultima thule* of these latest investigations that I believe is yet to be found the wave-length corresponding to the maximum ordinate in the heat spectrum of the surface of our planet.

I may make my meaning plainer, if I say that if the mean surface temperature of the globe be taken at 15° or 16° C., unless we could form an energy spectrum from heat of some such degree as *this*, and determine the maximum ordinate of its curve, we should, in my view, be unable to determine the approximate wave-length of the heat in question, to which I believe our atmosphere is sensibly impermeable; yet it is on the retaining of this extremely low *kind* of heat that the organic life of our globe, it seems to me, largely depends.

On the return to Allegheny, with the linear bolometers, and especially the one whose single strip was ⅓ of a millimeter in width and less than $\frac{1}{500}$ of a millimeter in thickness, a distinct special exploration of the invisible spectrum went on, as I have said, with particular reference to the mapping of these unknown cold bands. The publication of these results will be found in part in a memoir in the American Journal of Science for March, 1883, in which, by the permission of the Chief Signal Officer, certain of the present results are incorporated. I will in turn borrow from that memoir* the chart which includes the great band Ω discovered on Mount Whitney, which the reader will find in the lower curve of plates XI and XII. He will find, also, in the Appendix No. 2 an account of the method of determining the wave-lengths in these charts, from a memoir communicated to the National Academy of Sciences in April, 1883, which may perhaps receive a separate publication elsewhere later.

---

* The maximum ordinate on the normal curve, as deduced from the prismatic one, should fall in the same place as that observed directly in the heat spectrum of a grating, and actually does so in our charts with gratifying exactness. In the illustration first given in the American Journal of Science and other journals, owing to a slight error in the preparation of the drawing for the engraver, the maximum ordinate of the normal curve is somewhat nearer the red than it should be, and the descent of the curve toward the violet somewhat less abrupt.

CHAPTER XII.

BOLOMETER OBSERVATIONS ON THE SOLAR DIFFRACTION SPECTRUM MADE DURING THE MOUNT WHITNEY EXPEDITION.

With the spectro-bolometer, employing the large Rutherfurd grating of 681 lines to the milli meter, and other subsidiary apparatus already fully described, high and low sun observations were made in the diffraction spectrum at Lone Pine on August 11, 12, and 14, and at the Mountain Camp on Mount Whitney on September 1, 2, and 3. Owing to the difficulties incident to such an expedition, these days were the only ones on which observations of any value were secured with the bolometer.

Among the difficulties encountered at the lower station, one of the most serious arose from the derangement of the apparatus by high winds, the instruments being inclosed and protected by a tent only, where the hot wind found free entrance, every part of the instruments being covered by the desert sand and dust, while the heat of the darkened tent was often very near the limit of human endurance. Another difficulty arose from the rapid rise and fall of temperature in the morning and evening, which produced a galvanometer drift.

In spite of all care, the apparatus was too elaborate not to have suffered somewhat during transportation, and the number of other impediments to successful work was enough at the time to call out all the resources of our ingenuity and patience. In order to eliminate the effect of these changes, the conditions were assumed to remain constant only for the time required to complete a single series. The deflections were then approximately corrected for the effect of galvanometer drift, a further correction being made graphically by drawing smooth curves through the first irregular ones, and the numbers were so reduced that their sums should equal 1,000 for each series (thus representing one and the same sum of heat whether the previous absorption had been great or small). As the labor of separate reductions and comparisons of each day's observations would have been very great, and as the observations themselves were not such as to make such a course desirable, they were divided into groups, according to the hour-angle at which they were made. For the Lone Pine observations these groups are:

	h.	m.
Lone Pine I, including 8 series, average hour-angle	0	22
Lone Pine II, including 10 series, average hour-angle	4	23
Lone Pine III, including 4 series, average hour-angle	5	24

And for Mountain Camp observations:

	h.	m.
Mountain Camp I, including 6 series, average hour-angle	0	17
Mountain Camp Ia, including 4 series, average hour-angle	2	48
Mountain Camp II, including 7 series, average hour-angle	4	11
Mountain Camp III, including 3 series, average hour-angle	5	24

We give first the original observations, designating each series by a letter of the alphabet. What the reader has before him here in Table 108 is the observation as taken from the smooth curve. The wave-length is at the head of each column, and it is very obvious from the first that the higher the sun the greater is the proportionate heat in the short wave-lengths. Table 109 is elucidatory of Table 108, and needs no explanation.

TABLE 108.

Lone Pine observations with spectro-bolometer.

Date.	$\lambda=\mu.350$	$\lambda=\mu.375$	$\lambda=\mu.400$	$\lambda=\mu.450$	$\lambda=\mu.500$	$\lambda=\mu.600$	$\lambda=\mu.700$	$\lambda=\mu.800$	$\lambda=1\mu.000$	$\lambda=1\mu.200$	Sums.
Aug. 11, a	0	5	15	55	90	133	145	120	72	37	672
11, b	2	8	15	47	78	123	140	120	70	37	610
11, c	3	9	23	63	99	133	140	119	78	50	719
11, d	5	10	30	64	96	145	148	128	76	54	734
11, e	2	7	14	50	91	125	128	111	60	37	630
11, f	3	7	22	57	95	147	135	101	60	45	672
11, g	4	7	20	55	95	145	145	120	66	25	682
11, h	0	5		62	114	157	162	130	68	55	737
11, i	0	0	0	33	67	107	109	84	42	19	463
Aug. 12, a	2	7	29	82	132	177	172	140	81	42	855
12, b	3	8	24	65	110	180	194	170	98	45	897
12, c	2	6	20	75	122	170	160	131	88	51	823
12, d	0	0	5	45	80	128	130	110	79	56	633
12, e	0	0	4	41	74	126	142	121	73	49	635
Aug. 14, a	2	11	34	86	126	182	194	145	66	26	874
14, b	2	7	20	82	120	197	197	147	75	25	886
14, c	3	9	30	92	120	200	187	127	60	39	857
14, d	4	8	25	64	100	160	162	120	60	40	743
14, e	2	6	20	70	120	151	156	146	94	57	822
14, f	0	2	5	56	99	140	147	150	80	40	629
14, g	0	0	3	36	59	95	104	90	52	17	458
14, h	0	0	1	27	60	101	109	96	60	30	487

TABLE 109.

Lone Pine conditions during observations with spectro-bolometer.

Date.	Local time.	Sun's hour angle.	Zenith distance.	State of sky.	Wind.	Observers.	Remarks.
Aug. 11, a	6h 53m to 7h 17m a.m.	h. m. 5 09	69	Fair blue	Gentle.	L., K., and D.	Galvanometer drift +, moderate.
11, b	7 20 to 7 53 a.m.	4 27	62½	... do	... do	L., K., and D.	Do.
11, c	11 20 to 11 33 a.m.	0 38	23	Blue	Gentle.	L., K., and D.	Very little drift.
11, d	11 40 to 11 58 a.m.	0 16	22	... do	... do	L., K., and D.	Do.
11, e	12 05 to 12 10 p.m.	0 07	21½	... do	... do	L., K., and D.	Do.
11, f	12 19 to 12 37 p.m.	0 13	22	... do	... do	L., K., and D.	Do.
11, g	3 55 to 4 19 p.m.	4 02	57½	... do	... do	L., K., and D.	Do.
11, h	4 45 to 5 21 p.m.	4 58	68½	... do	Fresh	L., K., and D.	Drift +, greater than at noon.
11, i	5 40 to 6 03 p.m.	5 47	78	... do	... do	L., K., and D.	Temperature of galvanometer 105° F.
Aug 12, a	7 05 to 7 45 a.m.	4 30	65	... do	Gentle	L., K., and D.	Numerous small clouds, drift +, moderate.
12, b	7 45 to 8 14 a.m.	4 05	58	... do	... do	L., K., and D.	Do.
12, c	12 00 to 12 27 p.m.	0 09	22	Clear blue	Fresh	L., K., and D.	A few clouds at first, drift +, slight.
12, d	4 15 to 4 40 p.m.	4 73	62	... do	... do	L., K., and D.	Large clouds, drift +, moderately rapid.
12, e	4 30 to 4 46 p.m.	4 33	64	... do	... do	L., K., and D.	Heavy gale Aug. 13, prevented observation.
Aug 14, a	7 07 to 7 35 a.m.	4 43	66	Very fair blue	Calm	K. and D.	Drift +, rather rapid.
14, b	7 40 to 8 02 a.m.	4 13	60	... do	... do	K. and D.	Do.
14, c	11 13 to 11 40 a.m.	0 37		Uniform fair blue as in morning.	Gentle to fresh	K. and D.	Less drift.
14, d	12 09 to 12 31 p.m.	0 16	22½			K. and D.	Temperature of galvanometer 97° F.
14, e	12 35 to 12 56 p.m.	0 42	24			K. and D.	Very little drift.
14, f	4 00 to 4 19 p.m.	4 06	59	Same as at noon.	Gentle to fresh	K. and D.	Slow drift.
14, g	4 30 to 4 50 p.m.	4 54	64			K. and D.	Variable drift.
14, h	5 05 to 5 22 p.m.	5 10	71½			K. and D.	Packed up apparatus.

TABLE 110.

Mountain Camp observations with spectro-bolometer.

Date.	$\lambda=\mu.350$	$\lambda=\mu.375$	$\lambda=\mu.400$	$\lambda=\mu.450$	$\lambda=\mu.500$	$\lambda=\mu.600$	$\lambda=\mu.700$	$\lambda=\mu.800$	$\lambda=1\mu.000$	$\lambda=1\mu.200$	Sums.
Sept. 1, a	1	2	5	26	44	62	61	47	24	14	286*
1, b	0	1	3	14	26	37	35	28	18	10	171
1, c	1	2	6	23	34	44	37	26	17	10	200*
1, d	1	1	4	15	25	33	31	27	16	10	161*
1, e	0	1	3	26	43	56	60	50	30	18	282*
1, f	0	2	7	50	85	118	108	80	44	20	514
Sept. 2, a	3	5	27	106	135	155	142	114	62	30	773
2, b	2	5	28	105	145	165	148	108	58	30	794
2, c	7	22	55	200	263	288	263	193	109	60	1,460
2, d	9	27	75	189	270	296	268	200	111	65	1,580
2, e	5	14	30	144	221	250	225	158	92	49	1,200
2, f	2	10	34	150	225	250	221	165	90	42	1,109
2, g	1	4	25	110	160	200	200	150	85	40	975
2, h	0	4	22	90	140	200	209	171	92	42	970
Sept. 3, a	3	14	45	170	256	347	340	240	137	79	1,720
3, b	1	17	64	192	275	305	285	190	90	56	1,479
3, c	13	21	50	184	289	333	305	209	115	64	1,572
3, d	8	20	53	176	277	340	317	229	130	74	1,643
3, e	4	10	30	142	224	270	248	183	100	50	1,261
3, f	3	10	27	160	254	305	305	215	115	60	1,442

Table 110, above, shows the corresponding Mountain Camp observations. It will be remem-

* Observations made in weak first spectrum.

bered that the grating has one of its first spectra strong, the other weak, and that measures are customarily made in the former. The low readings on September 1 are due to an inadvertent reversal of the grating, causing the measures to be taken in the weak first spectrum.

The apparatus on September 2 and 3 was in all respects, save temperature, in the same condition as at Lone Pine, nor could the lower temperature of the bolometer and galvanometer possibly account for the great increase of the readings as shown in the table of sums. It is observable, too, that this increase has been made chiefly by the gain in the shorter wave-lengths, the wave-lengths 1,000, 1,200 (Table 110) not showing, for instance, the same proportionate increase over those in Table 108 that wave-lengths 0,500, 0,600 do. These last considerations show clearly that the change is not due to instrumental causes, but that the bolometer registers a very great increment of radiant heat in the visible part of the spectrum when on the mountain, while showing no corresponding gain in most infra-red radiations. It may even, it seems, indicate for some of these wave lengths greater *radiant* heat below in the valley than above on the mountain. This last effect is so slight that we must admit the probability of its being due to instrumental causes.* There is, however, a cause theoretically acting to produce such a result, and one not difficult to explain. In fact the column of air between the instrument and the sun which has absorbed heat of short wave-length must necessarily lose this heat in part by radiation toward the observer, and it will by the ordinary course (the degradation of heat) part with this by radiations of a greater wave-length than those it first absorbed, and it is not perhaps absolutely impossible that this *may* be sufficient in the case of some infra-red rays to mask the effect of the very slight atmospheric absorption of these rays themselves. It is to be observed also more generally, that whenever we measure the atmospheric absorption of any ray whatsoever, we should, according to what has just been said, strictly speaking make a correction for the heat of the wave-length under discussion radiated by the air column between the instrument and the sun, and that if we have not done so throughout this research it is not through any oversight but the occasion for correction, but because the whole amount of this correction is so small that its value is not determinable.

It may be remarked also that the mountain observations indicate an even greater facility of transmission for "dark heat" rays than has been inferred from the previous observations at Allegheny.

TABLE III.

Mountain Camp conditions during observations with spectro-bolometer.

[L. represents S. P. Langley. K., J. E. Keeler. D. W. C. Day.]

Date.	Local time.	Sun's hour angle.	Zenith distance.	State of sky.	Wind.	Observers.	Remarks.
		h. m.					
Sept 1, a	8 45 to 8 58 a. m.	3 31	78	Deep blue	Light breeze	K D	Observations made in weak 1st spec-
1, b	9 35 to 9 35 a. m.	2 35	61	do	do	L K D	trum
1, c	12 07 to 12 28 p. m.	0 18	38	do	Fresh	K D	Do
1, d	12 38 to 12 43 p. m.	0 36	39	do	do	L K D	Do
1, e	3 44 to 3 55 p. m.	3 18	57	do	Gentle	K D	Do
1, f	4 41 to 4 59 p. m.	4 12	69	do	do	K D	Grating turned, strong 1st spectrum.
Sept 2, a	8 49 to 8 59 a. m.	3 25	74	do	Light	L K D	Strong 1 drift. Changed bolometer.
2, b	3 51 to 3 59 p. m.	3 25	57	do	do	L K D	Strong 1 drift.
2, c	11 44 to 12 09 p. m.	0 03	38	do	High wind	L K D	But little drift
2, d	12 09 to 12 28 p. m.	0 40	39	do	do	L K D	Do
2, e	4 10 to 4 27 p. m.	3 40	65	do	do	L K D	Moderate drift.
2, f	4 27 to 4 46 p. m.	4 27	69	do	do	L K D	Do
2, g	5 05 to 5 25 p. m.	5 10	76	do	do	L K D	Do
Sept 3, a	3 50 to 5 46 p. m.	3 29	57	do	Fresh	L K D	Sky slightly milky about setting sun.
3, b	8 31 to 8 51 a. m.	1 15	73	do	High	L K D	
3, c	9 25 to 9 42 a. m.	2 20	61	do	do	L K D	
3, d	11 55 to 12 13 p. m.	0 05	39	do	do	L K D	
3, e	12 13 to 12 30 p. m.	0 22	39	do	do	L K D	
3, f	4 30 to 4 50 p. m.	4 10	71	do	do	L K D	A few cirrus clouds the first seen on
3, g	4 50 to 5 16 p. m.	5 08	75	do	do	L K D	the mountain

* In fact, observations of the battery galvanometer showed that, owing to some internal change in the battery, the current employed on the mountain was somewhat smaller than that used at Lone Pine. When an allowance is made for this change and the Mountain Camp results are increased to correspond with those obtained at Lone Pine, this discrepancy disappears, as will be shown in the chapter on hygrometric observations.

The following tables (112 and 113) exhibit all the bolometric observations made on the expedition, reduced so that the sum of all the deflections for each series equals 1,000, and divided into groups so as to show the distribution of energy in the spectrum at four different times of day, when the sun had the following hour-angles:

I, meridian observations, or hour-angle zero:

Ia, hour-angle of about 2½ hours;

II, hour-angle of 4+ hours;

III, hour-angle of 5+ hours.

TABLE 112.

LONE PINE OBSERVATIONS.

Deflections reduced to represent a constant radiant heat.

I.

Hour-angle.	0μ.350	0μ.375	0μ.400	0μ.450	0μ.500	0μ.600	0μ.700	0μ.800	1μ.000	1μ.200	Sums.
0°36″	4	12	32	91	157	185	198	165	108	70	1000
0 16	7	13	41	87	154	198	203	171	103	46	1000
0 07	3	11	22	79	147	198	203	175	108	58	1000
0 13	5	10	33	85	142	218	240	159	90	67	1000
0 09	2	7	24	91	148	207	195	160	104	62	1000
0 37	4	10	23	107	140	237	218	148	70	45	1000
0 16	5	11	24	80	135	216	217	161	81	54	1000
0 42	3	7	24	85	146	185	180	178	114	80	1000
0 22	4.1	10.1	29.1	88.9	146.2	205.0	202.8	163.5	97.3	58.9	1000

II.

	0μ.350	0μ.375	0μ.400	0μ.450	0μ.500	0μ.600	0μ.700	0μ.800	1μ.000	1μ.200	Sums.
4°27″	3	12	23	75	122	193	219	188	109	58	1000
4 42	6	10	28	83	178	213	213	176	97	37	1000
4 33	2	8	23	96	154	208	201	164	95	49	1000
4 05	3	9	27	72	129	201	216	190	105	50	1000
4 24	0	0	8	71	126	193	206	183	125	88	1000
4 33	0	0	9	64	116	199	224	196	118	77	1000
4 43	2	14	29	98	144	208	222	166	78	30	1000
4 13	2	8	23	80	146	224	224	167	85	28	1000
4 06	0	3	6	80	142	201	211	186	114	57	1000
4 34	0	8	11	78	129	204	228	196	113	37	1000
4 23	1.8	6.1	19.5	80.6	131.0	204.8	216.4	181.2	104.3	51.1	1000

III.

	0μ.350	0μ.375	0μ.400	0μ.450	0μ.500	0μ.600	0μ.700	0μ.800	1μ.000	1μ.200	Sums.
5°00″	0	7	22	82	134	198	216	179	107	55	1000
4 58	0	0	7	84	155	213	221	177	95	48	1000
5 47	0	0	0	76	145	231	235	181	91	41	1000
5 10	0	0	8	55	123	208	224	197	123	62	1000
5 14	0	1.8	9.3	74.2	135.2	213.0	224.0	183.5	103.5	51.5	1000

TABLE 113.

OBSERVATIONS AT MOUNTAIN CAMP.

Deflections reduced to represent a constant radiant heat.

I.

Hour-angle.	0μ.350	0μ.375	0μ.400	0μ.450	0μ.500	0μ.600	0μ.700	0μ.800	1μ.000	1μ.200	Sums.
0°18″	4	10	29	114	173	222	187	128	84	49	1000
0 36	1	9	24	90	156	210	192	161	97	61	1000
0 03	5	15	28	137	180	197	180	133	73	41	1000
0 19	6	18	30	126	180	197	173	132	74	37	1000
0 05	8	17	32	117	178	212	194	134	72	41	1000
0 22	5	12	32	107	169	219	193	139	79	45	1000
0 17	5.3	12.8	34.2	115.2	171.6	209.5	187.5	138.0	84.2	45.7	1000

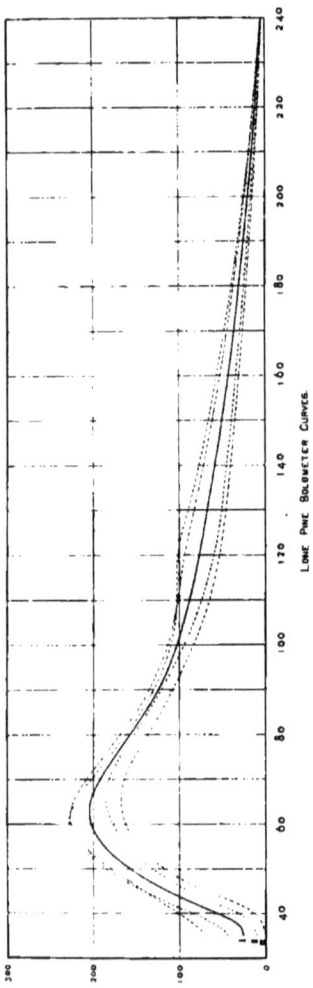

LONE PINE BOLOMETER CURVES.

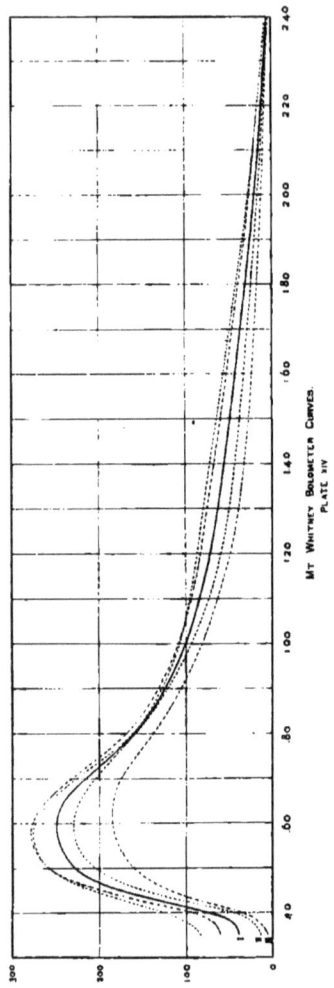

MT WHITNEY BOLOMETER CURVES.
PLATE XIV

OBSERVATIONS AT MOUNT CAMP- Continued.

Ia.

Hour angle.	0^m.250	0^m.375	0^m.400	0^m.450	0^m.500	0^m.600	0^m.700	0^m.800	1^m.000	1^m.200	Sums.
2^h 05^m	1	1	15	82	146	217	205	165	146	54	1000
2 50	5	6	35	102	183	208	186	136	73	58	1000
3 15	2	8	48	99	149	202	203	162	91	46	1000
2 26	2	10	46	130	186	206	185	128	61	28	1000
2 48	2.0	7.6	36.5	110.8	166.0	208.2	196.8	147.8	82.7	45.2	1000

II.

3^h 04^m	3	6	18	91	152	218	232	166	85	19	1000
3 18	1	3	11	92	141	201	211	178	106	55	1000
4 42	0	2	14	97	165	220	211	155	26	23	1000
3 25	4	6	25	129	171	201	184	118	80	40	1000
4 19	4	12	25	130	181	215	190	132	77	41	1000
4 37	2	8	29	128	189	210	186	139	76	15	1000
4 49	3	8	24	113	178	211	196	145	77	19	1000
4 41	2.4	6.6	22.5	109.7	168.1	212.7	198.6	151.9	81.1	42.3	1000

III.

5^h 16^m	1	4	26	113	164	205	205	154	87	41	1000
5 30	0	1	23	93	144	230	216	176	95	15	1000
5 08	2	7	19	111	177	212	201	149	80	12	1000
5 21	1.0	3.0	22.7	105.7	161.7	207.6	207.3	159.7	87.4	42.0	1000

The part of the spectrum thus far considered extends as far towards the lower limit as the grating which was used permitted, with due regard to the overlapping spectra, but a considerable part of the total heat, namely, that of the infra-red below wave-length, 1.2, is not represented in these first tables, while the actinometers, with which they are later to be compared, represent in the readings of their thermometers the aggregate effect of all heat of all wave-lengths. The curves (Plate XIV) have therefore been completed by extending them through the aid of subsequent measures at Allegheny to λ = 2$^\mu$.4 (the heat beyond this point being neglected), making the intermediate ordinates bear the same proportion to each other as in the Allegheny normal curve, determined after the return of the expedition.

This is shown in Fig. 12, where the dotted line represents the part of the curve supplied from the Allegheny observations.

Corrections having been applied for the angle of diffraction and for the selective absorption of the metallic reflecting surfaces employed, the curves were extended in the way just now explained, and the area of each curve was determined. For this purpose an Amsler's polar planimeter was used, to get by direct measurement the area above λ = 1$^\mu$.200. The section-paper on which the curves were drawn is divided into square inches and hundredths of square inches. On the axis of abscissæ Δλ = 0$^\mu$.1 = 1 inch, and on the axis of ordinates a deflection of 50 scale divisions = 1 inch. One square inch therefore = 2,500 in units of the tables. Below λ = 1$^\mu$.200, the areas of the curves were determined by means of a formula for approximate areas,

$$A = h \left(\frac{y_1}{2} + y_2 + y_3 + \ldots \ldots \frac{y_{n+1}}{2} \right)$$

y_1, y_2, y_3 &c., being the heights of ordinates separated by distances each equal to h.

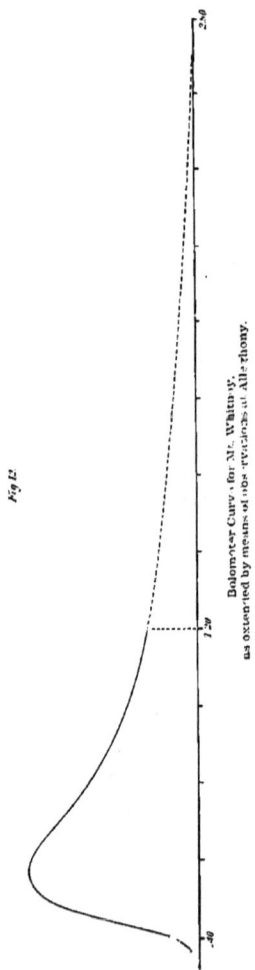

Fig 42

Bolometer Curve for Mt. Whitney,
as extended by means of observations at Allegheny.

TABLE 114.

Mean results of Tables 112 and 113 corrected for metallic absorption and angle of diffraction, extended to wave-length 2ᵘ.4, and again reduced to represent one and the same amount of heat for each series.

MEAN RESULTS OF TABLES MULTIPLIED BY FACTORS FOR SELECTIVE ABSORPTION OF METALLIC SURFACES AND ANGLE OF DIFFRACTION.

	Lone Pine.			Mount Whitney.			
	I.	II.	III	I.	Ia.	II.	III
A. 0ᵘ.350	23.1	11.6	0	35.7	13.7	16.5	6.7
B .375	28.4	18.1	5.2	39.2	21.5	20.6	15.9
0 .400	50.1	54.3	16.3	64.0	63.1	42.6	43.7
0 .450	110.6	102.1	93.7	155.8	151.0	151.0	116.8
0 .500	155.9	138.9	151.9	201.8	199.1	200.6	198.2
0 .600	207.0	204.7	211.8	225.3	223.2	230.7	227.4
0 .700	191.1	207.7	214.1	192.1	202.8	207.0	215.2
0 .800	155.3	175.5	177.9	142.6	153.5	169.7	169.5
1 .000	100.2	109.1	108.0	59.8	93.2	95.8	100.5
1 .200	76.1	67.6	67.7	64.5	64.1	60.7	60.9

EXTENDED BY SUBSEQUENT MEASURES.

Sums from ᵘ.35 to 1ᵘ.20	1092.3	1080.9	1018.7	1231.8	1185.2	1190.2	1148.0
λ = 1ᵘ.400	58.0	51.4	51.1	50.1	16.9	44.3	40.5
1 .600	42.5	35.9	37.7	38.2	35.1	33.5	29.9
1 .800	29.6	26.5	26.4	27.5	25.5	24.0	21.3
2 .000	18.9	17.0	16.9	18.2	17.1	16.0	14.2
2 .200	10.1	9.2	9.1	10.2	9.8	9.1	8.2
2 .400	3.2	2.9	2.9	3.4	3.3	3.3	3.2
Area above 1ᵘ.2	55.125	57.300	57.375	57.200	58.150	59.100	60.375
Area below 1ᵘ.2	19.875	17.700	17.625	17.800	16.850	15.900	14.625
Total area	75.000	75.000	75.000	75.000	75.000	75.000	75.000

Under I are the proportionate energies for each wave-length at noon. Under II are the proportionate energies for each wave-length at afternoon. Under III are the proportionate energies for each wave-length at late afternoon observations. The corrections for selective absorption in the apparatus, and for the angle of diffraction having been applied, the numbers now represent the true distribution of energy for each series.

Having now obtained the areas of the curves, we are prepared to still further improve the results directly obtainable from the observations made with the spectro-bolometer by combining the latter with the still more numerous results of actinometer observations. For this purpose we have merely to reduce the total areas of all the spectrum curves until the relation between them is the same as that of the solar radiation measured under like conditions by the actinometer.

We therefore determine reduction factors which, multiplied into the ordinates of the curves in Table 114, shall reduce the areas of the curves to correspond to the actinometer readings.

TABLE 115.

Factors for reducing bolometric to the standard of actinometric results.

Bolometer observations.	Calories by actinometers.*	Reduction factors, ratio of calories to Lone Pine I.
Lone Pine I	1.765	1.000
Lone Pine II	1.526	.865
Lone Pine III	1.310	.759
Mountain Camp I	1.972	1.118
Mountain Camp Ia	1.898	1.075
Mountain Camp II	1.745	.989
Mountain Camp III	1.423	.807

*These values differ so slightly from those on pp. 94 and 96 that it was not considered necessary to repeat these calculations, using the more exact values there given.

The next table gives the finally adopted ordinates of bolometric curves, the areas of which correspond to the readings of the actinometer at the same time. It was obtained from Table 114 by applying the above reduction factors.

TABLE 116.

Adopted values of bolometric energy in the spectrum at high and low measurements reduced to agree with corresponding actinometer readings.

		Lone Pine			Mountain Camp.			
		I.	II.	III.	I.	Ia.	II.	III.
A.	.350	28.1	16.0	0	39.9	14.8	16.3	5.4
	.375	28.1	15.7	4.0	44.8	23.2	20.3	12.8
	.400	30.1	29.6	12.4	71.5	68.0	42.1	35.3
	.450	119.6	88.3	71.1	174.2	162.5	149.3	118.5
	.500	151.9	120.5	117.6	229.0	214.4	203.2	160.0
	.600	201.0	177.0	160.7	249.7	240.4	228.1	183.5
	.700	191.1	180.7	162.5	214.8	218.3	204.6	176.1
	.800	155.5	151.7	131.3	169.1	165.3	157.9	136.9
	1.000	109.2	99.6	82.1	109.4	104.1	94.7	81.1
	1.250	76.1	55.5	51.1	72.1	69.1	68.0	49.1
	1.500	25.0	11.1	78.8	56.1	50.5	43.8	32.7
	1.750	12.5	32.7	28.6	42.7	37.8	33.1	24.1
	1.800	26.0	32.9	28.0	30.8	27.5	23.8	17.2
	2.000	18.0	14.7	12.9	20.1	18.7	13.8	11.4
	2.200	10.1	7.9	6.9	11.4	10.5	9.0	6.6
	2.400	3.2	2.5	2.2	3.8	3.6	3.2	2.6
Area above 1.2		55,125	49,550	43,540	63,350	62,540	58,450	48,720
Area below 1.2		19,875	15,300	13,360	19,920	18,130	15,700	11,790
Total area		75,000	64,850	56,900	83,850	80,670	74,150	60,510

All observations at Lone Pine are supposed to be made with a declination of $+14^\circ 30'$, and all at Mountain Camp with a declination of $+ 7^\circ 30'$. The actual declinations of the sun were: For August 11, a. m., $15^\circ 00'$; for August 14, p. m., $14^\circ 00'$; for September 1, a. m., $8^\circ 01'$; for September 3, p. m., $7^\circ 00'$.

The zenith distances in the following table were computed by the formula—

$$\cos z = \cos(\delta - \Phi) - 2 \cos \Phi \cos \delta \sin^2 \tfrac{1}{2} h$$

where $z =$ sun's zenith distance, $\delta =$ sun's declination, $\Phi =$ latitude $= 36^\circ 35'$ (for both stations), $h =$ hour-angle. The barometric readings (β) were obtained from the average of the corrected readings on the days of observation. M was determined from the formula,

$$M = \frac{.0174 \times \text{tabular refraction}}{\cos \text{apparent altitude}}$$

TABLE 117.

Air-masses traversed by the solar rays at each observation.

	Sun's hour-angle.	Sun's declination.	Sun's zenith distance.	M.	Barometer. (β)	Air-mass. (Mβ)
					P. m.	
Lone Pine, I	0 22ˢ	$-14 30$	22 58	1.083	6.62	7.17
Lone Pine, II	1 27	....	62 04	2.125	....	14.14
Lone Pine, III	5 14	....	72 27	3.25	....	21.52
Mount Whitney, I	0 17	$+ 7 50$	29 23	1.147	4.99	5.72
Mount Whitney, Ia	2 48	....	47 39	1.491	....	7.46
Mount Whitney, II	4 43	....	63 16	2.261	....	11.28
Mount Whitney, III	5 21	....	77 44	1.61	....	23.00

We now know (and largely from the present observations) that as we ascend in the atmosphere we find in general that the coefficient of transmission for any ray grows larger and larger *for like air-masses*, so that even for any given homogeneous ray there is a different coefficient for every stratum in the atmosphere and we could only represent the actual state of affairs by using as our coefficient of transmission the result of an integration of all these "differential" coefficients, an integration which we have no sufficient data for making. The more of these independent coefficients we can actually determine the more accurate will be the result, and as we employ fewer our result will increase, being largest when we employ but one (which is the usual practice). Since we have succeeded in determining but two independent coefficients for any given ray, the assumption (which we are forced to make by our limited observation) that these represent accurately the result that such an integration would give us, must obviously give a result somewhat too large, though, as obviously, we get a closer approximation to the truth than if this assumption had not been made. So far as depends on these considerations, then, the value of the constant thus obtained is too great.

In the observations which we now proceed to discuss, we go on to deal with separate rays, each of which, we may infer by analogy, will have a smaller coefficient of transmission determined by the comparison of the mountain and Lone Pine observations than that determined by observation through the air-mass above the mountain. The actual determination, however, of two distinct coefficients for each spectral ray is practically impossible, and even if we limit our study to a small number of typical rays taken in the infra-red, the yellow, and the violet or ultra violet parts of the spectrum, we shall find that with many of these it is impossible to do more than state certain general results. In regard to the infra-red rays, for instance, whose coefficients of transmission are in every case, save that of actual absorption bands, very near unity, minute errors of observation, such as are sure to present themselves, will be, relatively, great enough to prevent us from distinguishing between the coefficient above and the coefficient below. But we may expect to find a sensible difference, if anywhere, in those rays belonging to the shorter wavelengths, whose coefficients are small, and here we may hope that the differences between the upper and lower transmissibilities will not be proportionally affected by the errors of observation to be expected in circumstances of such difficulty.

The task, in fact, is a wholly untried one, and such results as we may obtain are to be regarded only as useful approximations.

We proceed as follows: We have already found that at 9.30 a. m. and at 2.30 p. m., on the mountain, the air masses were, approximately, the same as at noon in the valley. We select a number of typical points in the spectrum whose wave-lengths are 0.355, 0.375, 0.4, &c. We find for each one of these rays, by the use of the formula $\lg a_x^m = d$ the coefficients of transmission for the air-mass above the mountain by comparison of high and low sun; and again, by direct comparison of simultaneous observations at Lone Pine and Mountain Camp, we derive a second set of coefficients special to the air-mass between the two stations. We follow, in other words, in the case of each of the above mentioned rays the same method of procedure, which we have already explained in the "summary of actinometer results."

The results are given in the following table, whose data are, we believe, new:

TABLE IIS.

λ	Coefficients of transmission of the entire atmosphere of the quality of that above Mountain Camp	Coefficients of transmission of the entire atmosphere of the quality of that between the stations
λ = 0.375	0.35	0.10
0.400	0.48	0.15
0.450	0.61	0.29
0.500	0.83	0.42
0.600	0.88	0.72
0.700	0.91	0.74
0.800	0.99	0.84
1.000	0.92	0.89
1.200	0.97	0.95

The difference between the results of the order theory and of direct observation are here presented, and the discrepancy is most remarkable. Theory bids us determine with confidence the coefficients of transmission from the comparison of high and low sun at any single station, and this we have done at Allegheny, at Lone Pine, and the mountain with, on the whole, fairly concordant results. But an actual ascent above much of the absorbing mass of the atmosphere gives the coefficients just exhibited, which, as far as 1.200, represent the transmissibility of the atmosphere between the mountain and valley for each ray of the spectrum from immediate observation, and have nothing hypothetical about them. This remark applies to the coefficient b, special to the air-mass between Mountain Camp and Lone Pine, and if we calculate the value of the heat before absorption in each ray by these coefficients alone, we shall obtain enormous values for those in the green and blue, and a resulting value of the solar constant between 4 and 5 calories. (See column 6, Table 120.)

Our observed values here appear to us to bring novel and important data, but they are still not complete enough to enable us to determine the rate at which the transmissibility for each ray increases for like air-masses as we ascend in the atmosphere, though they show here how enormously this transmissibility diminishes between one ray and another as we pass from red to violet.

We may expect, then, that the use of these double coefficients, while giving us a smaller value than that obtained with a single coefficient by the customary formula, will give us an E_λ, or energy outside the atmosphere for each ray, which, so far as it depends on this process, will be rather in excess of the truth than within it, and we shall represent our results thus obtained by a curve whose area shows the energy before absorption, as obtained in this way. (See upper curve, Plate XV.) The value of a unit of area in this curve in calories is determined by a comparison with the like curves for the mountain and Lone Pine, where the value of this unit is known from simultaneous observations with the actinometer. We have, however, in what has preceded, neglected the consideration of the portion of the curve below 1°.0, where the coefficients, as obtained by the comparison of mountain and valley observations, are sensibly equal to unity.

With these values the column 7, in Table 120, is prepared, the value (E_λ) being the energy in each ray outside the atmosphere as thus determined.

With these values of E_λ as ordinates, we proceed to construct upon the normal scale the curve of energy outside the atmosphere shown in Plate XV by the upper line. (No. IV.) The lower (No. I) curve in the same plate is that of the normal spectrum already given, originally drawn to represent the distribution of energy in the spectrum of the high sun at Allegheny, and the area of which within the continued sinuous line closely represents 1.7 calories. The area of the curve (No. IV) outside the atmosphere, obtained by the process just described, is 3.505 calories, and 3.5 calories we regard, then, as a *maximum* value of the solar constant.

We now proceed to determine from our bolometer observations a value which we may believe, from considerations analogous to those just presented, to be a *minimum* of the solar constant, and one within the probable truth. All the evidence we possess shows, as we have already stated, that the atmosphere grows more transmissible as we ascend, or that for equal weights of air the transmissibility increases (and probably continuously) as we go up higher. In finding our minimum value we proceed as follows; still dealing with rays which are as approximately homogeneous as we can experimentally obtain them. Let us take one of these rays as an example, and let it be the one whose wave-length is 0.6 and which caused a deflection at Lone Pine of 201. (See table of adopted values, &c.) The coefficient of transmission for this ray, as determined by high and low sun at Lone Pine and referred to the vertical air-mass between Lone Pine and Mountain Camp, is .976. From the observations at Lone Pine, then, the heat of this ray upon the mountain should have been

$$201 \times \frac{1000}{976} = 206.0$$

but the heat in this ray actually observed on the mountain was 249.7. Therefore, multiplying the value for the energy of this ray outside the atmosphere calculated from Mountain Camp high and low sun observations (275) by the ratio $\frac{2497}{2060}$ we have 333.3, where 333.3 represents the energy in this ray outside the atmosphere as determined by this second process. In like manner we proceed to deal with the rays already used, thus forming column 8 in table 120.

With the values thus obtained as ordinates, we again construct a curve on the normal scale, whose area represents the solar constant on this hypothesis, so that the area of this curve gives a minimum value. The area thus obtained is found to be equal to 2.63 calories, the curve itself (No. II) being given by the line with the contour (—.—.—.—.—.—.—.—.) on plate.

Finally we draw the curve indicated by the line (—— —— —— —— ——), whose ordinates are intermediate between these two. (See Table 120, column 9.) The area of this last curve is 3.07 calories.

We have, then, upon Plate XV four curves. The lower curve (I) represents the actual observation of the solar spectrum, including its principal absorption bands on the normal scale near

sea-level, as determined at Allegheny, its area within the irregular line corresponding to an actually observed value of 1.7 calories. We have, by our subsequent long-continued observations at Allegheny, determined at leisure, and with all the accuracy we can at present command, the coefficients of transmission in this extreme infra-red portion, where we find them, in general, slightly less than unity. They are given in Table 123. It is by the use of these subsequent Allegheny observations, below $\lambda = 1^{\mu}.2$ (to which point the measures on the expedition were limited), that we have extended the upper curves (II, III, IV) shown in Plate XV to $2^{\mu}.7$ near the extremest limit of our latest observations. The calculated ordinates for curves II, III, IV from near $0^{\mu}.8$ to $1^{\mu}.2$ coincide very nearly with the Allegheny ordinates in the same region. In other words, we obtain in this part of the infra-red nearly the same heat at Allegheny as outside the atmosphere, these rays being transmitted almost unabsorbed (always with the exception of the cold bands).

We shall take, when necessary, a mean of these values and of those determined for the same rays by comparison of Lone Pine and Mountain Camp, and employ this single mean coefficient for each ray, slightly modifying the contour of curves II, III, IV near $\lambda = 1^{\mu}$ so that they may not be discontinuous with the curve showing the results in this region as derived from Allegheny. This modification, it will be understood, is made only to prevent a confusion of lines which the eye could not follow in the plate, the four lines here blending nearly into one. The measures of the areas have been made before this is done. The actual areas and the curves as presented, however, will be found to be almost exactly in the ratios already given.

In examining these curves, the reader's attention is directed to the fact that since the large depressions in the lower curve corresponding to $\lambda = 0^{\mu}.94$, $1^{\mu}.43$, $1^{\mu}.36$ to $1^{\mu}.37$, and $1^{\mu}.84$ to $1^{\mu}.87$, are deemed by us to be most probably due to telluric absorption, these depressions disappear in the curve representing energy before absorption. Although the coefficients of transmission for this extreme infra-red heat, derived from observations on the intervals between the cold bands, are very nearly unity, the undetermined coefficients of transmission in the cold bands themselves are much nearer zero. Accordingly, if we measure the area of the lower curve below $0^{\mu}.76$ (i. e, the whole dark-heat curve) following the sinuosities, we obtain an area for this part alone, which is less in proportion to the corresponding area in curve three; so that, roughly speaking, the coefficient of transmission of the dark-heat, considered as a whole, is less than might be inferred from a hasty consideration of the coefficients of transmission obtained from the little absorbed portions, as given in our preceding tables.

It will have been seen from what has preceded that there are at least four modes of combining the observations.

I. We may compare high and low sun observations at a given place. In so doing we assume that the diathermancy of the air has suffered no change in the interval between the observations, a supposition which is very improbable; but apart from this we know by direct comparison of the values thus deduced for the heat on Whitney from Lone Pine observations, or the heat at Lone Pine from Whitney observations, that this method gives results so erroneous that no dependence should be placed on it, when we have better at command.

II. We may combine the simultaneous observations at top and bottom of the mountain, assuming the transmission by the air above the mountain to be the same as that of the intermediate stratum between top and bottom, and this method is better than the preceding, but we know it to be incorrect by direct observation, and it gives so large a value that we cannot make use of it with safety.*

III. We may endeavor to use our knowledge that the atmosphere is made up of strata having different coefficients of transmission by the following process:

Consider the atmosphere divided into two strata having coefficients of transmission, which we will call a for the upper and b for the lower layer. First, using the customary exponential formula applied separately to our approximately homogeneous rays, determine the coefficient of transmission b_x for the atmospheric stratum between top and bottom of the mountain by comparison of Mount Whitney and Lone Pine noon observations. These values we may accept with considerable

---

* This was the method of Forbes, and he found the transmissibility of the upper air the same as that of the lower. Owing to his use of it, more than to any other cause, he obtained the result he did, which was far larger than he would have found by the legitimate use of his actinometer alone.

confidence, since they are quite independent of any hypotheses as to the manner in which the intermediate atmosphere exercises its absorption. Second, let the coefficients of transmission of the upper air, a^z, be determined by comparison of Mount Whitney high and low sun observations. Third, multiply the numbers denoting outside energy, obtained from method II, by the factor $\frac{b_\lambda}{a_\lambda}$.

An example will make the subject clearer. Fig. 43 represents the homogeneous atmosphere divided into two strata of equal thickness, but having different coefficients of transmission, namely, a for the upper and b for the lower stratum. The slanting lines represent rays from the sun, whose zenith distance is $60°$, so that sec $z=2$. Then if $A=$ true solar energy outside the atmosphere and $M=$ relative air-mass traversed $=$ sec $z=1$ and 2 in the above cases, the quantities written on the left of the upright line, or rays for zenith sun, are the energies of the ray from zenith sun, found at the points where written. Those on the right of the line are the energies found there when $z=60°$.

The following are modifications of the formulæ used in computing the energy of a ray before entering the atmosphere in reduction of bolometric observations:

$$(1) \quad\quad\quad\quad\quad\quad\quad\quad\quad\quad\quad t^{(M_2 - M_0)} = \frac{d_2}{d_1}$$

$$(2) \quad\quad\quad\quad\quad\quad\quad\quad\quad\quad\quad E = \frac{d_1}{t^{M_1}}$$

Their truth is apparent if the law of transmission of a homogeneous atmosphere is that assumed above. In these formulæ

    $d_2 =$ registered energy of a ray when air mass $= M_2$.
    $d_1 =$ registered energy of a ray when air-mass $= M_1$.
    $t =$ coefficient of transmission for zenith depth of atmosphere.
    $E =$ energy outside the atmosphere, computed by the formula (2).
    $M_2 =$ is taken greater than M_1, and consequently d_2 is less than d_1.

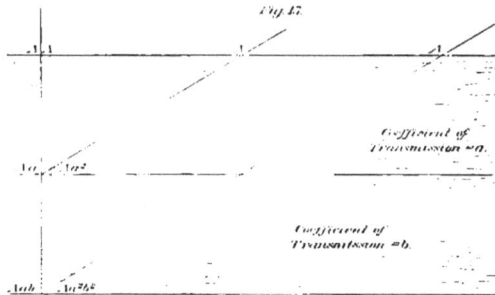

Fig. 43.

Illustrating Atmospheric Absorption.

For measurements taken at the upper and lower limits of the lower stratum the following examples are typical of the first three methods:

TABLE 119.

Comparison of high and low sun observations at upper station.	Comparison of high and low sun observations at lower station.	Comparison of high sun observations at upper and lower stations.
In this case we have—	In this case we have—	In this case we have—
$d_2 = Aa^4$ $d_1 = Aa^2$ $M_2 - M_1 = 1$	$d_2 = Aab$ $d_1 = Aa^2b^2$ $M_2 - M_1 = 1$	$d_2 = Aa^4$ $d_1 = Aab$ $M_2 - M_1 = 1$
Substituting in formula (1)—	Substituting in formula (1)—	Substituting in formula (1)—
$t = \frac{Aa^4}{Aa^2} = a$	$t = \frac{Aa^2b^2}{Aab} = ab$	$t = \frac{Aa^4}{Aa} = b$
Substituting in formula (2)—	Substituting in formula (2)—	Substituting in formula (2)—
$E = \frac{Aa^2}{a} = A$	$E = \frac{Aab}{ab} = A$	$E = \frac{Aa^4}{b} = A\frac{a}{b}$

The meaning to be drawn from the first two of the above examples is that, *if the absorption were invariable at all hours of the day and for all parts of the earth situated at the same height above the sea and subjected to the same air pressure,* the comparison of high and low sun observations would give us the true energy outside the air, whether the absorbent material were distributed uniformly throughout the atmosphere or were gathered into horizontal layers superposed according to any law whatever. The character of the atmosphere interposed between us and the sun, however, is constantly varying through the day, and even if it were at rest, a vertical section would have a different composition from that made at a very great inclination, which would necessarily pass over portions of the earth's surface subjected to conditions very different from those existing at the place of observation. But besides these unavoidable variations of the atmosphere, we have to consider the effect of the actual non homogeneity of the pencils observed on, and other objections already referred to, particularly that there seems to be a progressive change in the atmospheric transmission for the same air-mass dependent on the altitude of the sun, whose rays are continually affecting the condition and distribution of at least one of its constituents, atmospheric moisture. That these variations exist is conclusively shown by the want of agreement between the results for outside energy from the combination of different series, and notably by the different results from high and low sun observations at the summit and at the base of the mountain. The first two methods, then, while theoretically correct, are so only in the case of theoretical atmospheric conditions which nature never really presents us.'

In the third case, however, the formula gives a value for the energy outside the atmosphere which is certainly incorrect, or one which is only true when $a = b$. If $a > b$, as in the case in observations made on Mount Whitney, $E > A$; but since from the last equation $A = E \frac{b}{a}$, we could find

A, or the true value required, if b and a were known. As, however, we can only determine a by a method already shown to be objectionable, the determination by method III is still doubtless theoretically imperfect, though less so than that by method II. For reasons given later we may expect that the value found by it is more likely to be in excess than in defect.

There remains the fourth method which we have described as furnishing the most trustworthy minimum value; *i. e.,* from the Lone Pine high and low sun observations of each ray, compute what should have been observed at Mountain Camp, and then multiply the value representing the constant for the ray obtained from high and low sun observations on the mountain by the fraction

$$\frac{\text{Value actually observed on mountain}}{\text{Value computed from Lone Pine observations}}$$

We thus obtain the values in column 8. The following are the combinations which have been made:

No. 1. Lone Pine I and Lone Pine II.
No. 2. Lone Pine I and Lone Pine III.
No. 3. Mountain Camp I and Mountain Camp II.
No. 4. Mountain Camp I and Mountain Camp III.
No. 5. Mountain Camp Ia and Mountain Camp III.
No. 6. Lone Pine I and Mountain Camp I.
No. 7. Lone Pine I and Mountain Camp I (using 2 coefficients of transmission).
No. 8. Mountain Camp I and II multiplied by factor from Lone Pine observations.
No. 9. The mean of 8 and 7.

The results follow in the next table. Nos. 1 to 5 are obtained by method I, No. 6 by method II, No. 7 by method III, and No. 8 by method IV, while No. 9 is the mean of 8 and 7.

TABLE 120.

ENERGY OUTSIDE THE ATMOSPHERE.

Wave length.	No. 1.	No. 2.	No. 3.	No. 4.	No. 5.	No. 6.	No. 7.	No. 8.	No. 9.
0".350	64.8	7.	109.2	77.4	24.0	248.2	207.0	122.5	164.2
0.375	52.2	75.6	96.7	65.8	30.9	242.1	186.6	110.0	154.3
0.400	86.1	100.7	123.3	90.3	93.1	290.7	242.9	139.1	190.7
0.450	179.3	138.0	204.0	197.9	189.0	1046.0	583.2	195.5	544.4
0.500	184.0	175.0	258.8	237.8	246.8	1090.0	852.9	374.1	619.5
0.600	239.0	234.9	274.2	256.4	274.5	587.8	511.7	333.0	428.8
0.700	202.1	205.1	228.8	229.4	242.0	349.6	317.7	255.4	286.5
0.800	170.1	167.2	161.0	167.0	181.0	175.7	173.9	167.3	170.6
1.000	109.3	110.0	106.7	107.7	112.5	104.2	102.3	105.0	104.7
1.200	100.3	96.1	87.0	81.9	81.5	57.3	61.3	58.2	60.8
1.400	76.5	71.0	72.4	67.3	62.3	49.3	52.2	65.1	58.7
1.600	55.6	51.8	55.5	51.6	46.9	44.5	45.0	48.0	46.5
1.800	48.6	45.0	49.2	37.4	34.4	90.0	36.4	39.2	37.8
2.000	24.5	22.9	28.5	24.7	23.7	27.5	25.4	29.1	28.1
2.200	13.0	12.2	14.5	14.7	13.1	18.1	17.5	19.1	18.5
2.400	4.3	3.9	4.5	4.3	4.2	7.5	6.8	7.9	6.9

AREA OF OUTSIDE CURVE.

	No. 1.	No. 2.	No. 3.	No. 4.	No. 5.	No. 6.	No. 7.	No. 8.	No. 9.
Above λ = 1μ.2	61,900	61,075	70,625	70,500	70,590	152,950	127,830	69,688	98,759
Below λ = 1μ.2	26,030	24,240	25,490	23,760	22,320	20,710	21,230	26,245	23,708
Total area	87,930	85,315	96,115	94,260	92,800	173,660	149,060	95,933	122,167
Solar constant (calories)	2.067	2.005	2.267	2.217	2.184	4.084	3.505	2.666	3.068

The first five values are those obtained by combining high and low sun observations at a single station. They are presented here only for their use in determining the seventh and eighth, and as showing the errors from this mode of observation, since they are demonstrably too small. The sixth value is that obtained by the hypothesis of Forbes. It is demonstrably too large. We need concern ourselves only with the values given in columns 7 and 8. Column 9 is the mean of the last two, and represents a value nearly accordant with that finally adopted by us. The values below $\lambda = 1\mu.2$ are of comparatively small importance in their effect on the sun. They are obtained by aid of the Allegheny observations.

CHAPTER XIII.

SPECTRO-BOLOMETER OBSERVATIONS TAKEN AT ALLEGHENY, IN 1882, WITH FLINT-GLASS PRISM.

The following are the dates on which measurements of atmospheric transmission were attempted. Those which had to be rejected for the present purpose are indicated by an asterisk. 1882: February 15, March 3,* March 4,* March 23, March 29, March 31, April 3,* April 4,* April 17,* April 24, May 1, May 2, May 3,* May 19,* May 24,* May 29, June 22, September 1,* September 12, September 15, November 25.

The following table contains a record of high and low sun observations available for determinations of atmospheric transmission.

The same notation is used as in the previous article, except that on some days there are several sets of low-sun observations, which are distinguished by subscript figures. The noon deflection is, in all cases, denoted by d_1, and the afternoon measures by d_1, d_2, &c., in the order in which they were taken.

TABLE 121.

Deviation	52 00'	51 00'	50 00'	49 30'	49 00'	48 00'	47 30'	46 45'	46 30'	45 42'	45 30'	45 28'	44 30'
1882.													
Feb. 15 d_1						163	274	349	...	364	...	264	57
d_2						3	16	115		231		165	27
Mar. 23 d_1					...	104	183		363	424	337	212	34
d_2					...	...	156		360	415	335	203	27
d_1						104	183		364	428	337	212	34
d_3						30	61		241	300	279	176	24
d_1						104	183		363	424	337	212	34
d_4						14	31		163	178	127	124	14
d_1						104	183		363	428	337	212	34
d_5						6	9		89	125	106	107	18
d_1						104	183	...	360	425	337	212	34
d_7						4	7	...	60	91	70	80	12
Mar. 29 d_1						104	155	299	340	348	295	164	24
d_2						37	56	130	176	177	162	90	13
d_1						104	155	299	340	348	295	164	24
d_3						22	37	96	116	135	105	84	12
Mar. 31 d_1						112	173	306	350	535	267	176	18
d_2						79	49	148	170	191	169	114	13
Apr. 24 d_1	9.76	2.71	11.67	23	38	100	140			292	296	177	33
d_2	0.11	0.70	4.26	10	21	57	97			274	235	167	22
d_1				23	38	100	140			292		177	33
d_3				4	8	34	64			229		149	21
May 1 d_1		2.20	11.50		38	102	158	324	348	369	366	202	31
d_2		0.58	3.29		16.5	50	67.5	212	241	287	226	164	23.5
May 2 d_1	0.54	2.20	10.80	18.3	28								
d_2	0.15	1.33	7.15	13.23	24.71								
d_1		2.20	10.80	18.3	28	74	110	223	226	247	299	136	17
d_1		0.61	3.92	7.42	15.20	33	62	173	207	225	192	124	21
d_1		2.20	10.80	18.3	28								
d_1		0.18	1.85	3.87	8.37								

TABLE 121—Continued.

Deviation.		52° 00'	51° 00'	50° 00'	49° 30'	49° 00'	48° 00'	47° 30'	46° 45'	46° 30'	46° 12'	45° 53'	45° 28'	44° 30'
1882.														
May 29	{ d_1	2.6	9.0	16.6	23.5	78	126	272	235	201	202	141	17.5
	{ d_2	0.23	1.70	3.55	8	33.5	76	218	227	248	193	126	15
	{ d_1	2.6	9.0	16.6	23.5	78	126	222	235	202	202	141	17.5
	{ d_2	0.07	0.86	2.18	5	31.5	61	180	199	224	175	125	16.5
June 22	{ d_1	0.95	4.5			43	72	154		209	175	122	17
	{ d_2	0.03	0.39		3.0	12.5	38	109		134	107	79	11
Sept. 12	{ d_1	4.52	23.1		64.5	158	248	324	554	604	502	322	43
	{ d_2	0.73	5.5		26.8	69	113	341	402	472	381	269	38
	{ d_1				64.5	158	248	521	558	604	502	322	43
	{ d_2				3.7	29.6	62.5	230	286	363	304	229	33.6
Sept. 15	{ d_1	2.22	8.98		24.6	59	96	223	257	273	206	133	19.8
	{ d_2	0.80	3.10		12.9	39.5	69	178	210	228	170	119	19
	{ d_1		8.98		24.6	59	96	223	257	273	206	133	19.8
	{ d_2		1.73		7.7	27	51	99	189	217	162	111	18.5
	{ d_1				24.6	59	96	223	237	273	206	133	19.8
	{ d_2				2.3	12.5	30	119	138	186	130	99.5	14.8
Nov. 23	{ d_1			13	21	50	81	185	211	290	214	136	21
	{ d_2			0.3	1.5	7.5	22	76	99	137	103	96	18

The next table gives the sun's position, and the corresponding air-mass for each series in the previous table.

TABLE 122.

Date of observation.	High sun.				Low sun.			
	Sun's hour angle.	Sun's zenith distance.	Barometer (β_i).	Air mass ($M_i\beta_i$).	Sun's hour angle.	Sun's zenith distance.	Barometer (β_o).	Air mass ($M_o\beta_o$).
1882.			d. m.	d. m			d. m.	d. m.
Feb. 15	0°21′′	53°30′	7.36	12.37	4°53′′	85°48′	7.36	83.78
Mar. 23	1 45	45 56	7.38	10.61	3 31	61 43	7.37	15.57
23					5 06	79 00	7.36	37.76
23					5 33	84 04	7.38	54.11
23					5 43	86 00	7.35	48.66
23					5 50	87 20	7.35	111.72
Mar. 29	0 30	38 31	7.35	9.27	4 51	74 34	7.35	27.34
29					5 34	82 42	7.34	54 10
Mar. 31	0 26	38 39	7.37	9.14	4 51	73 55	7.37	26.82
Apr. 24	0.12a to 1 32	27 39′ to 33 65	7.37	8 36 to 8 89	4 48a to 5 26	67 55 to 74 46	7.37	19 53 to 29 42
24					5 46 to 6 01	78 33 to 81 44	7.37	37 54 to 48 64
May 1	1 34	32 32	7.36	8.73	5 15	71 46	7.36	23.39
May 2	6 37	26 15	7.43	8.25	4 45 to 5 01	65 52 to 68 53	7.40	18. 06 to 20. 51
2					5 26 to 5 39	73 37 to 76 05	7.40	26. 07 to 30. 25
2					5 50	78 26	7.40	36. 10
May 29	0 51 to 1 10	21 42 to 23 54	7.34	7.90 to 8.03	5 48 to 6 02	73 57 to 76 31	7.34	26. 22 to 31. 01
29					6 16 to 6 26	79 01 to 80 40	7.34	37. 95 to 44. 76
June 22	0 13	17 10	7.39	7.74	6 23	79 04	7.39	38. 27
Sept. 12	0 22	36 46	7.37	9.24	4 37	71 40	7.37	23.29
12					5 23	80 10	7.37	42 01
Sept. 15	1 02	40 07	7.38	9.62	4 38	72 28	7.36	24.36
15					5 09	77 30	7.36	33.41
15					5 33	83 00	7.36	55. 20
Nov. 25	0 58	62 46	7.37	16.07	4 00	82 50	7.37	54.83

From these data the coefficients of transmission through a stratum of air, capable of supporting 1^{dm} of mercury, have been deduced.

TABLE 123.

Coefficients of transmission.

Deviation ...	52 00	51 00	50 00	49 30	49 00	48 00	47 00	46 47	46 30	46 12	45	45 28	44 30
λ	0ᴹ 358	0ᴹ 581	0ᴹ 116	0ᴹ 440	0ᴹ 468	0ᴹ 550	0ᴹ 6.5	0ᴹ.781	0ᴹ.870	1ᴹ.01	1ᴹ 20	1ᴹ 50	2ᴹ.29
Feb. 15						.946	.961	.965	.....	.964	.....	.992	.990
Mar. 20							.971	.....	.966	.964	.969	.964	.955
20						.946	.966	.....	.962	.967	.992	.962	.987
21						.958	.967	.....	.975	.9.9	.961	.960	.984
24						.961	.964	.....	.981	.984	.964	.994	.990
25						.967	.966	.....	.982	.965	.965	.990	.964
Mar 29						.915	.94.5	.955	.961	.964	.967	.967	.962
29						.966	.9.9	.975	.976	.979	.977	.965	.965
Mar 31						.966	.960	.955	.939	.964	.974	.965	.961
Apr. 24	.964	.914	.950	.954	.961	.967	.975	.....	.....	.965	.960	.965	.965
24			.950	.964	.967	.976	.....	.....	.992	.994	.964	.....	
May 1		.913	.91s	.....	.915	.951	.961	.972	.975	.981	.980	.987	.967
May 2	.961	.960	.967	.951	.980								
2		.961	.915	.850	.946	.964	.974	.969	.986	.986	.986	.986	1.004
2		.911	.959	.946	9s	.....							
May 29		.960	.959	.965	.964	.95s	.973	.969	.966	.997	.997	.964	.984
29		.967	.915	.916	.979	.971	.976	.961	.915ᵃ	.965	.965	.966	.966
June 22		.925	.921	.....	.951	.960	.979	.966	.969	.961	.966	.966	.946
Sept 12		.869	.962	.....	.951	.961	.954	.971	.977	.961	.961	.967	.961
12					.920	.961	.959	.975	.980	.985	.965	.960	.962
Sept 15		.935	.908	.....	.957	.971	.972	.965	.946	.966	.961	.992	.967
15			.901	.....	.952	.968	.976	.967	.962	.966	.966	.962	.962
24					.919	.961	.975	.946	.961	.962	.960	.964	.964
Nov. 25				.967	.964	.952	.967	.957	.967	.961	.961	.961	.960

Adopted a⁷ˢ	.960	.920	.965	.942	.950	.960	.965	.978	.982	.965	.967	.968	.960
d	.449	.551	.660	.854	.677	.751	.761	.641	.671	.991	.905	.919	.926

By a comparison of the above table with Table 6, p. 25, it will be seen that the values of the coefficients of transmission determined in 1882, after the return of the expedition, are somewhat greater than those found in 1881. This may be due in part to the different seasons in which they were obtained, but the last values are entitled to more weight, because they rest on observations covering an entire year, and made with matured experience and improved apparatus.

In all, whether in 1881 or 1882, one fact is salient within the range of our experiments, that if we except the cold bands—*the transmissibility increases with the wave length, so that the "dark" heat is more transmissible than the "light,"* a conclusion directly opposed to the at present accepted belief.

The following noon observations, obtained on the clearest days during the year 1882, have been selected to give an average high sun curve. The battery current was not measured. The prism and lenses employed were of a specially diathermanous glass. As it was not absolutely so, a correction should be made, but this has not yet been applied. Its effect would be to relatively increase the values corresponding to the greater wave-lengths.

TABLE 124.

Deviation.	52 00	52 00	52 0	36 00	19 30	19 00	18 00	47	46 45	46 30	46 12	4 50	45 28	44 30
λ	0ᴹ.129	0ᴹ.358	0ᴹ.363	0ᴹ 116	0ᴹ.440	0ᴹ.464	0ᴹ.550	0ᴹ 645	0ᴹ.781	0ᴹ 870	1ᴹ.01	1ᴹ.20	1ᴹ.50	2ᴹ.29
April 24		0 76	2 71	11 67	21.0	.58	100	140	.....	.....	.292	260	177	33
May 1	.....	0 57	2 20	11 59	.....	.58	102	15s	.320	348	.369	306	202	31
May 2	0 16	0 54	2 20	10 88	18 3	28	71	119	224	226	247	209	136	17
May 3	.....	0 57	1 94	9 96	16 6	30	82	1.1	250	241	301	268	175	25
May 19	0 13	0 41	1 27	6 86	12 5	25	60	115	233	228	312	253	179	32
May 21	.....	0 41	1 55	7 96	16 4	18	56	94	178	191	193	174	129	20
May 29	0.09	0 71	2 56	9 00	16.6	21	74	126	222	215	264	202	141	18
Sept. 1	0.29	2 28	7 98	23 00	.....	64	159	211	128	211	351	251	170	24
Sept 12	.....	0 61	2 26	11 70	.....	12	79	124	267	278	302	251	161	22
Sept 15	.....	.....	2 22	8 99	.....	25	56	96	223	257	274	206	113	20
Nov 25	.....	0 41	1 89	7 00	11 0	21	50	81	184	244	280	214	136	21

| Mean deflection | 0 15 | 0 72 | 2 60 | 11 41 | 16.8 | .52 | 81 | 121 | 244 | 268 | 290 | 236 | 158 | 24 |

The corrections for the selective absorption of silver and glass have not been introduced in the above values.

The condition of the sky during these observations was as follows: April 24, excellent; May 1, fair blue, occasional clouds; May 2, blue, with passing clouds; May 3, faint haze, in irregular wisps; May 19, very thickly milky, with some clouds; May 24, rather irregular milky blue, with clouds; May 29, good blue; September 4, milky blue, with cumulus clouds; September 12, milky blue; September 15, milky blue, with fracto cumuli; November 25, blue, with clouds.

The air-masses at the time of observation are given in the next table.

TABLE 125.

Date of observation.	Sun's hour angle.	Sun's zenith distance.	Barometer. (S)	Air-mass (M β)
		° ′	d. m.	d. m.
April 24	0 54ᵐ	30 10	7.37	8.51
May 1	1 34	32 34	7.36	8.73
May 2	0 37	26 15	7.40	8.25
May 3	0 48	26 45	7.38	8.27
May 19	0 38	22 07	7.38	7.90
May 24	0 07	19 11	7.37	7.83
May 29	1 01	22 48	7.34	7.96
September 4	1 39	35 55	7.37	9.10
September 12	0 52	36 40	7.57	9.21
September 15	1 02	40 07	7.36	9.62
November 25	0 54	62 46	7.37	16.07
Mean air mass				9.23

CHAPTER XIV.

THE TRANSMISSIBILITY OF OUR ATMOSPHERE FOR LIGHT.

The absorption by our atmosphere of the heat in any ray must always be proportional to that of the light in the same ray, since light and heat are but names given to different manifestations of the same energy. It was of evident interest, then, to determine the transmissibility of the atmosphere at Mount Whitney for light; but as this was merely an adjunct to the study of the heat, no special photometers had been provided.

There is, however, one method of determination which requires none.

In 1877, Prof. E. C. Pickering suggested to the writer, who was then about to visit Mount Etna, that an estimate of the transparency of the atmosphere there might be made without any photometric apparatus, by comparing the light of two stars, one near the zenith, the other near the horizon, at the moment when they appear equally bright; for the absolute brightness being known from the magnitudes in the Star Catalogue, and the masses of air traversed by the rays being computable subsequently from the times of observation, we have all the data demanded by the usual formula. This method was therefore used on Mount Whitney in the following observations, made under my instruction by Mr. J. E. Keeler, assisted by Mr. W. C. Day.

Let it be understood that for our present purpose we use the word "light" as synonymous with the expression "light between the wave-lengths $0^s.4$ and $0^s.7$," and let this light of the sun or a star before absorption by our earth's atmosphere be denoted by L. If the celestial body be viewed in the zenith, a portion of the light will have been absorbed and a portion transmitted when it reaches an observer at the sea-level. The fraction expressing the percentage transmitted to the sea-level is called the *coefficient of transmission*, and denoted by l, so that the original light L becomes Ll after absorption by one such stratum, and would, if the light were homogeneous, become Ll^2 after absorption by n such strata. The light is not really homogeneous, but we shall first (in accordance with custom) here consider it as such, and shall afterward point out the consequences of this incorrect assumption. The absorption depends not upon the length of the path, but on the mass of air traversed,[*] and we may choose as the unit of mass anything we please. Since the weight of the mass of air in a vertical column above us equals at the sea-level that of 760 mm. of mercury, $l^{\frac{1}{76}}$ will represent the transmission by a mass $\frac{1}{76}$ as great, or to that of the mass corresponding to one decimeter of mercury (always on the assumption that the light is homogeneous, and the law of extinction such that the percentage transmitted by any one unit stratum is the same as by another, or that l is a constant). When the words "coefficient of transmission" are used without qualification, the transmission for the entire atmosphere (l) is referred to.

I arranged the subjoined form for the few observations made on Mount Etna,[†] which gave a transmission of 90 per cent. at that station, where the barometer was 660 mm. ($l^{\frac{66}{76}}=90$), whence by reduction to the sea-level, $l=0.88$. The early observations by Bouguer[‡] give, when expressed in the same terms as ours, $l=0.812$; those by Seidel[§] give $l=.794$; the recent ones by Professor Pritch-

---

[*] This statement is usually treated as axiomatic, and we do not here discuss it; but plausible reasons may be offered for thinking that the same mass may not exert the same absorption under different densities.

[†] See American Journal of Science, July, 1880, page 3.

[‡] Bouguer, "Traité d'optique," Paris, 1760.

[§] "Untersuchungen über die Extinction des Lichtes." Potsdam Obs. Rep., Vol. III, No. IV.

and, at Cairo,* give a higher value (l=.843); those at Oxford give l=.791; the still more recent observations by Müller† give l=0.825. Other values might be cited, giving abundant testimony that the absorption of light at the sea-level, in the opinion of the most trusted observers, is about 20 per cent. for a zenith star. In nearly all these observations, so far as is known, the same formula that we here employ provisionally, and under caution, has been used without reserve.

In all of them but the first, the determination has been made by photometric apparatus, but it is the peculiarity of the present that none is absolutely demanded; for let L be the original light: if l = the coefficient of transmission of light for a zenith depth of atmosphere, assuming no selective absorption of light, the light from this star in the zenith would become Ll.

If m = the mass of air traversed by the rays from the star at the zenith distance, z, the light reaching the observer, is $l^m L$.

If we select two stars at different altitudes, which appear equally bright to the eye, we have a means for determining l, provided the altitudes and magnitudes of the stars are known; for, since the apparent amount of light from each is the same,

$$l^{m_1} L_1 = l^{m_2} L_2 \qquad\qquad m_1 \log l + \log L_1 = m_2 \log l + \log L_2$$

from which

$$\log l = \frac{\log L_2 - \log L_1}{m_1 - m_2}$$

We shall here assume that the magnitudes of stars bear a relation to their light expressible by the formula $\log L = -.4\, M$, where L is the light and M the magnitude of the star.

The observations (which can be made by one person) have in this case, in order to eliminate personal peculiarities, been made by two observers, each of whom independently selects two stars, one not far from the zenith, and one only so far from the horizon as not to involve any sensible error in taking the air-mass proportional to sec. z. The two stars so selected must be of apparently equal magnitudes; and since the lower star, whose light has suffered greater absorption, must really be the brighter, the amount of this absorption is determined by the formula just given, which is that used implicitly or explicitly by all the observers just cited. Having selected the stars, the two observers then confer with each other and unite on what they deem the most perfect match, a search and comparison usually occupying some time, during which many comparison pairs are observed and rejected before one entirely satisfactory is found. When this is found, the time is noted to the nearest minute, and from this and the latitude, the secant of the zenith distance is obtained by subsequent computation. The tables which follow will be intelligible without further explanation.

The magnitudes of the stars used for the following comparisons have been furnished by the kindness of Prof. E. C. Pickering, director of the Harvard College Observatory.

---

* Memoirs of the Royal Astronomical Society, vol. XLVII. p. 116.
† Publications of Potsdam Observatory, 1883.

TABLE 126.

[Station, Mountain Camp. Observers, J. E. K. and W. C. D.]

Date.	Mean time.	Stars matched.	ζ	Mag.	Log L.	Sec ζ_2 − Sec ζ_1	Log L_1 − Log L_2	Log l [illegible]	l [illegible]
1881.									
Aug. 28	8h 16m	δ Herculis	25	3.28	−1.31				
		α Libræ	77	3.02	−1.20	+3.25	−0.11	−0.044	0.93
28	8 47	γ Cygni	12	2.31	−0.92				
		β Libræ	73	2.74	−1.09	+2.40	+0.17	+0.071	1.18
28	8 47	γ Draconis	22	2.35	−0.94				
		β Libræ	73	2.74	−1.09	+2.34	+0.15	+0.064	1.16
Aug. 29	8 17	δ Bootis	45	3.50	−1.39				
		ε Virginis	82½	3.00	−1.19	+6.18	−0.20	−0.032	0.93
29	9 12	γ Draconis	25	2.35	−0.94				
		β Libræ	79½	2.74	−1.09	+4.39	+0.15	+0.034	1.08
29	9 23	β Draconis	31	2.97	−1.18				
		ε Bootis	66½	2.56	−1.02	+1.34	−0.16	−0.119	0.76
29	9 23	η Cygni	1	4.01	−1.61				
		ε Bootis	66½	2.56	−1.02	+1.51	−0.59	−0.391	0.41
Aug. 30	8 48	ε Cygni	22	4.14	−1.77				
		β Scorpii	74	2.91	−1.16	+2.55	−0.61	−0.239	0.58
30	8 48	ζ Herculis	35	3.09	−1.23				
		β Scorpii	74	2.91	−1.16	+2.41	−0.07	−0.029	0.94
30	9 23	δ Cygni	11½	2.99	−1.19				
		η Bootis	81	2.89	−1.15	+5.37	−0.04	−0.007	0.98
30	9 23	γ Aquilæ	26½	2.77	−1.10				
		η Bootis	81	2.89	−1.15	+5.28	+0.05	+0.009	1.02
30	9 28	γ Delphini	22½	4.16	−1.66				
		ζ Bootis	76	3.84	−1.53	+3.05	−0.13	−0.043	0.91
30	9 58	β Lyræ	22½	3.57	−1.42				
		μ Serpentis	78	3.48	−1.39	+3.73	−0.03	−0.008	0.98
Aug. 31	10 25	ε Lyræ	29	3.84	−1.52				
		μ Serpentis	81	3.48	−1.39	+8.14	−0.13	−0.015	0.97
31	10 25	ε Delphini	27	4.11	−1.64				
		μ Serpentis	81	3.48	−1.39	+8.15	−0.25	−0.030	0.92
31	10 31	γ Cygni	12	2.31	−0.92				
		α Serpentis	82	2.71	−1.08	+6.16	+0.16	+0.026	1.06
31	10 34	ζ Cygni	7	3.48	−1.39				
		α Serpentis	82	2.71	−1.08	+6.18	−0.31	−0.050	0.89
Sept. 9	7 30	γ Draconis	17½	2.35	−0.94				
		δ Ursæ majoris	72½	2.60	−1.03	+2.28	+0.09	+0.040	1.10
9	7 41	γ Cygni	16	2.31	−0.92				
		α Ursæ majoris	70½	1.96	−0.78	+1.99	−0.11	−0.070	0.85
9	7 51	γ Aquilæ	46½	2.77	−1.10				
		β Scorpii	70¼	2.91	−1.16	+1.50	+0.06	+0.040	1.10
9	8 00	γ Lyræ	34	3.22	−1.28				
		α Can. Venat	72	3.00	−1.19	+2.31	−0.09	−0.039	0.91
9	8 05	γ Draconis	21½	2.35	−0.94				
		γ Ursæ majoris	72½	2.56	−1.02	+2.25	+0.08	+0.036	1.09
9	8 10	δ Cygni	9½	3.05	−1.21				
		ε Ursæ majoris	75	2.56	−1.02	+2.41	−0.19	−0.079	0.84
9	8 15	γ Cygni	12	2.31	−0.92				
		α Ursæ majoris	72½	1.96	−0.78	+2.30	−0.14	−0.061	0.87
Sept. 10	8 17	β Lyræ	11	3.57	−1.42				
		β Scorpii	78½	2.91	−1.16	+3.19	−0.26	−0.082	0.83
10	8 22	γ Draconis	21½	2.35	−0.94				
		α Ursæ majoris	73½	1.96	−0.78	+2.37	−0.16	−0.068	0.86
10	8 25	ε Herculis	26½	3.92	−1.56				
		α Can. Venat	77½	3.00	−1.19	+3.41	−0.37	−0.109	0.78
10	8 30	ζ Herculis	40	3.09	−1.23				
		η Bootis	79½	2.89	−1.15	+4.06	−0.08	−0.020	0.96
10	8 42	γ Cygni	4½	2.31	−0.92				
		α Ursæ majoris	75	1.96	−0.78	+2.86	−0.14	−0.049	0.89
10	8 53	δ Cygni	13	2.99	−1.19				
		ζ Ophiuchi	70½	2.84	−1.13	+1.97	−0.06	−0.030	0.93

During these observations the sky was constantly clear. On one occasion, September 9, the horizon was rather bright, owing to the sunset on one side and the rising moon on the other.

The mean from these thirty pairs of high and low stars at the Mountain Camp gives for the value of atmospheric transmission of stellar light l [illegible] = 0.92 ± .02, whence l = .88. A series of observations (not here given) made as the party was returning across the Inyo Desert, at a mean altitude of less than 4,000 feet, gave a smaller value for l than those on the mountain. I do not, regarding the considerable probable error, attach great weight to the value .88 above given, or to its coincidence with that obtained on Etna. It is certain that to obtain an entirely trustworthy value the exclusive time of the whole party during our stay on the mountain would have been insufficient; and it is doubtful whether there might not, even then, remain some systematic error tending to affect the results, as the conditions favoring systematic error are all present.

Professor Pickering, whom I consulted with reference to his own experience as to the value of high and low star comparison, writes me as follows:

HARVARD COLLEGE OBSERVATORY,
Cambridge, U. S., October 15, 1883.

* * * We have found here that the comparisons of high and low stars give a result apparently affected by a systematic error, according to which the lower star seems too bright, and the resulting coefficient of transmission will be made too large. The coefficient of transmission resulting from the observations of this kind made here has not been determined, but the observations give residuals from Seidel's tables of absorptions, ranging from one-tenth of a magnitude at a zenith distance of 65 to eight or nine-tenths close to the horizon. Seidel's table for extreme zenith distances does not agree precisely with that for ordinary zenith distances at the junction of the two, and at about 86 the comparisons between high and low stars made here give results exceeding those of Seidel for difference of absorption; but the general result is the reverse of this, as already stated. * * * It may be a general rule among observers that the lower of two stars seems comparatively bright. * * *

Very truly, yours,   EDWARD C. PICKERING.

We may observe that, whether there be a systematic error or not, we may be confident of our ability to draw the entirely legitimate conclusion that, at any rate, the air at such a site as Mount Whitney is not only clearer than at the sea-level, owing to its rarity, but intrinsically clearer, and in a very marked degree—clearer, that is, when equal masses of air from the mountain and sea-level are compared.

I have already remarked that, since we cannot actually observe the radiation of either sun or star before absorption, we cannot determine the coefficient of transmission except by employing some hypothesis; and that the ordinary assumption made implicitly by all the observers cited (the assumption that the original brightness and the coefficient can both be determined from the simple exponential formula used here) is erroneous, and not only theoretically so, but that it leads to sensible error in practice when we neglect, as we do in using this formula, the effects of selective absorption. I have alluded elsewhere to this fact in this same connection (see American Journal of Science, July, 1880, p. 37), and more recently I have demonstrated,* though not with the greatest generality possible, that when we neglect the effects of selective absorption the exponential formula not only gives erroneous results, but results which always err in one direction only, and under all circumstances make the calculated value of the original energy, light, or heat too small. I have also stated that the absolute value of the error introduced may be very considerable indeed, and that in the case of previous investigators the use of this formula by them has given a value for the solar heat essentially smaller than my subsequent determinations, in which the selective absorption is taken into account. These general views have already been developed in this volume in the chapter on the theory of the spectro-bolometer. I repeat them here with special application to photometric determinations.

Let us now consider whether they may not enable us to account in part for the fact suspected by us, and confirmed by Professor Pickering, that "the comparison of high and low stars gives a result apparently affected by a systematic error, according to which * * * the resulting coefficient of transmission will be made too large."

Having found that the actual energy (whether shown as heat or light) of the sun, as determined by the ordinary method, is, so far as it depends upon this method, always too *small*, and that the error is always (when referred to the same unit air-mass) greater when the absorption is greater, or when the sun is nearer the horizon, let us now apply these considerations, which belong equally to the stars, to the case of two stars, one of which, near the zenith, has suffered but slight absorption, the other, near the horizon, has suffered a greater one, and let us suppose that the real light before absorption in the two stars is approximately known. It appears, then, since what has already been demonstrated (loc. cit.) as regards heat applies equally to light, that the coefficient of transmission, which would be found by separate photometric observations, is larger (as reduced to the same unit stratum) when obtained from the lower star than from the other; and as the coefficient derived by the exponential formula from the direct comparison of the two stars will be intermediate between these two values, of which the least is itself too large, the coefficient of transmission obtained from the comparison of high and low stars will always be too great.

The above demonstration does not tell us in how great a degree this coefficient is too large, and, for aught we have here demonstrated, the error may be practically negligible. We have

* Comptes Rendus, tome 92, p. 701.

already stated that as a matter of fact, however, it is not negligible; and we are prepared to assert that the error is far greater than has been supposed. For it may be observed, in general terms, that since the rays with large coefficients are represented by diminishing geometric progressions, whose common ratio is near unity, these rays will persist, while others with small coefficients are very early extinguished; and something like this was shown by Biot, at the time when Melloni's first observations on the transmission of heat through successive strata attracted attention. But what we desire now further to point out is, that according as the difference of these coefficients of transmission for the different portions of the light of the same star is greater, so will the error of the result in treating them as equal be larger, a consequence so obvious that it is only necessary to make the statement in order to have its truth recognized.

Since it has now been demonstrated that the formula ordinarily employed leads to too small results, it might properly be left to those who still employ it to show that their error is negligible; but this has never been done. There is possibly an impression that if there were any considerable error its results would become apparent in such numerous observations as have been made all over the world in stellar photometry during this century. But it is, in my opinion, a fallacy to think so; and I believe, as I have elsewhere tried to show, that the error *might* be enormous—that the actual absorption[*] *might* be twice what it is customarily taken, or 40 per cent, instead of 20 per cent, without the errors being detected by such observations as are now made.

It is true that this error affects all magnitudes nearly alike, and consequently is not of the great importance in stellar photometry (which deals chiefly with relative magnitudes) that it is in solar work. All of those who, while admitting the sufficiency of the foregoing demonstration that error of a definite kind exists, continue to use the erroneous formula, may, however, be invited to consider whether the burden of proof does not properly lie with them, and asked to demonstrate that the continued use of methods and formulæ certainly in some unknown degree erroneous does *not* involve an error equal to the entire amount of the absorption[*] in question.

Nothing in what has preceded is calculated to disprove the observations made both by Professor Pickering and the writer, to the effect that in the special method of high and low stars there is *also* a systematic error calculated to give too large a coefficient of transmission, as compared with the ordinary method. What I have demonstrated is that both this and the ordinary method necessarily give too small a result for the absorption.

[*] The word "absorption" is used, it will be remembered, in a general sense here for every process in our atmosphere by which the light is prevented from reaching us, such as its scattering by dust particles, &c.

at the top of the page (continuation):

The heat sent in all directions from the sky is diffused or reflected or radiated sun heat, and the original intensity of the sun's radiation has evidently been diminished by this amount.

CHAPTER XV.

SKY RADIATION.

The heat sent in all directions from the sky is diffused or reflected or radiated sun heat, and the original intensity of the sun's radiation has evidently been diminished by this amount.

If we consider the analogy of light in the simple case of a single white cloud in an otherwise clear sky, we observe a greater amount of light from the cloud than from the adjacent parts of the heavens; but in this case we gain the added light at the expense of that portion of the earth in the cloud's shadow. If the sky is absolutely cloudless, however, it is clear that every point in our horizon is receiving sensibly the same amount of light which we do at our own station, and in this case we must admit that if we could rise to the upper limit of our atmosphere we should find the sun brighter (1) by the amount which the whole sky sent us at our station, and (2) by the amount which the sky diffuses, reflects, and radiates *away* from that station. If we add to the direct solar radiation only that directly observed from the clear sky, we obtain, then, an amount certainly *less* than that representing the radiation before absorption, a statement which seems to me incontrovertible, though advantage does not seem to have before been taken of this fact to determine a minimum value for the solar constant by a method perhaps the most trustworthy of all in our possession.

It is very desirable, then, to get the relation between sunlight and skylight in a clear sky; but I shall have to depend here largely upon evidence from other sources than our own direct observation on Mount Whitney, as the apparatus fitted for this end (such as the Marie-Davy thermometers) either failed to reach us in time for use, or was, as in the case of the solar comparator, not well adapted to the purpose, and as the already overburdened observers had no time to organize other experiments during the brief stay of the expedition.

It may be well to remark that, besides observation, we are not without the aid of theory upon this subject, which has been treated by the distinguished physicist, R. Clausius,* with great thoroughness from a theoretical standpoint.

It may be premised as almost self-evident that as the direct sunbeam is diminished in traversing a greater depth of atmosphere, the relative sky radiation will be greater. We first present a part of the result of the investigations of M. Clausius on this point in the following table, where the ratio of sky radiation to sun radiation for different altitudes of the sun is given first from the theory of M. Clausius, and second from direct observation at various stations.†

TABLE 127.

Height of the sun.	Theory of Clausius.	Heidelberg.	Cheetham.	Manchester.	Kew.	Lisbon.
20	0.38	0.45	0.19	0.10	0.36	0.37
25	0.55	0.48	0.20	0.11	0.47	0.44
30	0.71	0.65	0.23	....	0.57	0.51
35	0.86	0.82	0.26	....	0.65	0.64
45	1.09	1.00	....	....	0.75	0.82
50	1.21	1.37	....	0.28	....	1.02
60	1.43	1.60	....	....	....	1.43

* See Poggendorf, Annalen, Vol. 129, p. 239, 1866.
† See also Radau's "Radiations chimiques du soleil."

We have here, for instance, for a height of the sun of 60°, the statement that at Heidelberg the direct radiation of the sun was 1.6 that from a clear sky.

It is to be observed that the radiation considered in these experiments is not the total radiation of the entire spectrum, but of a sun emitting energy of a certain wave-length and transmissibility. This wave-length can only be obtained inferentially. It is that probably not far from 0μ.35 to 0μ.4, or that near the border of the invisible "actinic" spectrum, where the radiations appear to have been most efficient on the whole for the various chemical means here employed in noting it, and it corresponds in the theory of M. Clausius to the ray whose coefficient of transmission is about 0.56.

The following table has been calculated by Clausius on the assumption that the coefficient of atmospheric transmission $(p) = 0.75$ by the formula

$$C = Z \cos z (1 - p^z)$$

where

$C =$ light reflected and diffused by the sky,
$I =$ intensity of sunlight upon a surface exposed normally to its rays,
$N =$ intensity of sunlight upon a horizontal surface,
$z =$ zenith distance of sun,
$t =$ secant of zenith distance of sun,
$\cos z (1 - p^t) =$ the loss undergone by the direct light of the sun,
$Z =$ ratio of portion lost to that reflected and diffused by the sky, as determined by Clausius's theory of the diffusion of light.

TABLE 126.

Height of sun.	Z.	Sun.		Sky C	Total light	
		I.	S		C S	C I
10°	80°	0.45	0.19	0.03	0.07	0.10	0.30
15	75	0.51	0.33	0.09	0.09	0.18	0.42
20	70	0.55	0.43	0.15	0.11	0.26	0.51
25	65	0.61	0.51	0.21	0.13	0.34	0.63
30	60	0.65	0.56	0.28	0.14	0.42	0.70
35	55	0.64	0.61	0.35	0.15	0.50	0.75
40	50	0.67	0.64	0.41	0.16	0.57	0.80
50	40	0.70	0.69	0.51	0.17	0.69	0.86
60	30	0.72	0.72	0.62	0.18	0.80	0.90
70	20	0.73	0.74	0.69	0.18	0.87	0.92
80	10	0.74	0.75	0.74	0.18	0.92	0.93
90	0	0.74	0.75	0.75	0.19	0.94	0.94

According to this table it will be seen that when the sun is at an altitude of 40° the light given by the entire sky is 25 per cent. of that received upon a surface exposed normally to the sun's rays.

Photometric comparisons of sunlight and sky-light can seldom be made at Allegheny, on account of the rarity of a satisfactory blue sky. A single day's experiment [*] there (October 30, 1883), on an exceptionally fine day, gave the ratio of total sky-light to sunlight $\frac{1}{1000}$, the altitude of the sun being 38° and the light coming from the *central part of the solar disk*, and being therefore richer in blue rays than the average solar light. The comparison was made by a Bunsen photometer disk. The sunlight was reflected by the siderostat mirror through a hole 0.635 mm. in diameter (area 0.314 mm.), placed at a distance of 25.2 m. from the disk, forming there an image of the sun 0.24 m. in diameter. The sky-light was reflected by a similar mirror through an aperture 5 cm. square, situated at the other extremity of a long darkened passage, and distant 3.3 m. from the Bunsen disk when equality of lights was produced. For equal areas the ratio of intensities of sunlight (*central* part of *solar* disk) to zenith sky-light was consequently

$$\frac{0.314 \times (3.3)}{2,500 \times (25.2)} = 0.00000215$$

and the ratio of areas of the solar disk to the entire sky being 0.00001107, the ratio of total sky-light to sunlight would be $\frac{215}{1.107} = 0.19$, if the average light of the solar disk were the same as that

---

[*] The experiment was conducted by Mr. F. W. Very.

at the center. From measurements made at Allegheny, but not given here, we have found that the average brightness of the solar disk is 0.8 of that at the center, whence the ratio of total sky-light to average solar light would be $\frac{.19}{.8}=0.24$, which is in close agreement with the calculations of Clausius, based on the assumption that the average coefficient of transmission for the luminous rays between A and H is 0.75.

Professor Tyndall attributes, as is well known,[*] the blue color of the sky to selective reflection from fine particles of dimensions comparable with the wave-lengths of the more refrangible rays, a conclusion which we need hardly say we accept, since our whole theory of selective absorption [†] in the present work rests upon the belief that the heat is reflected and diffused by particles of various sizes, the grosser ones exercising a general absorption, the finer ones a partially selective one, the finest a purely selective one. (See pp. 13 and 14.)

Messrs. Bunsen and Roscoe have measured the direct effect of sunlight and sky-light upon a mixture of equal volumes of chlorine and hydrogen, in which, when exposed to a moderate radiation, chlohydric acid is generally formed.

We quote the table which they give, the effect being stated in "photo-chemical degrees." [‡]

TABLE 129.

Height of sun, h.	Direct rays.		Sky, C	Total light, I+S	Total light, I+C
	Normal effect, I.	Vertical effect, S.			
0	0.0	0.0	3.1	3.1	3.1
10	2.6	0.5	15.1	15.6	17.7
20	27.9	9.5	24.7	34.2	52.6
30	60.2	30.1	31.7	61.8	91.9
40	86.7	56.0	36.1	92.1	122.8
50	107.4	82.2	38.1	120.3	145.5
60	121.6	105.4	39.1	144.5	160.7
70	131.2	123.3	39.6	162.9	170.8
80	136.7	134.6	39.7	174.3	176.4
90	138.4	138.4	39.7	178.1	178.1

If in place of the mixture of chlorine and hydrogen, which is sensitive to the violet and ultra-violet rays, we choose a substance which is more powerfully acted upon by still shorter waves, the effect of sky-light will appear relatively greater. Thus it will be seen by reference to the table just given that the effect of normal radiation from the sun is equal to that from the entire sky when the altitude of the sun is about 18°; but when photographic paper is used in place of the hydrogen and chlorine mixture, the effect of normally received solar radiation does not equal that of sky-light until an altitude of 40° is attained by the sun. It is evident, then, that the extreme ultra-violet rays, which chiefly affect the photographic plate, are reflected by the sky in still greater proportion than the violet and blue rays, whose predominance gives the characteristic sky color, a conclusion which confirms our observation of the enormous absorption of these ultra-violet rays.

We need farther experiments, but with our present knowledge we may say, in reference to what has preceded, that under an *exceptionally pure* blue sky, when the sun's altitude is not far from 60° (which is about that in the mean of the Mount Whitney noon observations), the light (meaning by "light" all radiations between 0°.4 and 0°.7) from the sky is about ¼ that from the sun. If there is the slightest perceptible haze or milkiness in the blue, this value becomes greater. The amount sent upward and in other directions than toward us is, according to the estimates of M. Pouillet and M. Clausius, from 7 to 10 per cent.

We conclude, then, that if we added to the effect of the observed *light* (i. e., total light) of the sun not more than one-half its amount, we should get, according to our own observations, very nearly the *light* of the sun outside the atmosphere. The mean of the total "heat" radiation is,

---

[*] Proceedings of the Royal Society, No. 108, 1869.
[†] Mr. Koyl (Johns Hopkins University circular, August, 1883) suggests that the name "selective absorption" is misleading, and that we should rather speak altogether of "selective reflection."
[‡] Quoted from Radau's "Radiations chimiques du soleil."

according to our observation, somewhat more transmissible than that of the part between wave-lengths 0ᵘ.4 and 0ᵘ.7. We have found that the lower spectral radiation, as far at least as over 2ᵘ.0, is, except in the actual case of absorption bands, more transmissible than the luminous. Considering this ratio of total transmissibility to that of "light transmissibility" to be approximately as 80 to 75, and that the ratios of sky-light and sky-heat are nearly in the same proportion, we draw the final inference, from all that has preceded, that in the case of our highest actual observations of heat, taken in the purest sky, at an altitude of the sun of a little over 60°, the total sky-heat reflected, diffused, or radiated both toward and away from the observer is somewhat over ¼ the directly observed solar radiation.

CHAPTER XVI.

NOCTURNAL RADIATION.

The experiments of many physicists, notably Wells and Melloni, prove that during calm and cloudless nights bodies exposed freely in the open air lose a portion of their heat by radiation, the amount radiated varying with the nature of the surface and with the atmospheric permeability. That this transfer of heat takes place between the body at the earth's surface and the highest regions of the air or the celestial spaces beyond is rendered certain by the fact that the effect is obliterated by the interposition of a cloud. Some experiments in this direction were made on the expedition, the disposition of the thermometers employed being altogether similar to that used by Melloni.[*]

In his memoir on nocturnal cooling, Melloni, studying the subject with a view of improving on the work of Pouillet and Wells, employs three thermometers. His object is to find how much a thermometer, radiating freely toward space, falls below the temperature of the surrounding air. The radiating thermometer either has a black bulb or has its ordinary bulb covered with a black-ened thimble. The temperature of the surrounding air is to be determined by a thermometer which itself radiates as little as possible. This second or air thermometer, then, has a metal-covered bulb, or has its bulb inclosed in a bright silver thimble. This silver itself radiates in a minute degree. To allow for this Melloni takes a third thermometer, also covered with silver, and compares the action of the two latter when one is free to radiate toward the sky and the other is shielded. This third thermometer, then, is merely to obtain the correction for the slight radiating power of silver, whose effects we wish to eliminate.

These experiments are absolutely dependent for success on the calmness of the night; and those undertaken on the expedition could so seldom be made under favorable conditions in this respect that the results are of but moderate value. All the observations made at Lone Pine by Sergeant Dobbins prove, from this or other causes, to be useless; and a series made under circumstances of great difficulty at the peak of Whitney, by Captain Michaelis, who volunteered this trying service, are, though interesting, unfortunately prevented by the same cause from giving the results which might be expected under such otherwise uniquely favorable circumstances. Here, then, only a few of the considerable series of observations will be given.

DESCRIPTION OF APPARATUS.

Three thermometers, divided to 0.1° C., were provided. The first of these (Green, 4581) had a clear bulb, with black thimble for the radiation. The second (Green, 4583) was a clear-bulb thermometer, with silver thimble for measuring the temperature of the surrounding air. A black-bulb thermometer (Green, 4582) was also used as a radiation thermometer for comparison with the first, but its results were less satisfactory than where the blackened thimble was used, and they are here omitted. Our tables, therefore, give the comparison of thermometers 4581 (radiation) and 4582 (air temperature). Each thermometer was passed through a small cylinder of cork near its bulb. On this cork was fitted the thimble, in the one case covered with lamp-black, in the other of polished silver; and each thermometer was placed horizontally about 1 inches from the rock support, with its reservoir at the bottom of an inverted, truncated, tin cone, whose upper diameter (the aperture directed to the zenith) was 14 cm., and whose altitude was 9 cm. Each cone had a movable tin cover. According to Melloni, the action of such a cone is to almost exactly double

[*] See Memoir on Nocturnal Cooling, &c. Annales de Chimie et de Physique. February, 1848.

the effect of the radiation. The thermometer stems were themselves covered with tin cases, except at the moment of reading.

In general, the apparatus was brought out an hour before it was to be used, and after that time was alternately covered and exposed for intervals varying from 15 to 70 minutes, the thermometers being read once at the end of each interval. On one occasion (September 2, 1881) readings were taken every three minutes.

All the observations which have been thought worth preserving follow. In these tables the first column gives the date; the second, the time; the third, the readings of the black-thimble (i. e., radiation) thermometer; the fourth column (D_1) gives the absolute fall of this thermometer through radiation; the fifth column gives the silver-thimble initial and final readings; the sixth column, the difference of the silver-thimble readings (D_2). Were the silver an absolute non-radiator, and the lamp-black a perfect radiator, $D_1 - D_2$ would represent the effect of radiation. A correction has been introduced for the actual slight radiation from silver. This has been determined by comparing the readings of the silver-thimble covered and uncovered; and, on its being uncovered, we have in column 7 the fall due to the temperature of the air (D_3). Consequently, $D_1 - D_3$ is the effect due to radiation alone, always on the hypothesis that the night is *absolutely* calm, a condition that was only approximately obtained even on the nights of August 30 and 31.

OBSERVATIONS ON NOCTURNAL RADIATION.

TABLE 130.

[Station, Mountain Camp, Mount Whitney, California. Date August 28 and 29 1881. Weather, clear sky, wind light at 8h 15m p. m.; no observations recorded after this hour. Mean relative humidity 16 per cent.; mean force of vapor 2.2 mm. The relative humidity and force of vapor are given for the average of the times of observation. Corrections for instrumental errors have been applied. Instrument mounted on box on Table.]

Date.	Time.	Black thimble. (458.)	D_1	Silver thimble. (457.)	D_2	D_3	$D - D_3$	Exposure.
August 28	9 00 ? p. m.	2° 37′		3 4				Covered.
28	15m later.	−1 9	1 87	1 6	1 80	1 19	3 58	Uncovered.
28	10 08 p. m.	2 9		2 1				Covered.
28	10 31 p. m.	−0 5	3 40	2 4	0 10	0 41	3 81	Uncovered.
28	11 05 p. m.	1 90		3 4				Covered.
August 29	12 15 a. m.	−2 3	4 24	0 4	1 40	1 09	1 15	Uncovered.
	Mean nocturnal radiation	...	...	...			3 47	

TABLE 131.

[Station Mountain Camp, Mount Whitney California. Observer, Capt. O. E. Michaels. Date August 29 1881. Weather clear sky calm at 8h 15m p. m.; later, wind light. Mean relative humidity 67 per cent.; mean force of vapor 3.3 mm.]

Date	Time.	Black thimble. (458.)	D_1	Silver thimble. (457.)	D_2	D_3	$D_1 - D_3$	Exposure.
August 29	9 00 p. m.	1 2		1 2				Covered.
29	9 25 p. m.	−1 90	3 ??	1 3	−0 50	−0 61	3 84	Uncovered.
29	10 05 p. m.	2 7		2 8				Covered.
29	10 30 p. m.	−1 9	1 80	1 4	1 40	1 69	3 51	Uncovered.
	Mean nocturnal radiation	..	...	...			3 68	

TABLE 132.

[Station Mountain Camp, Mount Whitney, California. Observer, Capt. O. E. Michaels. Date August 30 1881. Weather, clear sky, wind light at 8h 15m p. m.; subsequently "night very still." Mean relative humidity 38 per cent.; mean force of vapor 1.8 mm.]

Date	Time.	Black thimble. (458.)	D_1	Silver thimble. (453.)	D_2	D_3	$D_1 - D_1$	Exposure.
August 30	8 50e p. m.	3 8		1 5				Covered.
30	9 10 p. m.	−0 6	4 40	2 5	1 30	0 70	61	Uncovered.
30	9 30 p. m.	4 4		3 3				Covered.
30	10 04 p. m.	−1 4	4 70	2 5	0 20	0 19	4 21	Uncovered.
30	10 25 p. m.	2 9		2 3				Covered.
30	11 10 p. m.	−1 8	4 70	1 ? 5	1 20	0 89	3 81	Uncovered.
	Mean nocturnal radiation	....	...	...			3 88	

* Recorded with a mark of interrogation after the "1."

TABLE 133.

[Station, Mountain Camp, Mount Whitney, California. Observer, Capt O E Michaelis. Date, August 31, 1881. Weather, clear sky, calm at 8h 15m p m., subsequently "very calm night." Mean relative humidity, 40 per cent.; mean force of vapor = 2.4 mm.]

Date	Time	Black thimble. 4581.	D_1	Silver thimble. 4583.	D_2	D_3	D_1-D_2	Exposure.
August 31	7h 55m p m	4.4		5.0				Covered.
31	8 25 p.m.	−0.5	4.99	4.1	0° 90	0° 50	4.31	Uncovered.
31	9 45 p.m.	2.7		4.1				Covered.
31	9 15 p.m.	−0.6	4.30	3.8	0.70	−0.01	4.31	Uncovered.
31	9 40 p.m.	4.0		4.2				Covered.
31	10 10 p.m.	−0.5	4.50	3.7	0.50	0.11	4.39	Uncovered.
31	10 40 p.m.	3.5		3.53				Covered.
31	11 10 p.m.	−0.4	3.90	3.5	0.03	−0.28	4.18	Uncovered.
Mean nocturnal radiation							4.30	

TABLE 134.

[Station, peak of Mount Whitney. Observer, Capt O.E Michaelis. Date September 2, 1881. Weather, clear sky "high wind." Mean relative humidity, 14 per cent., mean force of vapor = 1.6 mm. Instrument mounted between two rocks and protected from wind as much as practicable.]

Time.	Black thimble, 4581. Covered.	Uncovered.	Silver thimble. 4583 constantly covered.	4583−4581.	D_1	D_2	D_1-D_2
7h 45m p m	−3.0		−2.8	0.2			
8 00 p m		−3.5	−2.9	2.6	2.5	0.1	2.4
8 03 p m	−4.0		−2.9	2.0			
8 06 p m	−4.0		−2.9	1.1			
8 09 p m	−3.6		−2.9	0.7			
8 12 p m	−3.5		(2.7)*	0.5*			
8 15 p m	−3.4		(−2.5)	0.4			
8 18 p m		−4.7	−3.0	1.7			
8 21 p m		−5.2	−3.0	2.2			
8 24 p m		−5.6	−3.1	2.5			
8 27 p m		−5.7	−3.15	2.55			
8 30 p m		−5.83	−3.3	2.53			
8 33 p m		−5.8	−3.8	2.6	2.4	0.2	2.2
8 36 p m	−4.0		−3.2	1.7			
8 39 p m	−4.1		−3.15	0.95			
8 42 p m	−3.8		−3.2	0.6			
8 45 p m	−3.7		−3.2	0.5			
8 48 p m	−3.6		−3.2	0.4			
8 51 p m		−4.9	−3.2	1.7			
8 54 p m		−5.5	−3.25	2.25			
8 57 p m		−5.7	−3.25	2.45			
9 00 p m		−5.9	−3.3	2.6	2.3	0.1	2.2
Mean nocturnal radiation =							2.27

The numbers in parentheses have been rejected in favor of the following reading, viz.−3.0.

TABLE 135.

[Station, peak of Mount Whitney. Observer, Capt O E Michaelis. Date, September 3 and 4, 1881. Weather, clear sky, at first calm but afterward "slight north wind." Silver thimble kept constantly shielded from sky radiation No correction is therefore required to D_2, which indicates the change of temperature of the air. Consequently $D_1-D_2=$ nocturnal radiation.]

Date.	Time.	Black thimble. 4581.	D_1	Silver thimble. 4583.	D_2	D_1-D_2	Exposure (black thimble).
September 3	10h 00m p m	−6.0		−6.7			Covered.
3	10 15 p.m.	−9.6	2° 70	−6.9	0°.20	2.50	Uncovered.
3	10 30 p.m.	−7.1		−6.6			Covered.
3	10 45 p.m.	−10.3	3.20	−6.6	0.00	3.20	Uncovered.
3	11 00 p.m.	−6.5		−6.6			Covered.
3	11 15 p.m.	−8.8	2.30	−6.4	0.20	2.50	Uncovered.
3	11 30 p m	−7.0		−6.9			Covered.
3	11 45 p.m.	−8.0	1.00	7.1	0.20	(0.80)*	Uncovered.
September 4	12 00 m.	−7.8		7.6			Covered.
4	12 15 a.m.	−10.0	2.20	7.9	0.30	1.90	Uncovered.
Mean nocturnal radiation =						2.33	

* Rejected from the mean.

TABLE 136.

[Station, peak of Mount Whitney. Observer, Capt. O. F. Michaelis. Date, September 5 to 16, 1881. Weather clear sky, wind variable.]

Date.	Time	Black thermometer, ° C. Covered | Uncovered	Silver thermometer, ° C. constantly covered	D.	D.	D.	D.
September 5	7.25 p. m.		1.2				
5	7.35 p. m.		2.6	1.5			
5	7.45 p. m.		2.2	0.4		0.4	
5	8.15 p. m.	1.0	2.2	0.5			
5	8.25 p. m.		2.2	0.5			
5	9.00 p. m.		1.6	0.5	2.6		1
5	9.25 p. m.			1.2			
5	9.45 p. m.		1.1	1.1	.7	0.5	.6
5	10.25 p. m.		1.2	1.1		0.3	2.0
5	11.05 p. m.		1.8				
5	11.25 p. m.		1.5	1.7	2.5	0.3	.6
5	12.00 p. m.		1.4	1.0			
September 6	6.00 a. m.		.5	2.2	2.6	0.4	
6	6.10 a. m.						
Mean nocturnal radiation							

RESULTS.

The mean of four determinations made on the calmest night, August 31, gives us here as the nocturnal radiation from lamp black 1°.39 C. (in a dry air and at an altitude of nearly 12,000 feet); and there is reason to believe that with absolute calm the result would have been greater. Melloni's result, obtained by the same means October 9, 1846, in the clear air of Southern Italy, on a very calm night, was 3°.58 C.

Pouillet's values (see *Comptes Rendus*, July 9, 1838) are nearly double this, though obtained at Paris, near the sea-level. They are probably exaggerated by the defects of his actinometer, and through his apparent habit of placing his radiation thermometer near the earth and his air thermometer at some distance above it, a practice which usually insures too high readings for the latter, as Melloni has shown.

Pouillet uses his results on nocturnal radiation indirectly in obtaining his celebrated value (−142° C.) for the "temperature of space."

All the present writer's observations point to the conclusion that this "temperature of space" is little above that of the absolute zero, or, in other words, that all the heat the earth receives from all external sources, excepting the sun, but including that radiated from the stars, the dark heat from invisible bodies in space, the heat communicated dynamically by the contact of meteorites, &c., are collectively negligible, or nearly so. The ground for this conclusion cannot be given here,[*] but an extension of these experiments on nocturnal radiation would furnish a method of testing them, did we know the permeability of the different strata of our atmosphere to heat rays of low refrangibility. We should in this case need to place the bulb of our radiation thermometer in a suitable inclosure, so that it could radiate to only a limited portion of the sky at once, and then inclining it to successive portions from the zenith to the horizon, and noting its indications, we should obtain data for determining the amount of heat radiated earthward by this atmosphere alone, for the heat coming from the celestial spaces (if any) must be nearly constant at every inclination, while that from the atmosphere must ordinarily increase as we approach the horizon. If, then, we know the law of this increase, we can determine whether a constant term is to be added (to the expression for atmospheric radiation) for the "temperature of space," and if so, its amount. The actual experiment, it need hardly be said, would be a delicate one, but it does not seem beyond the reach of effort. An attempt to realize it was actually made on Mount Whitney, but was defeated by the presence of wind and the imperfection of the apparatus.

* See, however, the foot-note on p. 122.

CHAPTER XVII.

"HOT BOX" AND SOLAR RADIATION THERMOMETERS.

A hastily devised and constructed apparatus was taken with the expedition to observe the temperature which could be attained by the unconcentrated solar rays on the mountain.

A nest of shallow boxes, alternately of wood and of blackened copper, separated by air spaces or by loose cotton packing, and covered by sheets of common American window-glass, formed a sort of "hot box" (shown in section in Fig. 14). Within the inner vessel of thin spun copper, 4 cm. deep and 16½ cm. in diameter, was placed the bulb of a thermometer for registering the air temperature of the inclosure. This copper vessel had a glass cover, and was placed in a wooden vessel, which in turn rested on the bottom of the larger copper one, 8 cm. deep and 32 cm. in diameter, itself covered by a layer of glass and protected as much as possible from loss of heat by an outer envelope of wood, with loose cotton packing. The whole was inclined so as to receive the sun's rays normally to its glass face.

At the Mountain Camp the result of the best trial was on the 9th of September, 1881, when at 1ʰ 40ᵐ p. m. a temperature of 143½° C. was attained, the shade temperature at the time being 142.8, and the excess in the inner compartment of hot box 98°.5 C.

The following table gives the details of the observation on the mountain:

Fig. 14.

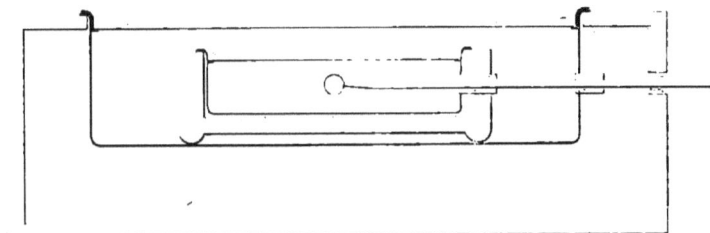

Section of Hot Box.

107

TABLE 137.

[Station: Mountain Camp, Observer O. E. M. Date: September 9, 1881. Thermometers corrected for instrumental errors.]

"HOT BOX."

Time	Sun thermometer.	Shade thermometer.	Sun thermometer.	Shade thermometer.	Difference	Remarks.
	Fahr.	*Fahr.*	*Cent.*	*Cent.*	*Cent.*	
11ʰ 30ᵐ a. m.	191.2	....	88.44	....	....	
11 35 a. m.	196.2	....	91.22	....	....	
11 40 a. m.	201.4	....	94.10	....	....	
11 45 a. m.	206.1	....	96.72	....	....	
11 50 a. m.	210.0	....	98.88	....	....	
11 55 a. m.	214.8	....	101.55	....	....	
12 00 m.	216.7	60.1	102.61	15.78	86.81	
12 05 p. m.	219.0	60.4	103.89	15.78	88.11	
12 10 p. m.	221.0	60.7	105.00	15.94	89.06	
12 15 p. m.	222.7	61.9	105.94	16.61	89.33	
12 20 p. m.	224.9	60.2	107.17	15.67	91.50	
12 25 p. m.	226.4	60.0	108.00	15.56	92.44	
12 30 p. m.	228.1	59.7	108.94	15.39	93.55	
12 35 p. m.	229.2	59.7	109.55	15.39	94.16	
12 40 p. m.	230.2	60.7	110.11	15.94	94.17	
12 45 p. m.	231.2	62.7	110.67	17.06	93.61	
12 50 p. m.	232.1	62.2	111.17	16.78	94.39	
12 55 p. m.	232.8	61.1	111.55	16.17	95.38	
1 00 p. m.	233.3	59.8	111.83	15.44	96.39	
1 05 p. m.	233.2	60.0	109.55	15.56	93.99	Out of adjustment
1 10 p. m.	232.3	59.9	111.28	15.50	95.78	
1 15 p. m.	233.3	61.0	111.83	16.11	95.72	
1 20 p. m.	234.5	60.7	112.50	15.94	96.56	
1 25 p. m.	235.2	....	112.87	....	....	
1 30 p. m.	235.7	....	113.17	....	....	
1 35 p. m.	236.0	60.8	113.33	16.00	97.33	
1 40 p. m.	236.0	59.7	113.33	11.83	98.50	
1 45 p. m.	235.2	58.0	112.89	14.44	98.45	
1 50 p. m.	233.8	57.9	112.41	14.39	97.72	

SOLAR RADIATION THERMOMETERS.

There was provided a pair of "conjugate thermometers," *i. e.*, one having a blackened bulb, the other a bright one, each in vacuo, not connected, but always placed together and read together. The following observations were made at Lone Pine:

TABLE 138.

[Solar conjugate thermometers. August and September 1881. Station: Lone Pine. Observer A. C. D.]

Date.	Time	Fahrenheit.			Centigrade.		
		Bright bulb	Black bulb	Difference	Bright bulb	Black bulb	Difference
Aug. 8	7ʰ 00ᵐ a. m.	102°.7	119°.5	16°.8	39°.28	48°.61	9.33
8	8 00 a. m.	102.6	134.5	31.9	39.22	56.95	17.73
8	12 00 m.	136.4	175.7	49.3	57.44	79.83	27.39
9	8 00 a. m.	108.6	137.2	28.6	42.56	58.44	15.88
9	12 00 m.	106.5	158.0	51.5	41.39	70.00	28.61
9	5 00 p. m.	90.7	94.5	4.2	32.39	34.72	2.33
10	7 00 a. m.	89.0	117.5	28.5	31.67	47.50	15.83
10	8 00 a. m.	105.5	141.0	35.5	40.83	60.56	19.73
10	12 00 m.	124.4	163.7	40.3	51.33	73.72	22.39
10	4 00 p. m.	113.8	136.5	22.7	45.44	58.06	12.62
10	5 00 p. m.	109.1	139.5	30.4	42.83	59.72	16.89
11	7 30 a. m.	95.1	129.0	33.9	35.06	53.89	18.83
11	8 00 a. m.	102.8	127.3	27.5	39.33	53.06	19.73
11	12 00 m.	120.0	162.9	42.9	48.89	72.74	23.89
11	4 00 p. m.	116.3	153.5	37.2	46.83	67.50	20.67
11	5 15 a. m.	104.0	144.4	40.4	40.00	62.44	22.44
12	12 00 m.	122.5	170.0	47.5	50.28	76.67	26.39
12	5 00 p. m.	92.0	141.0	19.0	33.33	44.89	10.56
13	7 00 a. m.	80.1	110.1	31.0	29.22	48.56	18.34
13	8 00 a. m.	99.2	143.5	44.3	37.33	61.95	24.62
13	12 00 m.	119.5	163.7	44.2	48.61	73.17	24.56
14	7 30 a. m.	83.6	115.0	31.4	28.67	46.11	17.44
14	8 00 a. m.	99.5	131.1	31.6	37.50	55.06	17.56
14	12 00 m.	119.5	136.2	26.7	48.61	58.00	20.39
14	4 00 p. m.	103.0	123.0	20.0	39.44	50.56	11.12
14	5 00 p. m.	102.4	110.0	8.8	39.89	43.33	1.44
15	7 10 a. m.	87.0	116.4	29.0	30.56	46.85	16.11
15	5 05 p. m.	94.8	107.5	12.7	34.89	41.94	7.05
16	8 30 a. m.	79.0	112.5	33.5	26.11	44.72	18.61
16	5 05 p. m.	99.7	121.0	21.3	37.61	49.44	11.83
17	6 33 a. m.	84.0	114.0	30.0	28.89	45.56	16.67
17	5 05 p. m.	86.1	102.0	15.9	30.06	38.89	8.83
18	6 55 a. m.	72.5	108.5	36.0	22.50	42.50	20.00
18	5 44 p. m.	89.2	104.0	14.8	31.78	40.00	8.22
19	8 55 a. m.	77.0	119.0	42.0	25.00	48.72	17.78
19	5 05 p. m.	89.7	96.0	6.3	32.06	35.56	3.50
20	8 53 a. m.	73.7	109.2	35.5	23.17	42.89	19.72
20	3 05 p. m.	94.2	110.0	15.8	34.56	43.33	8.77

TABLE 138—Continued.

Date	Time	Fahrenheit.			Centigrade.		
		Bright bulb.	Black bulb.	Difference.	Bright bulb.	Black bulb.	Difference.
Aug. 21	6 55 a. m.	77 .5	109 .5	32 .0	25 .28	43 .06	17 .78
21	12 05 p. m.	121 .3	168 .5	47 .2	49 .61	75 .84	20 .23
21	5 05 p. m.	94 .5	105 .6	11 .1	34 .72	40 .89	6 .17
22	8 05 a. m.	101 .0	145 .5	44 .5	38 .33	63 .06	24 .73
22	12 00 m.	118 .6	165 .5	46 .9	48 .11	74 .17	26 .06
22	5 05 p. m.	97 .8	113 .0	15 .2	36 .55	45 .00	8 .41
23	6 55 a. m.	79 .5	111 .5	32 .0	26 .30	44 .17	17 .74
23	12 00 m.	117 .2	163 .5	46 .3	47 .33	73 .06	25 .73
23	5 05 p. m.	91 .7	104 .0	12 .3	33 .17	40 .00	6 .83
24	7 00 a. m.	83 .5	120 .0	36 .5	28 .61	48 .89	20 .28
24	12 00 m.	115 .7	161 .8	46 .1	46 .50	72 .11	25 .61
24	5 05 p. m.	93 .0	105 .0	12 .0	33 .88	40 .56	6 .67
25	6 58 a. m.	73 .0	104 .0	35 .0	22 .78	42 .22	19 .44
25	12 00 m.	112 .5	158 .8	46 .3	44 .72	70 .44	25 .72
25	5 05 p. m.	96 .7	101 .1	10 .4	32 .61	38 .39	5 .78
'26	6 55 a. m.	74 .1	107 .0	33 .5	23 .56	42 .17	18 .61
27	12 00 m.	110 .3	156 .5	46 .2	43 .50	69 .17	25 .67
27	5 05 p. m.	80 .0	91 .0	11 .0	26 .67	32 .78	6 .11
28	6 55 a. m.	69 .2	101 .3	32 .1	20 .67	38 .50	17 .83
28	12 00 m.	113 .2	158 .3	45 .1	45 .11	70 .17	25 .06
29	6 55 a. m.	71 .2	101 .5	32 .3	21 .78	40 .72	17 .84
29	12 00 m.	112 .0	156 .3	44 .3	44 .44	69 .06	24 .62
30	6 05(?) p. m.	86 .6	112 .8	26 .2	30 .33	44 .89	14 .56
30	6 55 a. m.	73 .3	102 .3	29 .0	22 .91	39 .06	16 .12
30	12 00 m.	109 .5	153 .8	44 .0	43 .22	67 .60	24 .44
30	5 05 p. m.	82 .1	90 .0	7 .9	27 .83	32 .22	4 .39
31	6 55 a. m.	65 .1	98 .6	33 .5	18 .39	37 .00	18 .61
31	12 00 m.	109 .8	155 .0	45 .2	43 .22	68 .33	25 .11
31	5 05 p. m.	93 .5	109 .6	16 .1	34 .17	43 .11	8 .94
Sept. 1	12 00 m.	113 .5	159 .1	45 .8	45 .17	70 .62	25 .45
1	5 05(?) p m	92 .7	111 .7	19 .0	33 .72	44 .28	10 .56
2	6 55 a. m.	71 .5	102 .3	30 .0	21 .83	39 .06	17 .23
2	12 00 m.	114 .6	156 .3	41 .7	45 .89	69 .06	23 .17
2	5 05 p. m.	90 .8	112 .3	18 .5	34 .33	44 .61	10 .28
3							
4	6 55 a. m.	80 .0	108 .0	28 .0	26 .67	42 .22	15 .55
4	12 00 m.	112 .3	154 .2	38 .9	44 .61	66 .22	21 .61
5	6 55 a. m.	79 .0	109 .0	30 .0	26 .11	42 .78	16 .67
5	12 00 m.	122 .0	159 .1	37 .1	50 .00	70 .62	20 .62
5	5 05 p. m.	97 .4	119 .5	22 .1	36 .33	48 .61	12 .28

* Observations suspended on account of gale.   † On account of gale noon and afternoon readings were omitted.

TABLE 139.

Summary of observations showing differences of readings of bright and black bulbs in vacuo at Lone Pine.

(In degrees Centigrade.)

Date.	7 a. m.	8 a. m.	12 m.	4 p. m	5 p. m.
Aug. 8	9 .39	17 .74	27 .39	....	....
9	....	15 .58	28 .61	....	2 .33
10	15 .83	19 .73	22 .39	12 .62	16 .89
11	18 .83	19 .74	25 .89	20 .67	10 .72
12	....	22 .44	26 .39	....	10 .56
13	18 .34	24 .02	24 .56	....	....
14	17 .44	17 .56	20 .39	11 .12	4 .44
15	16 .11	....	....	....	7 .05
16	18 .61	....	....	....	11 .83
17	16 .67	....	....	....	8 .83
18	20 .00	....	....	....	8 .22
19	17 .78	....	....	....	3 .50
20	19 .72	....	....	....	8 .77
21	17 .78	....	26 .23	....	6 .17
22	24 .73	....	26 .06	....	8 .44
23	17 .78	....	25 .73	....	6 .83
24	20 .28	....	25 .61	....	6 .67
25	19 .44	....	25 .72	....	5 .78
26	18 .61	....	....	....	....
27	....	....	25 .67	....	6 .11
28	17 .83	....	25 .06	....	....
29	17 .84	....	24 .62	....	11 .56
30	16 .12	....	24 .11	....	4 .39
31	18 .61	....	25 .11	....	8 .94
Sept. 1	....	....	25 .45	....	10 .56
2	17 .23	....	23 .17	....	10 .28
3	15 .55	....	21 .61	....	....
5	16 .67	....	20 .62	....	12 .28
Mean	17 .80	19 .67	24 .70	14 .80	8 .44

From a comparison of fourteen observations of the solar-radiation thermometers at noon with corresponding readings of the shade temperature, it appears that on the average the bright bulb registered 162.85 C. and the black bulb 41°.50 C. above the shade thermometer.

It will be noticed that the radiation is always greater (for the same altitude of the sun) in the morning than in the afternoon, a fact deducible also from the actinometer curves (see Fig. 11, p. 119), which, however, present it in a less salient manner.

CHAPTER XVIII.

HYGROMETRIC OBSERVATIONS.

In any investigation of atmospheric absorption of radiation, the great importance of water as an absorbing agent is evident. In spite of numerous controversies as to its mode of action at different temperatures and in different physical states, there is no doubt that water in some form has much to do with the variations in atmospheric permeability to radiation.

Accordingly a large number of observations with the psychrometer and other hygrometrical instruments were made on the expedition, which are now to be reduced. It is believed that the long-continued tri-hourly observations at Lone Pine furnish a valuable record of the peculiarities of a desert climate. It was intended to make similar ones on the mountain, but the excessive fatigue and difficulty attendant on the latter observations, made it impossible for the overworked observers on the mountain to accomplish their share.

INSTRUMENTS USED FOR HYGROMETRICAL OBSERVATIONS AND CONDITION OF ENVIRONMENT.

At Lone Pine (elevation 3,760 feet) the psychrometer, in charge of Sergeant Dobbins, United States Signal Service, was in one of the galvanized iron frames provided by the Signal Service. It consisted of dry bulb thermometer S. S. 1037; wet-bulb thermometer S. S. 1015, covered with thick wicking instead of the ordinary standard thin muslin. [*]

At the lower camp on Mount Whitney (altitude 11,600 feet) the psychrometer, in charge of Sergeant Nanry, United States Signal Service, was hung in an extemporized chamber, formed of a box with lattice-work opening, looking towards the north, and as well shielded from air-currents as could be obtained. It consisted of dry-bulb thermometer S. S. ——; wet-bulb thermometer S. S. ——, covered with thick wicking. [*]

On the mountain it was impossible to obtain any deposit of dew with Regnault's hygrometer or dew-point apparatus in its ordinary treatment, and therefore no observations will be found recorded by Sergeant Nanry, who was in charge at the upper station. Subsequently, however, results were obtained on one day (September 9), by Captain Michaelis, by the use of a frigorific mixture. These observations are given and discussed further on.

The Regnault dew-point readings obtained by Sergeant Dobbins at Lone Pine were taken in the ordinary way, by blowing air through the ether in the "dew-point" flask. These observations are given and discussed on pages 171 and 172.

REDUCTION OF PSYCHROMETER OBSERVATIONS.

The Fahrenheit thermometer readings are from the original records kept by Sergeants Dobbins and Nanry. The readings of the dry bulb, t, and of the wet bulb, t', and their difference, $t-t'$, are reduced to the centigrade scale by Table A IV, "Smithsonian Meteorological and Physical Tables."

With these numbers expressing the values of t' and $t-t'$ in Centigrade degrees used as arguments, the value of x, the force of vapor, expressed in millimeters of pressure of mercury, is obtained from the table B II. This table is calculated by Regnault's formula for the psychrometer,

$$x = f - \frac{0.480}{610 - t'}(t - t')h$$

* There is some doubt about the nature of the covering used. The question is fully discussed in Appendix I.

(161)

in which h represents the height of the barometer, assumed equal to 755 mm., and f' the force of aqueous vapor in saturated air at a temperature equal to t'. (See Smithsonian Tables, B, page 12.) At the end of the table is a shorter one giving the "correction for the barometrical height."

EXAMPLE.

$$t' = 16°.50 \qquad t - t' = 8°.39 \qquad \text{Bar.} = 666 \text{ mm.}$$

In the table we find for $t' = 16°$ and $t - t' = 8°.2$,

Force of vapor = 8.53 mm.

For $t - t' = 8°.4$,

Force of vapor = 8.41 mm.

Whence, by interpolation, for $t - t' = 8.39$,

Force of vapor = 8.42 mm.

Correction for extra 0°.50 in value of $t' = +0.15$ mm.
Correction for barometer (666 mm.) = +0.60 mm.
Concluded force of vapor = 9.17 mm.

The fluctuations of the barometer being very slight, and the deviations in the barometer correction produced by neglecting them causing at the most only a change of a few hundredths of a millimeter in the resulting force of vapor, the following table was compiled from the Smithsonian table, by interpolation and extension, for a mean barometer of 660 mm., the height at Lone Pine, in order to expedite the calculation.

TABLE 140.

[Correction for reduction to mean barometer, 660mm, for each tenth of a degree in the value of $t - t'$ from 1° to 17°.9.]

$t - t'$

	0.0	0.1	0.2	0.3	0.4	0.5	0.6	0.7	0.8	0.9
	mm.	mm.	mm.	mm.	mm.	mm.	mm.	mm.	mm.	mm.
1	.08	.09	.09	.10	.11	.12	.12	.13	.14	.14
2	.15	.16	.17	.17	.18	.19	.20	.21	.21	.22
3	.23	.24	.24	.25	.26	.27	.27	.28	.29	.29
4	.30	.31	.32	.32	.33	.34	.35	.36	.36	.37
5	.38	.39	.40	.40	.41	.42	.43	.43	.44	.45
6	.46	.47	.47	.48	.49	.50	.50	.51	.52	.52
7	.53	.54	.55	.55	.56	.57	.58	.59	.59	.60
8	.61	.62	.62	.63	.64	.65	.65	.66	.67	.67
9	.68	.69	.70	.70	.71	.72	.73	.73	.74	.75
10	.76	.77	.78	.78	.79	.80	.81	.82	.83	.84
11	.84	.85	.85	.86	.87	.88	.88	.89	.90	.90
12	.91	.92	.93	.93	.94	.95	.96	.97	.97	.98
13	.99	1.00	1.00	1.01	1.02	1.03	1.03	1.04	1.05	1.05
14	1.06	1.07	1.08	1.08	1.09	1.10	1.11	1.12	1.12	1.14
15	1.14	1.15	1.16	1.16	1.17	1.18	1.19	1.20	1.20	1.21
16	1.22	1.23	1.23	1.24	1.25	1.26	1.26	1.27	1.28	1.28
17	1.29	1.30	1.31	1.31	1.32	1.33	1.34	1.35	1.35	1.36

For the reduction of observations at the Mountain Camp, a mean barometric pressure of 500 mm. was adopted, and the following corrections were deduced:

TABLE 141.

When the temperature of the wet bulb is below the freezing point, Regnault's formula becomes

$$x = f' - \frac{0.480}{689 - t'}\, \overline{t - t'}\, h$$

with which the following barometer corrections have been used, calculated for the lower camp, Mount Whitney:

TABLE 142.

$t - t'$

Correction for a difference of 1 do.	.27	.14	.21	.25	.15	.18
Correction for a difference of 1 c.m.	.1	.07	.11	.11	.17	.21
Correction for 650 mm	.25	.15	.22	.29	.4	.47	.50
Correction for Motion	.18	.26	.34	.71	.87	1.05	1.22

On account of the dryness of the air a considerable number of the observations fall outside the limit of the Smithsonian tables, which have not been calculated for so low a relative humidity, and which in consequence are probably less reliable in these parts, being founded on instrumental readings taken under different conditions. The values that have been deduced by extending the tables are distinguished in the following pages by an asterisk.

(For a discussion of the method employed in the reduction of the psychrometer observations see Appendix No. 1.)

TABLE 143.

Direct comparison of results by the psychrometer and by Regnault hygrometer or dew-point apparatus.

(Station, Lone Pine. Observer, A. C. D.)

Date.	Local time.	Temperature of air.	Dew-point			Sky.	Wind and classification according to wind velocity.	Force of vapor.		Relative humidity.	
			Psychrometer	Regnault hygrometer	Dew-point			Psychrometer	Regnault hygrometer	Psychrometer	Regnault hygrometer
		Cent.	Cent.	Cent.	Cent.			m.m.	m.m.	per ct.	per ct.
Aug. 8	8.45 a.m.	28.91	18.85	18.85	+7.94	Clear	Gentle A	16.15	9.72	54.7	32.9
8	8.15 p.m.	19.88	12.82	9.50	+1.12	Cloudy	Fresh B	10.85	8.87	63.0	51.4
9	8.15 a.m.	20.35	12.11	13.72	...	Clear	Calm A	10.65	11.70	53.0	58.6
10	8.15 p.m.	29.50	11.11	9.50	+1.91	Clear	Fresh B	10.08	8.53	18.9	41.2
11	8.15 a.m.	26.89	12.81	6.47	+7.44	Clear	Gentle A	11.81	7.90	41.0	26.8
11	8.15 p.m.	27.59	1.91	-1.50	+1.11	Clear	Fresh B	6.52	5.41	24.1	20.5
12	8.15 a.m.	27.01	11.57	8.11	+7.50	Clear	Gentle A	12.20	7.20	44.1	26.9
12	8.15 p.m.	27.72	14.27	...	...	Clear	Fresh B	12.10	8.23	64.7	31.3
13	8.15 a.m.	27.11	15.70	7.11	...	Clear	Gentle A	13.60	7.49	47.8	27.6
14	8.15 a.m.	25.55	11.20	8.11	-2.15	Clear	Calm A	9.88	8.08	44.0	32.2
14	8.15 p.m.	25.91	1.28	2.50	-1.28	Clear	Variable, Fresh to gentle B	5.94	5.51	20.3	22.2
Mean		25.88	12.18	7.79	1.91			11.30	8.07	45.6	33.1

The above observations were made during a short period of comparatively moist weather. Subsequently, however, the air became so excessively dry that the difficulties in the use of the Regnault hygrometer became almost insuperable with the limited resources at command.

The next table (144) repeats the previous results, classified according to the velocity of the wind.

10 59	8.87	63 1	5) 4	12.02	9.50	+3.12
10 05	8 57	46 9	41 3	11.44	9.50	+1.84
11 55	5 21	24 1	21 5	4 99	3.39	+1.54
12 10	8 21	60 7	41 3	14 25	8 39	+5 86
5 04	5 54	20 3	22 2	4 28	2.56	−1.28
Means 8 05	7 16	45 9	35 5	8 99	6 67	+2.24

As far as they go, these observations show a marked tendency toward a diminution of the difference between the indications of the two instruments as the wind increases.

TABLE 145.

Comparison of psychrometer and Regnault hygrometer.—Record of the results by the psychrometer.

[Station, Mountain Camp   Observer O. E. M.   Date, September 9, 1881.]

Local mean time.	Fahrenheit.			Reduced to Centigrade.			By Smithsonian tables (Regnault).	
	Dry.	Wet.	Difference.	Dry.	Wet.	Difference.	Force of vapor.	Dew-point.
							mm.	*°*
0 58 30	67 5	50 5	17 0	19 72	10 28	9 . 44	5. 41	+2.28
1 56 30	66 8	49 1	17 7	19 333	9 50	9 . 83	4 84	+0.71
1 06 50	69 6	52 2	17 2	20 56	11 00	9 56	6 01	+5.77
1 21 00	64 5	46 5	18 0	18 06	8 . 06	10 . 00	4 08	−1.60
1 56 50	65 1	46 7	18 4	18 39	8 . 17	10 22	4 03	−1.73
1 21 50	64 0	46 0	18 0	17 78	7 . 78	10 00	3 91	−2.15
1 20 5	64 0	46 5	18 0	18 22	8 . 22	10 00	4 17	−1.50
1 54 20	65 0	47 0	18 0	18 33	8 . 33	10 . 00	4 24	−1.09
1 56 50	62 0	44 0	18 0	16 67	6 . 67	10 . 00	3 38	−4 93
1 11 50	64 0	45 8	18 2	17 78	7 . 67	10 . 11	3 80	−2.52
1 46 20	67 8	49 8	28 0	19 89	9 . 89	10 . 00	5 12	+1.50
1 51 50	65 0	47 5	18 1	18 67	8 . 01	10 . 06	4 58	−0.46
Mean				18 42			4 15	−0.55

By the tables for tension of aqueous vapor, the dew-point, corresponding to a mean force of vapor of 4.15 mm., is−0°.55 which is adopted as the dew-point given by the psychrometer.

On the same day readings of the Regnault hygrometer were obtained by Captain Michaelis with the following result:

"Tried Regnault hygrometer. No success with ether. Made frigorific mixture of snow and salt. Obtained temperature of 13° and 15° F., but no deposition of moisture. By using the hygrometer (Signal) thimble a temperature of 7° was obtained with ice and salt, and the dew-point was reached with the following reading:

Regnault thermometers—dry, 61° Fahr.; wet, 7° Fahr. (shade); in the sun, copious dew at 7°, dry, 62°; at 12°, moisture began slowly to evaporate; at 13°, rapidly—dry-bulb, 61°.5."

The result of the comparison is therefore

Dew-point by Regnault hygrometer....   =7° to 12° F.
Mean dew-point by Regnault hygrometer.. = −12°.5 C
Mean dew-point by psychrometer ..   = −0.45 C.
Psychrometer − Regnault hygrometer ..... = + 12°.05 C.

This agrees with the previous measures, obtained at Lone Pine during light or gentle wind, in indicating that the psychrometer gave dew-points somewhat too high.

On several other occasions attempts were made to reach the dew-point, but always without success.

Tabular statements now follow of the original observations with the psychrometer (expressed both in the Fahrenheit and Centigrade scales), together with the results of their reduction.

TABLE 146.

Reduction of wet and dry bulb thermometer readings.

(A. C. D., Observer. Station Lone Pine. Latitude, 36° 36′. Longitude, 118° 05′ 47″. Elevation 3760 feet.)

Date.	Local time	Corrected Fahrenheit.			Reduced to Centigrade.			By Smithsonian Tables.			Wind and weather
		Dry s. No. 1037.	Wet s. No. 1045	Differ ence.	Dry.	Wet.	Differ ence	Force of vapor	Dew point.	Relative humidity	

RESEARCHES ON SOLAR HEAT.

TABLE 146—Continued.

Reduction of wet and dry bulb thermometer readings—Continued.

Date.	Local time.	Corrected Fahrenheit.			Reduced to Centigrade.			By Smithsonian tables.			Wind and weather.
		Dry a.s. No. 1037.	Wet a.s. No. 1045.	Differ-ence.	Wet.	Dry.	Differ-ence.	Force of vapor.	Dew-point.	Relative humidity.	
1881.						°	°	mm.	°Cent.	Per cent.	
Aug. 19	12 40 p. m.	88.8	60.5	28.3	31.56	15.63	15.73	·4.80	0.9	14.1	{ Gentle. { Clear.
19	8 15 p. m.	58.1	51.9	6.2	14.50	11.06	3.44	8.01	8.0	65.3	{ Gentle. { Clear.
20	8 15 a. m.	72.9	58.0	14.9	22.72	14.44	8.28	7.80	7.6	38.0	{ Gentle. { Clear.
20	12 35 p. m.	91.5	61.4	30.1	33.06	16.33	16.73	·4.70	0.6	12.7	{ Fresh. { Clear.
20	8 15 p. m	69.4	56.9	12.5	20.78	13.83	6.95	8.02	8.0	43.0	{ Gentle. { Clear.
21	8 15 a. m.	77.2	61.5	15.7	25.11	16.39	8.72	9.29	10.1	38.8	{ Light. { Clear.
21	12 35 p. m.	91.2	62.9	28.3	32.89	17.17	15.72	6.11	4.0	16.4	{ Fresh. { Clear.
21	8 15 p m	80.3	58.0	22.3	26.83	14.44	12.39	5.62	2.8	21.5	{ Brisk. { Clear.
22	8 15 a. m.	73.5	60.9	14.6	24.17	16.06	8.11	9.22	10.1	41.0	{ Light. { Clear.
22	12 35 p. m.	87.8	60.2	27.5	31.00	15.72	15.28	·5.07	1.4	15.2	{ Fresh. { Clear.
22	8 15 p. m.	63.0	53.0	10.6	17.56	11.07	5.89	7.12	6.3	47.6	{ Fresh. { Clear.
23	8 15 a. m.	77.0	61.5	15.5	25.00	16.39	8.61	9.24	10.1	39.3	{ Light. { Clear.
23	12 35 p. m.	88.7	59.6	29.1	31.50	15.33	16.17	·4.27	-1.0	12.4	{ Fresh. { Clear.
23	8 15 p. m.	71.7	56.5	15.2	22.06	13.61	8.45	7.98	6.2	35.8	{ Fresh. { Clear.
24	8 15 a m.	79.3	58.5	20.8	26.28	14.72	11.56	6.30	4.5	24.8	{ Fresh. { Clear.
24	12 35 p. m.	86.6	61.0	25.6	30.33	16.11	14.22	6.00	3.8	18.7	{ Fresh. { Clear.
24	8 15 p. m.	76.8	55.5	21.3	24.89	13.06	11.83	4.89	0.0	20.9	{ Brisk. { Clear.
25	8 15 a. m.	72.7	58.7	14.0	22.61	14.83	7.78	8.39	8.7	41.1	{ Calm. { Clear.
25	12 35 p. m.	82.4	57.7	24.7	28.00	14.28	13.72	4.78	0.5	17.0	{ Fresh. { Clear.
25	8 15 p. m.	68.7	52.6	16.1	20.39	11.44	8.95	5.34	2.1	30.0	{ Fresh. { Clear.
26	8 15 a. m.	74.2	59.4	14.8	23.44	15.22	8.22	8.50	8.9	39.7	{ Light. { Clear.
26	12 35 p. m.	86.3	57.5	28.8	30.17	14.17	16.00	3.54	-3.4	14.1	{ Brisk. { Clear.
26	8 15 p m	65.2	50.6	14.6	18.44	10.33	8.11	5.08	1.4	32.3	{ Fresh. { Clear.
27	8 15 a. m.	65.2	49.6	15.6	18.44	9.78	8.66	4.48	-0.4	28.4	{ Brisk. { Clear.
27	12 35 p. m.	76.4	55.5	20.9	24.67	13.06	11.61	5.03	1.3	21.7	{ Light. { Clear.
27	8 15 p. m.	53.2	45.2	8.0	11.78	7.33	4.45	5.31	2.1	51.6	{ Calm. { Clear.
28	8 15 a. m.	66.5	53.2	13.3	19.17	11.78	7.39	6.43	4.8	38.8	{ Light. { Clear.
28	12 35 p. m.	79.5	59.0	20.5	26.39	15.00	11.39	6.63	5.2	25.9	{ Fresh. { Clear.
28	8 15 p. m.	60.6	53.1	7.5	15.89	11.72	4.17	8.09	8.1	60.1	{ Gentle. { Clear.
29	8 15 a. m.	69.2	54.2	15.0	20.67	12.33	8.34	6.27	4.4	34.5	{ Light. { Clear.
29	12 35 p. m.	79.8	56.4	23.4	26.56	13.56	13.00	4.65	0.2	17.9	{ Fresh. { Clear.
29	8 15 p. m.	60.0	47.6	12.4	15.56	8.67	6.89	4.79	0.6	36.3	{ Gentle. { Clear.
30	8 15 a. m.	67.7	53.6	14.1	19.83	12.00	7.87	6.30	4.5	36.7	{ Fresh. { Clear.
30	12 35 p m	77.8	57.5	20.3	25.44	14.17	11.27	6.06	3.9	23.1	{ Fresh. { Clear.
30	8 15 p m	59.7	48.6	11.1	15.39	9.22	6.17	5.46	2.4	41.9	{ Calm. { Clear.
31	8 15 a. m.	66.7	54.4	12.3	19.28	12.44	6.84	7.15	6.3	42.0	{ Light. { Clear.
31	12 35 p m	89.8	56.5	24.3	27.11	13.61	13.50	4.43	-0.5	16.6	{ Fresh. { Clear.
31	8 15 p. m.	59.3	48.8	10.5	15.17	9.33	5.84	5.70	3.0	44.3	{ Calm. { Clear.
Sept. 1	8 15 a. m.	67.5	52.6	14.9	19.72	11.44	8.28	5.72	3.1	33.5	{ Calm. { Clear.
1	12 35 p. m.	80.6	59.0	21.6	27.00	15.00	12.00	6.30	4.5	23.8	{ Fresh. { Clear.
1	8 15 p. m.	66.1	52.0	14.1	18.94	11.11	7.83	5.72	3.1	35.2	{ Calm. { Clear.
2	8 15 a. m.	73.7	56.5	17.2	23.17	13.61	9.56	6.52	5.0	30.8	{ Calm. { Clear.
2	12 35 p. m.	87.8	58.5	29.3	31.00	14.72	16.28	·3.83	-2.4	11.5	{ Brisk. { Clear.

TABLE 146—Continued.

Reduction of wet and dry bulb thermometer readings.—Continued.

Date	Local time.	Corrected Fahrenheit.			Reduced to Centigrade.			By Smithsonian tables.			Wind and weather.
		Dry s.s. No. 1037.	Wet s.s. No. 1045.	Differ-ence.	Wet.	Dry.	Differ-ence.	Force of vapor.	Dew-point.	Relative humidity.	
								a. m.	Cent.	Per cent	
1883.											
Sept. 2	8 45 p.m.	73.3	52.1	21.2	22.94	11.17	11.77	3.69	-- 2.9	17.8	{ Brisk. { Clear.
3	8 45 a.m.	76.3	52.9	23.4	24.61	11.61	13.00	3.50	-- 4.5	14.4	{ Gale. { Clear.
3	12 35 p.m.	85.3	56.7	28.6	29.61	15.72	15.89	2.72	-- 6.8	8.8	{ Gale. { Clear.
3	8 45 p.m.	69.7	49.1	20.6	20.94	9.50	11.44	2.85	-- 6.2	15.5	{ Brisk. { Clear.
4	8 45 a.m.	67.7	49.2	18.5	19.83	9.56	10.27	3.50	-- 3.6	26.4	{ Brisk. { Clear.
4	11 27 p.m.	77.6	54.6	23.0	25.33	12.56	12.77	4.08	-- 1.6	17.0	{ Fresh. { Clear.
4	8 45 p.m.	74.2	51.2	23.0	23.44	10.67	12.77	2.89	6.0	11.5	{ Brisk. { Clear.
5	8 45 a.m.	72.2	54.5	17.7	22.33	12.50	9.83	5.20	2.8	27.9	{ Brisk. { Clear.
5	12 35 p.m.	81.0	59.5	21.5	27.22	15.28	11.94	6.55	3.0	21.4	{ Gentle. { Clear.
5	8 45 p.m.	72.9	52.6	20.3	22.72	11.44	11.28	4.34	-- 1.4	20.2	{ Gale. { Clear.
6	8 45 a.m.	65.7	50.4	15.3	18.72	10.22	8.50	4.84	0.6	26.0	{ Brisk. { Clear.
6	12 35 p.m.	....	....	....	....	....	....	....	....	....	
6	8 45 p.m.	52.7	47.5	5.2	11.50	8.61	2.89	6.87	5.7	67.9	{ Calm. { Clear.
7	8 45 a.m.	67.1	53.6	13.5	19.50	12.60	7.50	6.48	4.9	38.4	{ Light. { Clear.
7	12 35 p.m.	79.8	56.9	22.9	26.56	13.83	12.73	4.98	1.1	19.2	{ Fresh. { Clear.
7	8 45 p.m.	52.9	49.2	3.7	11.61	9.56	2.05	7.84	7.7	77.0	{ Calm. { Clear.
8	8 45 a.m.	68.5	56.0	13.5	20.83	13.33	7.50	7.85	7.7	12.9	{ Light. { Clear.
8	12 35 p.m.	84.3	59.5	24.8	29.06	15.28	13.78	5.55	2.7	18.6	{ Fresh. { Clear.
8	8 45 p.m.	....	....	....	....	....	....	....	....	....	

† Owing to the illness of the observer the readings at 12 35 were delayed until 1.27.
‡ Omitted on account of illness of Sergeant Dobbins.
§ Station closed by direction of Professor Langley.

TABLE 147.

Reduction of wet and dry bulb thermometer readings.

[J. J. N. observer. Latitude, 36° 3½′. Longitude, 118° 18′ 30″. Elevation, 11,600 feet.]

MOUNTAIN CAMP.

Date.	Local time.	Corrected Fahrenheit.			Reduced to Centigrade.			By Smithsonian tables.			Wind and weather.
		Dry s.s. No.	Wet s.s. No.	Differ- ence.	Dry.	Wet.	Differ- ence.	Force of vapor.	Dew- point.	Relative humidity.	
								mm.	Cent.	Per cent.	
1881. Aug. 22	12 35 p.m.	61.5	41.	20.5	16.4	5.0	11.4	2.04	−10.2	14.7	Fresh, NW., steady, clear.
22	8 15 p.m.	46.	37.	9.	7.8	2.8	5.0	3.66	−3.0	46.1	Light, NW., steady, clear.
23	8 15 a.m.	51.	32.	19.	10.6	0.0	10.6	0.42	−28.7	4.4	Light, NE., steady, clear.
23	12 35 p.m.	59.	39.	20.	15.0	3.9	11.1	1.60	−13.2	12.6	Light, NE., variable, clear.
23	8 15 p.m.	42.5	33.	9.5	5.8	0.6	5.2	2.75	−6.6	38.9	Light, SE., steady, clear.
24	8 15 a.m.	49.	33.	16.	9.4	0.6	8.8	1.35	−15.3	15.4	Light, E., steady, clear.
24	12 35 p.m.	59.	41.	18.	15.0	5.0	10.0	2.58	−7.4	20.3	Fresh, NW., variable, clear.
24	8 15 p.m.	44.	32.	12.	6.7	0.0	6.7	1.9a	−10.6	26.0	Light, NW., steady, clear.
25	8 15 a.m.	47.	32.	15.	8.3	0.0	8.3	1.47	−14.3	17.9	Light, NW., steady, clear.
25	12 35 p.m.	40.	18.		14.4	4.4	10.0	2.32	−8.7	19.0	Fresh, N., steady, clear.
25	8 15 p.m.	41.	30.	11.	5.0	−1.1	6.1	2.11	−9.8	22.3	Calm, clear.
26	8 15 a.m.	43.	32.	11.	6.1	0.0	8.1	2.34	−8.6	33.2	Light, NW., steady, clear.
26	12 35 p.m.	54.	38.	16.	12.2	3.3	8.9	2.31	−8.7	21.8	Fresh, NW., variable, clear.
26	8 15 p.m.	40.	27.5	12.5	4.4	−2.5	6.9	1.39	−15.0	22.1	Light, NW., steady, clear.
27	8 15 a.m.	37.	30.	7.	2.8	−1.1	3.9	2.88	−6.0	51.3	Light, NW., steady, clear.
27	12 35 p.m.	52.	37.	15.	11.1	2.8	8.3	2.36	−8.5	21.8	Fresh, NW., variable, clear.
27	8 15 p.m.	36.	30.	6.	2.2	−1.1	3.3	3.11	−5.1	57.6	Light, NW., steady, clear.
28	8 15 a.m.	41.	30.	11.	5.0	−1.1	6.1	2.11	−9.8	32.3	Light, N., steady, clear.
28	12 35 p.m.	32.5	37.8	5.	11.4	3.3	8.3	2.46	−8.0	24.5	Gentle, E., steady, clear.
28	8 15 p.m.	36.	28.	8.	2.2	−2.2	1.4	2.38	−8.1	44.3	Light, NW., steady, clear.
29	8 15 a.m.	44.5	32.5	12.	6.9	0.3	6.6	2.11	9.8	28.3	Gentle, E., steady, clear.
29	12 35 p.m.	53.	38.5	14.5	11.7	3.6	8.1	2.74	−6.7	26.7	Fresh, NW., steady, clear.
29	8 15 p.m.	36.	31.5	4.5	2.2	−0.3	2.5	3.67	−3.1	67.5	Calm, clear.
30	8 15 a.m.	45.	32.	13.	7.2	0.0	7.2	1.79	−11.9	23.6	Light, NW., steady, clear.
30	12 35 p.m.	56.	39.	17.	13.3	3.9	9.4	2.58	−8.5	26.7	Fresh, NW., steady, clear.
30	8 15 p.m.	39.	29.	10.	3.9	−1.7	5.6	2.08	−10.0	34.3	Light, NW., steady, clear.
31	8 15 a.m.	44.	37.	7.	6.7	2.8	3.9	4.10	−1.5	35.8	Light, NW., steady, clear.
31	12 35 p.m.	58.	39.5	18.5	14.4	4.2	10.2	2.16	9.5	17.7	Light, NW., steady, clear.
31	8 15 p.m.	29.	30.	9.	3.9	−1.1	5.0	2.48	−7.9	40.9	Calm, clear.
Sept. 1	8 15 a.m.	47.5	32.	15.5	8.6	0.0	8.6	1.23	−16.1	11.8	Light, NW., steady, clear.
1	12 35 p.m.	56.5	40.5	16.	13.6	4.7	8.9	2.87	−6.1	24.7	Fresh, W., steady, clear.
1	8 15 p.m.	58.	28.5		3.3	−1.9	5.2	2.17	9.5	37.3	Calm, clear.
2	8 15 a.m.	44.	32.	12.	6.7	0.0	6.7	1.98	−10.6	26.9	Light, NW., steady, clear.
2	12 35 p.m.	...	...	...	...	...	...	...	...	...	

MOUNT WHITNEY PEAK.

Sept. 2	6 00 p.m.	30.	21.	9.	−1.1	−6.1	5.0	1.30	−15.8	30.7	Gale, NW., variable, fair.
2	9 00 p.m.	27.	20.5	6.5	−2.8	−6.4	3.6	1.68	−12.6	45.2	Gale, NW., variable, clear.
2	Midnight.	25.5	18.	7.5	−3.6	−7.8	2.5	1.21	−16.6	34.6	Gale, NW., variable, clear.
3	3 00 a.m.	22.5	18.	4.5	−5.3	−7.8	2.5	1.74	−12.2	56.9	Gale, NW., variable, clear.
3	6 00 a.m.	22.	19.	3.	−5.6	−7.2	1.6	2.13	−9.7	71.4	Gale, NW., variable, clear.

[Instruments on the way to the summit.]

TABLE 148.

Reduction of special tri-hourly observations of wet and dry bulb thermometers.

[Station, Lone Pine. Observers, A. C. D. and H. L.]

Corrected Fahrenheit. Reduced to Centigrade. By Smithsonian tables.

Date.	Local time.	Observer.	Dry s. No. 1637.	Wet s. No. 1645.	Differ-ence	Dry.	Wet.	Differ-ence.	Force of va-por.	Dew-point	Rela-tive hu-midity.
									mm	°Cent.	Pr. ct.
1881											
Aug. 15	Noon	H. L.	80.3	61.0	25.3	31.28	16.11	15.72	5.24	1.8	19.0
15	3 p. m.	A. C. D.	89.8	60.0	29.8	32.11	15.56	16.55	4.42	-0.8	12.1
15	6 p. m.	A. C. D.	82.8	58.3	24.5	28.22	14.61	13.61	5.15	1.6	18.1
15	9 p. m.	A. C. D.	74.7	53.4	21.3	23.72	11.89	11.83	4.15	-1.4	19.0
15	Midnight	A. C. D	67.7	49.6	18.1	19.83	9.78	10.05	3.75	2.7	21.8
16	3 a. m.	H. L.	58.6	48.1	10.5	14.78	8.94	5.84	5.74	3.1	45.0
16	6 a. m.	A. C. D.	50.3	42.7	7.5	10.11	5.91	4.17	4.84	0.7	52.2
16	9 a. m.	A. C. D.	79.3	61.6	17.7	26.28	16.44	9.84	8.68	9.2	34.1
16	Noon	H. L.	81.8	58.0	23.8	29.33	15.00	14.33	5.05	1.3	16.7
16	3 p. m.	A. C. D.	89.8	59.5	20.3	31.11	15.28	15.83	4.47	-0.4	13.3
16	6 p. m.	A. C. D.	79.3	54.5	24.8	26.78	14.72	11.56	6.15	4.8	25.0
16	9 p. m.	A. C. D.	69.5	58.3	11.0	20.83	11.72	8.11	9.21	10.1	50.5
16	Midnight	A. C. D	62.1	49.0	12.8	16.72	9.78	6.94	5.58	2.2	38.0
17	3 a. m.	A. C. D	50.2	44.7	5.5	10.11	7.16	2.95	6.93	3.6	64.0
17	6 a. m.	A. C. D	49.7	46.2	3.5	9.83	7.89	1.94	6.92	4.4	76.8
17	9 a. m.	A. C. D.	78.2	58.0	20.2	25.67	14.44	11.23	6.28	4.4	25.6
17	Noon	H. L.	84.8	58.2	26.0	29.31	14.56	14.75	4.50	-0.3	14.9
17	3 p. m.	A. C. D	88.3	60.1	28.2	31.28	15.63	15.67	4.84	0.7	14.2
17	6 p. m.	A. C. D.	79.8	58.5	21.3	26.56	14.72	11.84	6.20	4.2	23.9
17	9 p. m.	A. C. D.	62.1	53.6	8.5	16.72	12.00	4.72	7.95	7.9	56.3
17	Midnight	A. C. D	57.8	47.0	10.8	14.33	8.33	6.00	5.13	1.8	42.3
18	3 a. m.	A. C. D	54.2	41.2	13.0	12.33	6.56	5.83	4.19	-1.2	39.3
18	6 a. m.	A. C. D.	52.7	44.7	8.0	11.50	7.06	4.44	5.19	1.7	51.3
18	9 a. m.	A. C. D.	78.0	59.1	18.8	25.56	15.06	10.50	7.16	6.3	29.4
18	Noon	A. C. D.	86.8	61.7	25.1	30.44	16.50	13.94	6.34	5.0	20.3
18	3 p. m.	A. C. D.	89.8	59.5	30.3	32.11	15.28	16.83	3.94	-2.0	11.1
18	6 p. m.	A. C. D.	79.8	61.2	18.6	26.56	16.22	10.34	8.21	8.4	31.7
18	9 p. m.	A. C. D.	54.7	51.3	3.4	12.61	10.61	2.00	8.44	8.8	78.0
18	Midnight	A. C. D.	50.1	47.6	2.5	13.39	8.67	4.72	5.94	3.6	51.9
19	3 a. m.	A. C. D.	51.7	44.5	7.2	10.94	6.94	4.00	5.37	2.2	55.3
19	6 a. m.	A. C. D.	52.2	44.7	7.5	11.22	7.06	4.16	5.34	2.1	53.9
19	9 a. m.	A. C. D.	79.8	61.1	18.7	26.56	16.17	10.39	8.14	8.2	31.4
19	Noon	A. C. D.	91.3	63.5	27.8	32.94	17.50	15.44	6.61	5.2	17.8
19	3 p. m.	A. C. D	92.3	62.5	29.8	33.50	16.94	16.56	5.55	2.7	14.4
19	6 p. m.	A. C. D	82.8	59.5	23.3	28.22	15.28	12.94	6.02	3.8	21.2
19	9 p. m.	A. C. D	57.6	42.6	15.0	14.22	5.94	8.28	4.63	0.1	71.8
19	Midnight	A. C. D	50.2	49.1	11.1	13.11	9.94	6.17	5.31	2.1	41.4
20	3 a. m.	A. C. D	50.2	40.7	9.5	10.11	4.83	5.28	3.67	-3.0	39.6
20	6 a. m.	A. C. D.	51.7	44.7	7.0	10.94	7.06	3.88	5.48	2.5	56.4
20	9 a. m.	A. C. D	80.0	61.5	18.5	28.67	16.39	10.28	8.89	9.7	32.2
20	Noon	A. C. D	92.5	63.5	29.0	33.61	17.50	16.11	6.24	4.4	16.2
20	3 p. m.	A. C. D	89.7	62.0	31.7	34.50	16.67	17.61	1.75	0.5	11.8
20	6 p. m.	A. C. D.	83.5	60.0	23.1	27.83	15.56	12.27	6.61	5.2	23.8
20	9 p. m.	A. C. D	72.0	56.5	15.5	22.22	13.61	8.61	7.01	6.0	35.2
20	Midnight	A. C. D	67.8	53.1	14.7	19.89	11.72	8.17	5.97	3.7	34.8
21	3 a. m.	A. C. D	56.6	49.3	8.7	11.78	9.94	4.84	6.58	5.1	52.5
21	6 a. m.	A. C. D.	58.6	48.6	9.0	14.78	9.22	5.50	6.41	4.7	51.1
21	9 a. m.	A. C. D	81.6	63.9	17.9	27.67	17.50	9.95	9.75	10.9	35.3
21	Noon	H. L.	91.8	63.0	28.8	33.22	17.22	16.00	6.07	3.9	16.0
21	3 p. m.	A. C. D	91.3	63.3	28.0	32.94	17.39	15.55	6.16	4.5	17.4
21	6 p. m.	A. C. D	83.8	61.5	23.3	22.83	16.39	12.44	7.54	5.8	22.1
21	9 p. m.	H. L.	72.2	57.5	14.7	22.33	11.17	8.16	7.70	7.4	39.4
21	Midnight										
22	3 a. m.	A. C. D	56.2	51.4	7.2	14.78	10.78	4.00	7.52	7.1	60.0
22	6 a. m.	A. C. D	53.7	46.6	7.1	14.67	8.11	3.05	6.01	3.8	57.1
22	9 a. m.	A. C. D	80.8	63.0	17.8	27.11	17.22	9.84	9.33	10.3	35.0
22	Noon	H. L.	90.5	63.0	27.3	33.28	17.22	15.47	6.52	5.0	18.1
22	3 p. m.	A. C. D	92.3	61.1	31.2	33.50	16.17	17.33	4.43	-0.5	11.5
22	6 p. m.	A. C. D.	81.3	60.0	21.3	27.39	15.56	11.83	6.64	5.7	25.2
22	9 p. m.	A. C. D.	56.7	51.6	5.1	13.72	10.89	2.84	8.19	8.3	70.1
22	Midnight	A. C. D.	58.2	51.0	6.8	14.56	10.89	4.67	7.56	7.5	62.7
23	3 a. m.	A. C. D	56.3	50.1	6.6	13.72	10.06	3.66	7.27	6.6	62.2
23	6 a. m.	A. C. D.	55.2	47.6	7.6	12.89	8.67	4.22	6.39	4.2	55.9
23	9 a. m.	A. C. D.	80.0	62.5	17.5	26.67	16.94	9.73	9.19	10.0	35.3
23	Noon	H. L.	89.6	62.0	27.6	32.00	16.67	15.33	5.95	3.6	16.4
23	3 p. m.	A. C. D	92.3	60.5	31.7	33.44	15.83	17.61	3.97	-2.0	10.4
23	6 p. m.	A. C. D.	82.6	60.0	22.6	27.11	15.56	11.55	6.29	0.9	17.0
23	9 p. m.	A. C. D.	66.7	53.2	13.5	19.28	11.78	7.50	6.38	4.7	38.3
24	Midnight	H. L.	59.5	50.9	8.6	15.28	10.50	4.78	6.35	4.6	56.7
24	3 a. m.	H. L.	55.7	48.1	7.6	13.17	8.94	4.23	6.35	4.6	58.1
24	6 a. m.	A. C. D.	52.6	48.6	4.0	11.44	9.22	2.22	7.54	7.1	75.0
24	9 a. m.	A. C. D.	81.5	59.0	22.5	27.50	15.00	12.50	6.04	3.9	22.1
24	Noon	H. L.	90.2	62.0	27.6	32.44	16.11	16.33	5.15	2.4	16.0
24	3 p. m.	H. L.	90.6	61.0	29.6	32.56	16.11	16.45	4.67	0.8	13.3
24	6 p. m.	A. C. D.	68.1	57.0	24.5	27.61	13.89	13.78	4.47	0.4	18.2
24	9 p. m.	A. C. D.	67.2	51.3	15.9	19.56	10.72	8.84	4.93	1.0	29.0
25	Midnight	A. C. D.	53.2	46.6	6.6	11.78	8.11	3.66	6.10	4.1	58.7
25	3 a. m.	A. C. D.	48.2	41.2	12.0	12.33	6.78	5.49	5.19	3.5	42.1
25	6 a. m.	A. C. D.	52.0	43.6	8.4	11.11	6.44	4.67	4.78	0.5	48.3
25	9 a. m.	A. C. D.	79.3	59.4	20.0	26.28	15.22	11.06	6.82	5.8	27.2
25	Noon	H. L.	81.0	56.5	24.5	27.67	13.61	13.72	5.29	2.6	19.4
25	3 p. m.	H. L.	88.2	59.5	28.7	31.22	15.28	15.94	4.42	-0.5	13.1
25	6 p. m.	A. C. D.	74.2	64.0	10.2	23.44	17.78	5.66	12.10	14.2	56.5
25	9 p. m.	A. C. D.	66.2	52.1	16.1	20.11	11.17	8.94	6.11	3.1	32.1
25	Midnight	A. C. D.	65.6	48.1	15.5	17.56	8.94	8.62	4.61	-1.7	27.0
26	3 a. m.	A. C. D.	50.7	44.2	6.5	10.39	6.78	3.61	5.49	3.5	58.2

‡ Observer failed to awaken.
‡ This observation appears doubtful.

RESEARCHES ON SOLAR HEAT.

TABLE 148—Continued.

Reduction of special tri-hourly observations of wet and dry bulb thermometers.

[Station, Lone Pine. Observers, A. C. D. and H. L.]

Corrected Fahrenheit. Reduced to Centigrade. By Smithsonian tables.

Date.	Local time	Observer.	Dry s. s. No. 1037.	Wet s. s. No. 1045.	Differ- ence.	Dry.	Wet.	Differ- ence.	Force of va- por.	Dew point.	Rela- tive hu midity
									mm.	Cent.	Pr. ct.
1884.											
Aug. 26	6 a. m.	A. C. D.	51.7	43.7	8.0	10.94	6.50	4.44	4.92	0.9	59.7
26	9 a. m.	A. C. D.	78.5	60.8	17.7	25.83	16.00	9.83	8.28	8.5	33.5
26	Noon	A. C. D.	87.8	58.0	29.8	31.00	14.44	16.56	3.45	3.8	10.3
26	3 p. m.	H. L.	87.3	57.5	30.0	30.72	14.06	16.66	3.10	5.1	9.5
26	6 p. m.	H. L.	80.1	54.6	25.5	26.72	12.56	14.16	3.34	4.2	12.8
26	9 p. m.	A. C. D.	65.7	50.1	15.6	18.72	10.06	8.66	4.63	0.1	28.8
26	Midnight	A. C. D.	65.2	46.5	19.0	18.44	7.80	10.55	2.41	6.6	15.3
27	3 a. m.	A. C. D.	58.1	43.7	14.4	14.50	6.50	8.00	5.06	5.3	24.9
27	6 a. m.	A. C. D	56.2	42.7	13.5	13.41	5.94	7.50	3.08	5.2	36.9
27	9 a.m.	A. C. D	68.0	50.7	17.3	20.00	10.39	9.61	4.32	0.8	26.8
27	Noon	H. L.	77.8	53.4	24.4	25.44	11.89	13.55	3.21	4.7	13.4
27	3 p. m.	A. C. D.	81.3	57.0	24.3	27.39	13.89	13.50	4.02	0.1	17.0
27	6 p. m.	A. C. D.	64.6	57.0	7.6	18.11	13.89	4.22	9.54	10.6	61.7
27	9 p. m.	H. L.	51.0	46.1	4.9	10.56	7.83	2.73	6.48	4.9	68.0
27	Midnight	H. L.	49.0	43.2	5.8	9.44	6.22	3.22	5.42	2.3	61.8
28	3 a. m										
28	6 a. m	H. L.	48.2	40.7	7.5	9.00	4.83	4.17	4.26	1.0	49.6
28	9 a. m.	A. C. D.	70.2	55.2	15.0	21.22	12.89	8.33	6.66	5.3	35.6
28	Noon	H. L.	82.1	57.7	24.4	27.83	14.28	13.55	4.02	0.9	17.7
28	3 p. m.	H. L.	84.3	57.8	26.5	29.06	14.33	14.73	4.33	0.8	14.5
28	6 p. m.	A. C. D.	76.3	56.5	19.8	24.61	13.61	11.00	5.76	3.2	23.1
28	9 p. m.	A. C. D.	56.5	48.6	8.7	14.61	9.78	4.83	9.40	4.0	52.4
28	Midnight	A. C. D.	54.4	43.7	10.7	12.44	6.50	5.91	4.14	1.4	34.6
29	3 a. m	A. C. D	45.4	40.7	4.7	7.44	4.82	2.61	5.06	1.3	65.8
29	6 a. m.	A. C. D.	49.7	41.7	8.0	9.85	5.39	4.44	4.41	0.6	50.0
29	9 a. m.	A. C. D.	72.7	56.1	16.6	22.61	13.39	9.22	6.54	5.0	32.0
29	Noon	H. L.	84.6	58.0	24.6	27.00	13.33	13.67	4.14	1.4	15.0
29	3 p. m.	A. C. D.	84.1	56.6	27.5	28.94	13.67	15.27	3.32	4.3	11.2
29	6 p. m.	A. C. D.	66.7	58.5	7.2	19.28	15.28	4.00	10.78	12.5	64.5
29	9 p. m.	H. L.	65.0	48.1	16.9	18.33	8.94	9.39	3.63	3.1	23.2
29	Midnight	H. L.	66.2	47.6	18.6	19.00	8.67	10.33	2.98	5.6	18.2
30	3 a. m.	H. L.	58.9	44.5	14.4	14.94	6.94	8.00	3.24	4.0	26.0
30	6 a. m.	H. L.	56.6	44.5	12.1	13.67	6.94	6.73	3.04	2.0	33.7
30	9 a. m.	A. C. D.	69.9	55.0	14.9	21.06	12.78	8.28	6.62	5.2	35.5
30	Noon	H. L.	80.0	59.5	20.5	26.67	15.28	11.39	6.85	5.7	26.3
30	3 p. m.	A. C. D.	82.5	57.0	25.5	28.06	13.89	14.17	4.26	1.0	15.1
30	6 p. m.	A. C. D.	76.7	53.4	17.3	21.50	11.89	5.22	2.0	27.8
30	9 p. m.	A. C. D.	53.6	47.4	8.2	13.11	8.56	4.53	3.97	3.7	53.1
30	Midnight	H. L.	49.2	44.5	4.7	9.56	6.94	2.62	6.10	4.0	68.3
31	3 a. m.	H. L.	48.0	41.2	6.8	8.89	5.11	3.78	4.62	0.1	54.1
31	6 a. m.	H. L.	44.6	39.0	5.6	7.00	3.89	3.11	4.43	0.5	59.4
31	9 a. m.	A. C. D.	73.7	56.0	17.7	23.17	13.33	9.84	8.16	4.1	29.1
31	Noon	A. C. D.	82.4	57.7	24.7	28.00	14.28	13.72	4.83	0.7	17.2
31	3 p. m.	A. C. D.	84.0	57.0	27.0	28.89	13.89	15.00	3.81	2.5	12.9
31	6 p. m.	A. C. D.	67.0	57.5	9.5	18.44	14.17	4.17	8.25	16.1	53.2
31	9 p. m.	A. C. D.	57.6	47.6	10.0	14.22	8.67	5.55	5.51	2.5	45.7
1	Midnight	H. L.	47.2	43.5	3.7	8.44	6.39	2.05	6.15	4.1	74.2
Sept. 1	3 a. m.	H. L.	48.7	41.0	7.7	9.28	5.00	4.28	4.29	0.9	54.8
1	6 a. m.	H. L.	48.2	40.0	8.2	9.00	4.44	4.56	3.90	2.2	45.6
1	9 a. m.	A. C. D.	73.8	53.0	20.8	23.22	11.67	11.55	4.16	1.3	19.7
1	Noon	A. C. D.	83.3	59.2	24.1	28.50	15.13	13.39	3.65	2.9	19.5
1	3 p. m.	A. C. D.	76.3	54.3	28.0	30.17	14.61	15.56	3.13	1.4	12.9
1	6 p. m.	A. C. D.	66.9	58.4	8.5	19.39	14.67	4.72	9.34	11.2	59.3
1	9 p. m.	A. C. D.	60.3	52.1	8.2	15.72	11.17	4.55	7.51	7.0	56.6
1	Midnight	A. C. D.	62.6	47.5	15.1	17.00	8.61	4.39	3.97	2.0	27.5
2	3 a. m.	A. C. D.	50.6	42.1	8.5	10.31	5.61	4.72	4.36	0.7	46.0
2	6 a. m.	A. C. D.	47.0	40.0	7.0	8.67	4.44	4.24	4.07	1.6	48.4
2	9 a. m.	A. C. D.	76.4	58.6	17.8	24.67	14.89	8.78	7.41	6.5	32.0
2	Noon	A. C. D.	87.8	57.9	29.9	31.00	14.39	16.61	3.39	4.3	9.9
2	3 p. m.	A. C. D.	89.6	58.5	31.1	32.00	14.72	17.28	4.29	4.4	9.3
2	6 p. m.	A. C. D.	69.9	53.6	16.3	21.06	12.00	9.06	5.05	2.9	30.3
2	9 p. m.	A. C. D.	71.7	50.9	20.8	22.06	10.50	11.56	3.36	4.1	17.0
2	Midnight	A. C. D.	70.1	50.0	20.1	21.17	10.00	11.17	4.33	4.5	17.8
3	3 a. m.	A. C. D.	69.7	48.2	21.5	20.94	9.00	11.94	2.28	8.9	12.4
3	6 a. m.	A. C. D.	68.7	49.0	16.1	18.72	9.7	8.94	2.40	10.0	27.0
3	9 a. m.	A. C. D.	66.3	53.5	25.7	25.72	11.67	14.05	2.81	6.3	14.6
3	Noon	A. C. D.	88.4	56.0	29.4	29.67	13.33	16.34	2.70	6.8	8.7
3	3 p. m.	A. C. D.	84.5	56.3	28.2	29.17	13.50	15.67	3.19	4.4	10.6
3	6 p. m.	H. L.	72.9	50.9	22.0	22.72	10.50	12.22	3.08	5.2	15.0
4	3 a. m.	A. C. D.	89.1	48.1	21.0	20.61	8.94	11.67	1.98	9.1	10.5
4	Midnight	A. C. D.	62.1	44.7	17.4	16.72	7.06	9.66	2.81	8.8	17.1
4	3 a. m.	A. C. D.	63.3	47.6	15.7	17.39	8.67	8.72	3.81	2.4	28.9
4	6 a. m.	A. C. D.	62.0	45.6	16.4	16.67	7.56	9.11	2.98	5.6	21.1
4	9 a. m.	A. C. D.	48.9	39.6	19.1	21.00	10.44	10.62	2.55	20.5	
4	Noon	A. C. D.	77.8	55.5	22.3	25.44	12.7	12.66	4.30	6.9	17.8
4	9 p. m.	A. C. D.	58.3	53.5	24.9	26.89	13.06	13.83	1.88	2.3	14.0
4	6 p. m.	A. C. D.	70.8	51.6	25.2	24.89	10.89	14.00	3.36	5.3	10.1
4	Midnight	A. C. D.	72.7	51.4	21.3	22.61	10.78	11.83	3.39	4.0	10.6
5	3 a. m.	A. C. D.	54.7	46.1	8.6	12.61	7.83	4.78	5.40	2.3	49.7
5	6 a. m.	A. C. D.	59.6	48.6	11.0	15.33	9.22	6.11	5.49	2.5	42.4
5	9 a. m.	A. C. D.	74.7	56.3	18.4	23.72	13.50	10.22	6.09	4.9	28.0
5	Noon	A. C. D.	82.8	59.8	22.9	28.22	15.44	12.78	6.24	4.3	22.0
5	3 p. m	A. C. D.	84.5	60.8	23.7	29.17	16.00	13.17	6.51	4.9	21.6

† Observer failed to awake.
* This observation appears doubtful.
‡ Omitted; observer sick
§ Discontinued by direction of Professor Langley, dated September 3, 1884.

TABLE 149.

Summary of special tri-hourly observations of force of aqueous vapor at Lone Pine, showing diurnal variation.

Date	3 a. m.	6 a. m.	9 a. m.	Noon.	3 p. m.	6 p. m.	9 p. m	Mid night.
	m m.	m m.	m m.	m m	m m.	m m.	m m	m m.
1881. Aug. 15	......	......	......	5.24	4.32	5.13	4 15	3.75
16	5 74	4.83	8 68	5.65	4 47	6 35	9.24	5 34
17	5 83	6 92	6 36	4.50	4.54	8 20	7 36	5 13
18	4 19	5.19	7 16	6 54	3 94	8 21	8 44	5 91
19	5.37	5.34	8 11	6.61	5.55	6 02	8.83	5 34
20	3 67	5 4*	8 39	6.26	4 75	6 61	7 91	5 37
21	4 58	6 41	9.75	6.07	8.16	6.82	7 59	
22	7 52	6.01	9 33	6 52	4 43	6.84	8 39	7 76
23	7 27	6.20	9.19	5.95	3 97	4 76	6.54	7 51
24	6 55	7.54	6.04	5.45	4 87	4.47	4 93	6 16
25	4 49	4.78	6.92	5.39	4.42	(12.10)*	5 18	4 94
26	5.49	4 92	8.2*	3.45	3.10	3.34	4 65	2.41
27	3.06	3 06	4.72	3 21	4 02	8.54	6 44	3.42
28	......	4 26	8 66	4 92	4.33	5.76	6 49	4.14
29	5 96	4 41	6 54	4 14	3 32	(10 78)*	3.63	2.98
30	3 28	5 94	6 62	6.85	4 26	5.32	5.97	6.10
31	4 62	4 43	6.16	4.83	3.81	9.25	5 51	6.13
Sept. 1	4.29	5.90	4.16	5.05	4 13	9.94	7 51	3 97
2	4.38	4 07	7.41	3.30	3 29	5.65	3.36	3.27
3	2.2*	4 33	2.*1	2.70	3 19	3.94	2 44	2 45
4	3 83	2.98	3 42	4.30	3.*6	2.36	3 39	
5	5 49	5.49	6.09	6.24	6 51			
Mean.	4.94	4.98	6 80	5.14	4.34	6.09	6.06	4.94

* Rejected.

TABLE 149a.

Summary of special tri-hourly observations of relative humidity at Lone Pine showing diurnal variation.

Date.	3 a. m	6 a m	9 a m	Noon.	3 p m	6 p. m.	9 p. m.	Mid night.
	Per ct.	Per ct.	Per ct.	Per ct	Per ct.	Per ct.	Per ct.	Per ct.
1881. Aug. 15	......	......	......	15.9	12 1	14 1	19 0	21 8
16	45 *	52.2	34 1	16.7	13.3	25.0	50.5	34 0
17	64 0	76.8	55.6	14 9	14 2	27 9	56 0	42 3
18	39 3	51.3	29 4	20 3	11.1	31.7	78 0	51 9
19	55.3	53.9	31.4	17 8	14.4	21.2	71.5	41 8
20	39.6	56.4	32 2	18.2	11.*	23.8	35 2	34 6
21	32.5	51 1	35.3	16.0	17 4	22.1	35 4	......
22	60.0	57.1	35.0	18.1	11.5	25 2	70.1	63.7
23	62.2	55.9	35 3	16.4	10 4	17 9	58 3	56 7
24	56 1	75.0	22 1	14.0	13.3	16 2	29 0	50 7
25	42.1	4* 3	27 2	18.1	13.1	*56 5	29 6	27.0
26	56 2	50.7	33.5	16.3	9.5	12.4	2* 8	13 3
27	24 8	36 9	24 8	13.3	17 0	65.7	68 0	61.8
28	......	49.8	35 6	17.7	14.5	23 1	32.4	34 6
29	65.8	50.0	32 0	15.6	11.2	*64 5	23 2	1* 2
30	28.0	33.7	35.5	20.3	18.1	27.0	53 1	64 3
31	54.1	59 4	28 7	11.2	12.9	55 2	45 7	74 2
Sept. 1	44 9	45 6	19.7	19.3	12 9	59.5	56. 6	27.5
2	46.9	48 4	32 0	9.9	8 3	36.3	17 0	17 5
3	12.4	27.0	11.4	8.7	10.6	15.0	13.5	17.3
4	28.9	21 1	29 5	17.8	14.6	10.1	16 6	......
5	49 7	42.4	28 0	22.0	21.6	......		
Mean ...	46.5	49.2	29.0	16.6	13.3	27.5	42.4	40.9

* These observations are rejected.

TABLE 150.

Summary of psychrometer determinations of force of aqueous vapor at Lone Pine showing influence of wind.

Calm.	Light.	Gentle.	Fresh.	Brisk.	Gale.
m m.	m m.	m m.	m m.	m m.	m m.
9 46	9.20	12 58	8.83	7 08	12 29	3.30
13.24	9.22	10 *1	6.92	6.00	4 71	2.72
16.65	9.24	13.99	5.04	6.00	5.94	4 14
11.61	8 50	10.0*	6.71	4.7*	4 13	......
9.08	5.03	13.00	4 89	5.34	3.82	......
7 32	6.13	4.09	3.76	4.0*	4 89	......
8.20	6.27	7.96	8.28	6.63	3.54	......
5.53	7.15	7.43	4 56	4.65	4 48	......
5 40	4 48	7.72	4.82	6.30	3 83	......
5 70	7.85	4.88	4.79	6.06	3 69	......
5.72	......	8.01	6.11	4 43	2.85	......
5.82	......	7.*0	5.07	6 30	3.50	......
6.52	......	*.02	7.12	4 0*	2.*9	......
6.*7	......	*.09	4.27	4.98	5.59	......
7.84	......	4.79	......	5.5*	4 81	......
	......	6.55	......	......	......	
Mean. 8.59	7.54	8 54	5.72	5.84	3.39	

*Note as to wind the * at end. Definions.*—In estimating the wind velocity the following terms have been used. Calm indicates no perceptible wind" light indicates a velocity of from 1 to 2 miles per hour, gentle, 3 to 5 miles, fresh, 6 to 14 miles, brisk, 15 to 21 miles, gale indicates a velocity of more than 21 miles per hour. It will be remembered that the velocity is only estimated and the word variable always refers to velocity and not to direction.

It is impossible to estimate how far the apparent reduction of atmospheric moisture with increasing wind, shown in Table 150, may be due to the influence of air currents about the instrument, and how far it may be owing to the coming of the ordinary winds from a very dry quarter, since there was no record kept of the direction of the wind at Lone Pine.

TABLE 151.

Summary of observations of atmospheric moisture made at the Mountain Camp.

Date.	Force of vapor			Relative humidity		
	8ʰ20ᵐ a. m.	12ʰ55ᵐ p. m.	8ʰ15ᵐ p. m.	8ʰ15ᵐ a. m.	12ʰ55ᵐ p. m.	8ʰ15ᵐ p. m.
1881.	m.m.	m.m.	m.m.	Per cent.	Per cent.	Per cent.
Aug. 22	....	2.01	3.66	....	14.7	46.1
23	0.42	1.60	2.75	4.4	12.6	39.9
24	1.35	2.58	1.98	15.4	20.3	28.9
25	1.47	2.32	2.11	17.9	19.0	32.3
26	2.34	2.31	1.39	33.2	21.8	22.1
27	2.58	2.56	3.11	51.3	23.8	57.8
28	2.41	2.46	2.34	32.3	24.5	44.3
29	2.11	2.74	3.63	28.2	26.7	67.5
30	1.79	2.36	2.08	23.6	29.7	34.3
31	4.10	2.16	2.48	55.8	17.7	40.9
Sept. 1	1.21	2.87	2.17	14.8	24.7	37.3
2	1.96	....	....	26.9	....	....
Mean ...	1.96	2.35	2.52	27.6	20.6	40.9

TABLE 152.

Effect of wind upon the force of vapor observed with the psychrometer upon the mountain.

[Station, Mount Whitney Camp. Date, from August 22 to September 2, 1881. Observer, J. J. N.]

	Variation with velocity of wind.				Variation with direction of wind.				
Calm.	Light.	Gentle.	Fresh.	Gale.	N.	N W.	W.	S E.	E.
m.m.	m.m.	m.m.	m.m.	m.m.	m.m.	m.m.	m.m.	m.m.	m.m.
2.11	2.66	2.46	2.08	....	2.32	2.01	2.87	0.42	1.35
3.63	0.42	2.31	2.58	....	2.11	1.66	....	1.60	2.11
2.48	1.60	....	2.52	....	2.68	2.58	....	2.75	....
2.17	2.75	....	3.11	....	....	1.98	....	....	....
....	1.35	....	3.66	....	....	1.47	....	....	....
....	1.39	....	2.74	....	....	2.31	....	....	....
....	1.47	....	2.06	....	....	2.21	....	....	....
....	2.34	....	2.87	....	....	1.39	....	....	....
....	2.11	....	....	....	....	2.88	....	....	....
....	2.58	....	....	....	....	2.56	....	....	....
....	3.11	....	....	....	....	3.11	....	....	....
....	2.11	....	....	....	....	2.48	....	....	....
....	2.38	....	....	....	....	2.36	....	....	....
....	1.79	....	....	....	....	2.74	....	....	....
....	2.08	....	....	....	....	1.79	....	....	....
....	4.10	....	....	....	....	2.36	....	....	....
....	2.16	....	....	....	....	4.10	....	....	....
....	1.21	....	....	....	....	2.16	....	....	....
....	1.96	....	....	....	....	1.23	....	....	....
....	....	....	....	....	....	1.96	....	....	....
Mean 2.60	2.18	2.29	2.48	....	2.17	2.37	2.87	1.59	1.73

Though the psychrometer observations at Lone Pine were sufficient, those on the mountain which are grouped together in Table 152, are too few, and the results too unequally distributed, to base any conclusions upon them as to the influence of wind upon the psychrometer. The most that can be said is this:

The Mount Whitney station was at the bottom of an immense amphitheater of rock, rising to the height of 1,200 feet on the east, on the north, and on the south, and only open on the west. A glance at the map and frontispiece will enable the reader to realize the influence which these surroundings must have exerted upon the direction of the wind. At the mountain camp, west of the summit of Mount Whitney, during the latter part of August, 1881, northwest winds prevailed, being generally light, of a force not exceeding "fresh," the strongest winds blowing always in the middle of the day, as may be seen from the table, where the noon observations are marked with a small *n*. It is probable that the large surface of rock, strongly heated by the sun, produced a

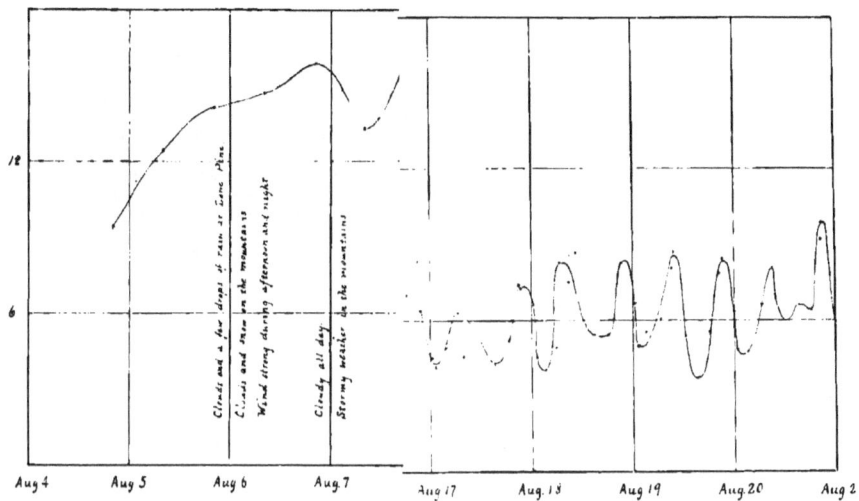

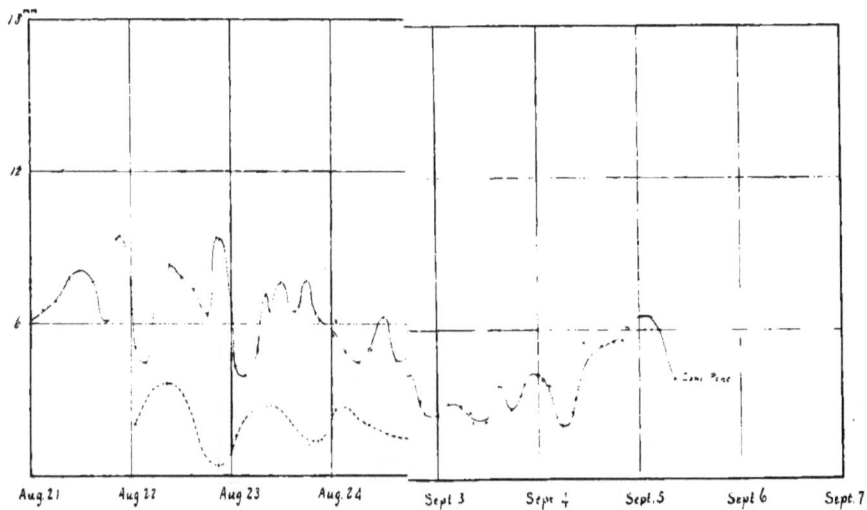

powerful indraught towards the summit of the mountain at mid-day. On the few occasions when southeast winds blew they seem to have been drier than those from the northwest.

It will be seen by an inspection of the tables and of the corresponding curves (Plate XVI, that the force of water vapor on the mountain usually increases during the day. Apparently, after sundown it begins to decrease, and has ordinarily fallen somewhat by the time of the evening observation at 8.15 p. m.

The relative humidity is lowest at the noon observation and highest in the evening. No observations were made before sunrise at the mountain camp, but the tri-hourly readings during a single night at the peak showed the highest relative humidity in the early morning. The curves of relative humidity (Fig. 15) at the upper and lower stations are in tolerably close agreement, both as to form and numerical value; but the *absolute* humidity, or force of vapor, is totally different in amount and in the law of its variation at the two stations.

Fig. 15.

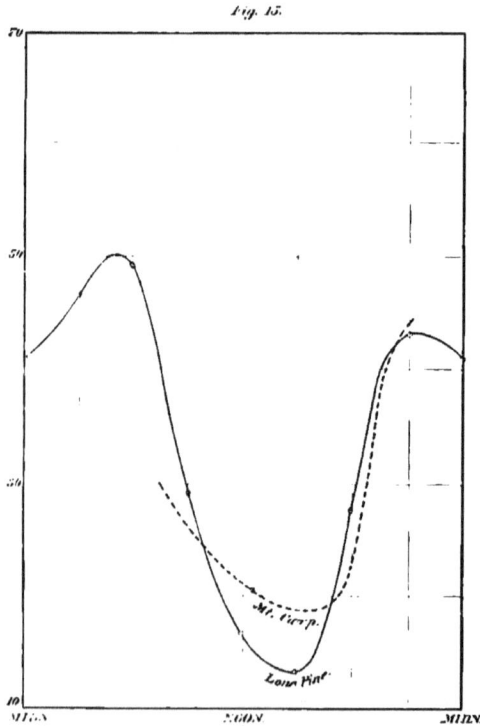

Diurnal Variation of Relative Humidity.

powerful indraught towards the summit of the mountain at mid-day. On the few occasions when southeast winds blew they seem to have been drier than those from the northwest.

It will be seen by an inspection of the tables and of the corresponding curves (Plate XVI), that the force of water vapor on the mountain usually increases during the day. Apparently, after sundown it begins to decrease, and has ordinarily fallen somewhat by the time of the evening observation at 8.15 p. m.

The relative humidity is lowest at the noon observation and highest in the evening. No observations were made before sunrise at the mountain camp, but the tri-hourly readings during a single night at the peak showed the highest relative humidity in the early morning. The curves of relative humidity (Fig. 15) at the upper and lower stations are in tolerably close agreement, both as to form and numerical value; but the *absolute* humidity, or force of vapor, is totally different in amount and in the law of its variation at the two stations.

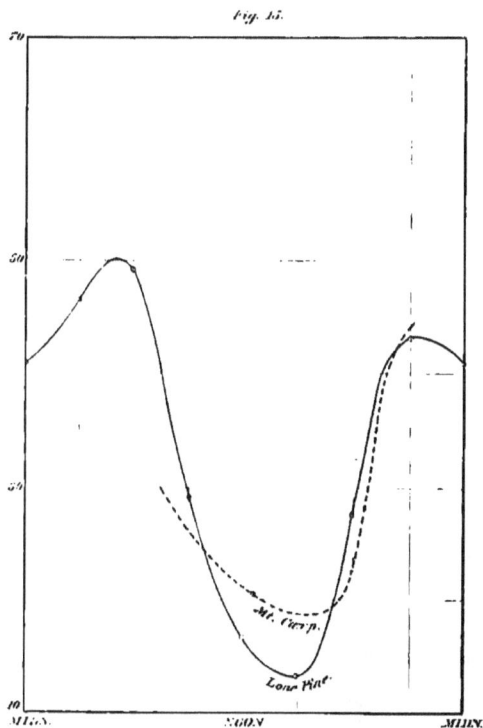

Fig. 15.

Diurnal Variation of Relative Humidity.

Attention is particularly called to the curve showing diurnal variation of moisture at Lone Pine (Fig. 16). It is the result of tri-hourly observations which, individually considered, are fairly concordant, and which extend over a period of three weeks, sufficient, it is imagined, to eliminate all abnormal variations. The curve should also be compared with that for each day (Plate XVI). It will be seen that during the night, from midnight to sunrise, the force of vapor remains nearly constant. Upon the rising of the sun it rapidly increases, usually attaining its greatest development about 9 a. m., after which it diminishes until the middle of the afternoon, becoming smallest at the hottest hour of the day. The moisture after this time progressively *increases*, reaching a second maximum after sundown, and then decreases again until midnight.

The curve, which is entirely different from the diurnal variation observed at sea or in moist climates, is confirmatory of the results obtained by other observers in hot and dry climates, and may be considered characteristic of such.*

Quite different is the curve of absolute humidity at the mountain station, which exhibits a continual rise during the hours of sunshine, if we may be permitted to draw conclusions from

Fig. 16

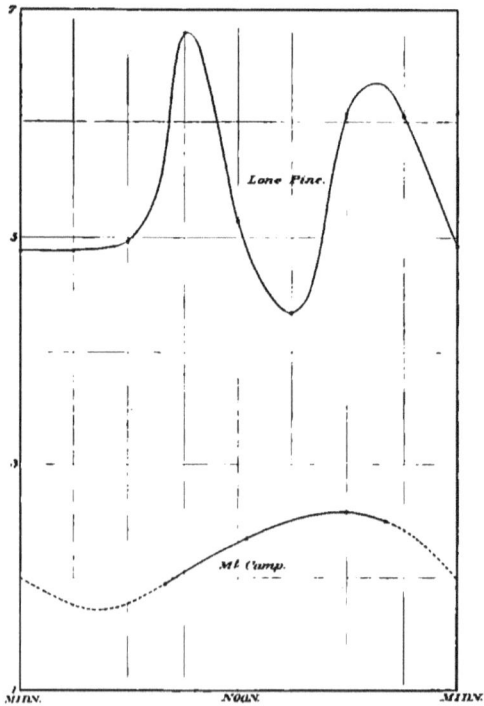

Diurnal Variation of Tension of Aqueous Vapor.

* Compare Blanford, " Indian Meteorologist's Vade-Mecum," page 110, where a very probable explanation of this effect is given. Quoted in this report, page 186.

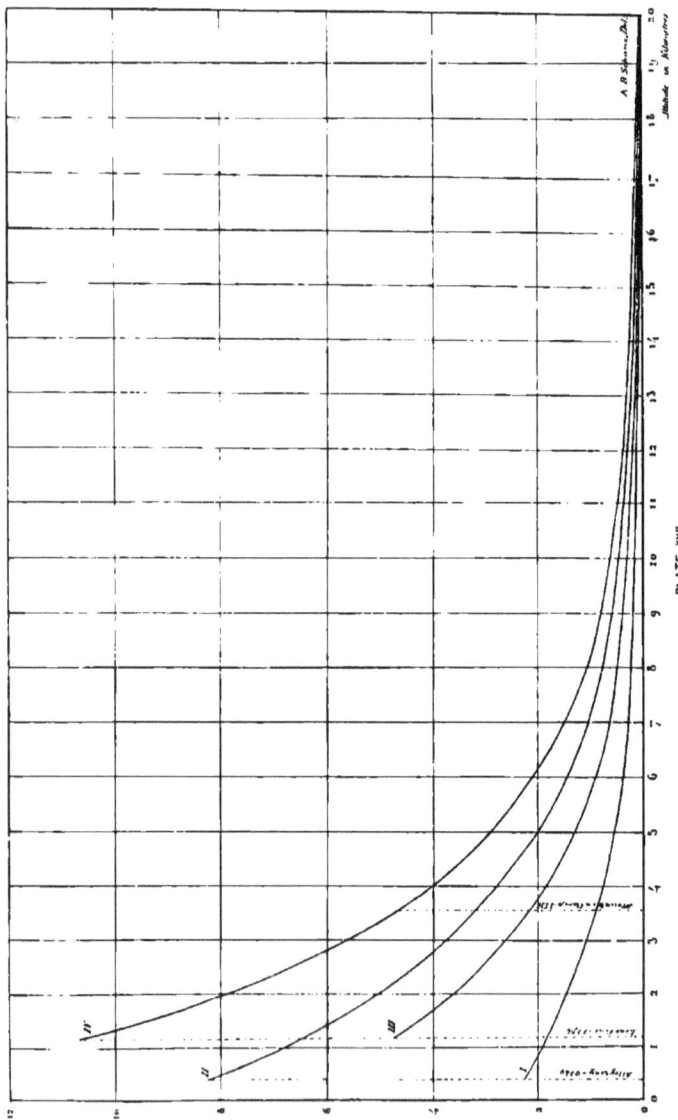

PLATE XVII.

AQUEOUS VAPOR IN THE ATMOSPHERE.

the comparatively small number of observations at our disposal. This result is probably to be explained by the transference of water vapor from the lower regions of the atmosphere to the higher by diffusion and convection—motions which are largely produced by the heating effect of the solar rays. If the relative humidity of the upper air does not also increase, it is because the source of supply at the surface of the desert plains is too limited to counteract the very great increment in the capacity of the air for retaining moisture, produced by its rapid rise of temperature. It will be noticed that the *range* of variation of relative humidity is smaller at the upper station than at the lower. In a moister climate this midday decrement of relative humidity at high altitudes tends to disappear, and may even be changed to an increment. The tendency to increased cloud-formation and rainfall in the afternoon has been noted by many meteorologists. It, no doubt, indicates a corresponding law of diurnal variation in the relative humidity of the cloud-bearing layers of the atmosphere. That no such law was observed on Mount Whitney is again to be attributed to the extraordinary dryness of the climate. The transference of moisture from the lower layers of the atmosphere to the higher by diffusion and convection, which, in a nearly saturated atmosphere, might produce cloud and precipitation, is here powerless to effect more than a slight lessening of the desiccation produced by the midday sun. It is worthy of note that, during the latter part of August and early in September, no such thing as a cloud-bearing stratum of air appears to have existed at any altitude. With the exception of smoke from forest fires, not the slightest visual obscuration of the sky could be detected. This fact is of importance in any estimation of the probable quantity of water existing as vapor in the atmosphere above Mount Whitney; for, since the air-layer between Lone Pine and Mount Whitney is but one-fourth of the entire air-mass above the lower station, we remain in comparative ignorance of the hygrometric condition of the larger part of the atmosphere. It is conceivable, therefore, that a layer of moist air might exist, unknown to us, superposed on the dry one; but when we know that for weeks not even the faintest streak of cirrus was visible, such an assumption appears most improbable, and we are justified in expecting a comparatively regular decrease of moisture with the increase of altitude. I have endeavored, on this hypothesis, to obtain an approximate notion of the amount of water contained in the air above Lone Pine and Mount Whitney at the time when observations of atmospheric transmission were being made with the spectro-bolometer. It will be seen by reference to Plate XVI that the atmospheric conditions were quite different at these two epochs. From August 5 to 14 a moist atmosphere prevailed. Clouds, and even a few drops of rain, were formed at Lone Pine, where, from the average of observations at noon on August 11, 12, and 14, the weight of vapor per cubic meter of air may be assumed equal to 10.48 grammes.

The greatest dryness occurred from September 2 to 4. During the bolometric observations on Mount Whitney of September 2 and 3 the average weight of vapor per cubic meter at noon was 2.05 grammes. Synchronous midday comparisons from August 22 to September 1 showed that the force of vapor at Lone Pine was usually about 2.2 times as great as at the lower camp, Mount Whitney. With these data the curves in Plate XVII were drawn, in which abscissæ represent altitudes, and the ordinates give the probable weight of vapor in grams per cubic meter at each altitude up to the height where the curves coincide with the axis (at about 20 kilometers) for the two epochs.* The areas included between the curves and the ordinates at any two altitudes are, then, proportional to the total moisture in the included atmospheric layers.

The most common way of expressing the amount of water vapor in the atmosphere is to state the pressure which it would exert upon a barometric column. As thus determined at a particular point in the atmosphere, it is independent of the quantity present in higher or lower strata. If, therefore, we would know the total quantity of water vapor in the atmosphere, it is necessary to

* It is believed that the allowance made for water vapor in the higher atmosphere is not excessive, as a very much slower decrement has been observed on several occasions by Glaisher in his balloon ascents. For example, in that of July 17, 1862, Glaisher found a dew-point of 21° Fahr., corresponding to a force of vapor of 3.27 mm, at a height of 6 kilometers, the dew-point at the surface of the ground being 55° Fahr. and the force of vapor 11.00 mm. In this instance, a current of warm moist air was superposed over a cold one, and for a short space the usual decrement was reversed, but there is reason to believe that this condition is not very unusual. The distribution of moisture during the Mount Whitney observations, however, was probably more nearly like that found by Glaisher in his third ascent September 5, 1862. (See "Report of the British Association for the Advancement of Science" for 1862, pages 463 and 464.)

make some assumption as to its distribution, which has been done in the above instances by assuming all moisture to cease at a height of 20 kilometers, and drawing smooth curves through the points of observation. The quantity of water present in the form of vapor in a layer of air of a given thickness can be best expressed by stating the depth of liquid water which would result from its condensation; and since a stratum of air 1 kilometer thick, containing, on an average, 1 gramme of water vapor per cubic meter, would give a liquid layer 1 mm. deep, if the water were all condensed, we may conveniently take this quantity as our unit.

Calling h''' the height above sea-level at which water vapor practically ceases (say, 20 kilometers),

h'' the height above sea-level of the upper station (Mountain Camp),

h' the height above sea-level of the lower station (Lone Pine),

w the average weight, in grammes, of water vapor per cubic meter = millimeters of water capable of being precipitated from a layer 1 kilometer thick,

we have, approximately,

Liquid layer from atmosphere above Lone Pine . . . . . . = w' $(h''' - h')$
Liquid layer from atmosphere above Mountain Camp = w'' $(h''' - h'')$

These quantities, as graphically determined by the method just described, are as follows :

Above Lone Pine, August 11 to 14 . . . . w' $(h''' - h')$ = 30.6 mm. precipitable water.
Above Mount Whitney Camp, September 2 to 3, w'' $(h''' - h'')$ = 6.0 mm. precipitable water.

THE CONNECTION BETWEEN ATMOSPHERIC MOISTURE AND GENERAL SELECTIVE ABSORPTION.[*]

We are led to believe that water vapor is an efficient agent in modifying the solar radiation by three classes of observation :

First, by comparisons of observations at different seasons.
Second, by comparisons of observations at different hours of the day.
Third, by comparisons of observations at different altitudes above the sea-level.

As an instance of the first, we take observations already made at Allegheny.

When we compare observations made in the winter with those of the spring, the sun being at the same altitude and the air-masses the same, we find certain rays most absorbed when the moisture is greatest. Apparently, therefore, these rays are especially cut off by the absorption of water vapor or by the action of some substance whose amount varies synchronously with the atmospheric moisture and in nearly the same proportion; or, what is exceedingly improbable, the composition of the solar radiation which we receive is itself variable within notable limits. The last hypothesis may be dismissed without further consideration, and we shall, for the sake of illustration and *in this preliminary consideration merely*, assume that this part of the atmospheric absorption, which exhibits a seasonal variation, is directly produced by water vapor, and that the law of extinction for this substance is the same as that deducible from Melloni's experiments on the transmission of lamp-radiation by liquid water. (See " La Thermochrose," Table IV, pp. 206, 207.) By applying to Melloni's figures a modification of Bouguer's formula for transmission, $y = p^x$, in which p still denotes the transmission by a layer having a thickness of unity, but in which x, instead of representing the number of such layers, is some function of the thickness, the following values of the exponents are obtained :

TABLE 153.

Thickness ...	1	2	3	4	5	6	7	8	9	10	50	100
Exponents	1.00	1.16	1.26	1.34	1.40	1.46	1.51	1.56	1.60	1.63	2.27	2.64

* We here use the word absorption in its most general sense, and intend it, in the absence of a better word, to cover every process of reflection, diffusion, or other interruption, by which the ray is hindered on its passage to us. Messrs. Lecher & Pernter assert that absorption of heat by our atmosphere is chiefly due to carbonic acid associated with the vapor of water, and not to the pure vapor itself. For our present purpose, we are not called on to affirm or deny their statement; for, by vapor of water, in atmospheric moisture, we here mean, it will be noticed, that whatever is at all times and places associated in nature with water vapor and varies synchronously with it (if there be such association), shall be held to be water vapor in our present sense.

A very similar law of extinction is deducible from Melloni's observations on the transmission of lamp radiation by rock-crystal, glass, and other substances; and although it would be desirable to repeat his experiments with homogeneous and solar radiations, and tubes filled with water vapor, these numbers will, nevertheless, serve to illustrate the method, which is now merely a tentative one.

Let the coefficient of transmission of homogeneous rays, of wave-length λ, by 1 kilometer of water vapor at an average density such that 1 cubic meter shall contain 1 gramme of water, be denoted by W_λ; h being the height of station above sea level in kilometers; h''' the height above sea-level (in kilometers) at which water vapor practically ceases; w_i, the depth of water (in millimeters) capable of being precipitated from a layer one kilometer thick, at the average humidity of the air column between h and h''' in winter; w_{ii} the corresponding depth in spring; and z_i, z_{ii}, the sun's zenith distances at these seasons. Then the seasonal change of transmission, represented by the ratio of galvanometer deflections,

spring (d_2)
winter (d_1)

if due to unequal prevalence of water vapor, may be expressed by

$$(W_\lambda)^{w_{ii}(h'''-h \cdot \sec z_{ii}) - w_i (h'''-h \cdot \sec z_i}$$

or (W_λ), raised to a power whose exponent is the difference between certain functions of the quantities in brackets, which are here taken from the above table (153). In order to eliminate the effects produced by other atmospheric absorbents, which vary as the air mass, it is desirable to combine measures made at nearly the same zenith distance, or for which z_i and z_{ii} are as nearly as possible identical. Hence we select spring observations, made when the sun was at a distance from the meridian, for comparison with winter ones taken at noon. Also, since the earth's distance from the sun has changed considerably in the interval, a correction must be applied to eliminate the effect of this variation, which is here accomplished by reduction to the earth's mean distance ($\rho = 1$).

We have the following data: Winter of 1881, average $z_i = 57° 02'$; sec $z_i = 1.84$. Spring of 1881, average $z_{ii} = 56° 13'$; sec $z_{ii} = 1.80$. Average force of vapor (winter) = 2 mm.; spring = 8 mm. Average weight of vapor per cubic meter (winter) = 2.20 grammes; spring = 8.23 grammes. Average depth of precipitable water (winter) = 7.3 mm.; spring = 26.3 mm.

The depths of precipitable water have been obtained by measuring the areas of the curves in Plate XVII, which involve assumptions as to the distribution of moisture in the upper air, which have already been alluded to.

The formula by which coefficients of transmission have been calculated, are

Winter of 1881, $(W_\lambda)^{w_i (7.3 \cdot 1.84)} = W_\lambda^{1.76}$ (by Table 153)

Spring of 1881, $(W_\lambda)^{w_{ii} (26.3 \cdot 1.80)} = W^{2.24}$ (by Table 153)

and

$$\frac{W_\lambda^{2.24}}{(W)^{1.76}} = (W_\lambda)^{0.54} = \frac{d_2 \text{ (spring)}}{d_1 \text{ (winter)}}$$

TABLE 154.

Transmission, probably in some way connected with atmospheric moisture.

[From observations at different seasons.]

λ	Winter d_1.	Spring d_2.	Winter reduced to $\mu=1$.	Spring reduced to $\mu=1$.	W
.375	192.6	71.5	187.5	72.5	.17
.400	301.4	119.8	353.9	121.4	.14
.450	279.3	275.6	501.1	279.6	.27
.500	767.8	309.1	747.8	374.3	.28
.600	724.9	439.9	705.8	445.2	.15
.700	547.9	433.9	534.0	410.0	.75
.800	336.3	298.5	326.5	302.7	.85
.900	215.1	191.1	209.7	191.2	.87
1.000	173.6	166.4	162.9	168.9	1.00

* This table is given in order that the reader may decide for himself how far atmospheric moisture influences the transmission of solar radiation.

The above values of an atmospheric absorption, dependent in some way upon water vapor, cannot be regarded as possessing an absolute significance; but there seems hardly room to doubt of the existence of such an absorption, or of the fact that it increases progressively from wave-length 1.°0 to 0.°4.

Secondly, we find on comparing actinometer observations made in the morning and afternoon with equal altitudes of the sun, and hence with nearly equal air-masses, that the measured radiation is greatest when there is least moisture.

The very numerous Allegheny observations taken with the same altitude of sun (morning and afternoon) are inconclusive on this special point, owing to the great irregularities of its sky. The independent ones in the clear atmosphere of Lone Pine and Mount Whitney agree in showing that, for the same altitude of the sun and the same air-mass traversed, the total heat absorption, as indicated by the actinometer, increases with the amount of water vapor in the air. (The extremely minute barometrical change between morning and afternoon evidently cannot account for the effect.)

The low relative humidity of the desert climate and the almost complete absence of haze or cloudiness at all hours of the day render these observations uncommonly well fitted to decide the question as to the influence of the absolute humidity upon atmospheric absorption, since they are freed from all complications which the capricious skies of moister climates introduce. It must, however, be remembered that we are here concerned with the entire quantity of moisture contained in the atmosphere above the place of observation.

Both at Lone Pine and Mount Whitney the morning readings with the actinometer surpassed those taken in the afternoon with the same altitude of the sun (see Fig. 14). The discrepancy was, however, relatively greater for the mountain observations. The same thing has been noticed by other observers, and it is presumed that the effect is due to the increase of the absolute quantity of moisture in the upper atmosphere, produced by evaporation at the earth's surface with subsequent ascension of the water vapor by diffusion or convection currents, an increase which goes on as long as the sun shines.

At the earth's surface the law of diurnal variation of atmospheric moisture is different for land and sea, and is affected by various local causes. Thus on the arid desert around Lone Pine the vapor tension rises for the first hours of the forenoon, attaining a maximum at about 9 a. m. It then decreases until the hottest part of the afternoon (about 3 p. m.), after which it again increases until sundown. During the night the water vapor decreases until midnight, after which it remains constant until the rising of the sun again sets in action the process of distillation and diffusion. This diurnal variation is characteristic of an arid climate on the plains, and, according to Blanford ("The Indian Meteorologist's Vade-Mecum," p. 110, "it probably depends on the ratio between the rate of production of vapor on the one hand and its rate of removal on the other; the rate of diffusion varies as the square of the absolute temperature, and therefore by diffusion alone the removal of vapor will be accelerated, at least in that proportion as the temperature rises; while from a dry land surface, with little vegetation, the production of vapor may not increase even directly as the temperature; nay, may even fall after the more superficial moisture has been dissipated."

In the upper atmosphere, on the other hand, there is usually a gradual increase of moisture during the day, as is evident from Table 151 and an inspection of the lower curve, Fig. 16, showing force of vapor on Mount Whitney. "That the humidity of the cloud-forming strata of the atmosphere and in all probability the tension of vapor at comparatively moderate heights do not follow the same law of diurnal variation as in that stratum which rests immediately on the earth's surface, may be inferred conclusively from the observed diurnal variation of the cloud proportion and of the frequency of rainfall." (Ibid, p. 110.)

At Calcutta rainfall is most frequent from 1 to 3 p. m. Loomis "found a decided diurnal inequality in the rainfall at Philadelphia, showing a maximum about 6 p. m." (American Journal of Science, vol. CXI, p. 7.) He quotes from Kreil the results of ten years' observations at Prague, showing a maximum about 4 p. m.

The psychrometer observations on Mount Whitney were not sufficiently numerous to give any very reliable information concerning the diurnal variation of moisture at that high altitude; but, so far as they go, they indicate an increase of water vapor throughout the day.

If the want of symmetry between the two halves of a diurnal curve of radiation is thus due to the increase of moisture in the upper air with the hour of the day, it should nearly disappear in the cold winter weather when the absolute quantity of moisture becomes minute, and the evaporating power of the sun is diminished. This is what M. Crova has found. He says:

"During the winter beautiful days may be encountered at Montpellier, in which a series of observations can be continued under good conditions from morning to evening; * * * in these cases the horary curves of calories may present a symmetry so nearly complete that we may, without sensible error, consider it as perfect."

On the other hand, the almost constant character for summer days, he says, is a want of symmetry. The curves are "hardly ever symmetrical with relation to the ordinate which passes through true solar noon; they are generally more regular after noon than in the morning; the maximum is attained before midday, and the tangent at true solar noon inclines toward the afternoon." (Mesure de l'intensité calorifique des radiations solaires," p. 50.)

The diminution in the solar radiation penetrating our atmosphere after midday cannot be accounted for by any assumption of instrumental error; let us then consider it to be the effect of increased absorption by the vapor of water, and, assuming that the mountain observations give a fair approximation to the variation of moisture for the entire atmosphere, let us calculate the absorption produced by water alone. If this value should be found to agree with that obtained by other processes, it would lend confirmation to these, though in itself it cannot pretend to the possession of great accuracy.

The following figures giving values of solar radiation and atmospheric moisture are taken from smooth curves representing average results at Mount Whitney.

Sun 4 hours from meridian; ☉ decln. = +10 ; 𝓏 zenith dist. = 60 ; sec 𝓏 = 2.

		A. M.	P. M
Solar radiation	calories	1.86	1.72
Force of vapor	millimeters	1.98	2.
Precipitable water	do	6.4	8.3

from these

$$(W)^{\frac{56.5 \cdot 2}{}} = \frac{1.72}{1.86} = .925$$

and determined as before the values of (by table 153)

$$\log W = -.0339 ; \quad W = \log W = -.0339 \div .068 = -.1985 \quad W = .6174.$$

Here W is the coefficient of transmission for the entire or complex solar radiation.

Thirdly, some agent present in the air between the top and bottom of the mountain causes a greater absorption of heat, for a given air-mass at the lower station, than is produced by the same air-mass above the mountain (see Fig. 10). We know of no conspicuous agents which make the constitution of the lower air differ from that of the upper except water vapor and dust, and we do know that there is more of both in the same air-mass below than above. We will confine our present attention to the water vapor, and proceed as if it were the only agent.

Considering for example the difference in the radiation observed at Lone Pine and Mountain Camp with an air-mass of unity we have the following data. Radiation (average): Lone Pine r 1.75 calories; Mountain Camp, r' 1.91 calories. Approximate depth of precipitable water: Lone Pine, 13.7 millimeters; Mountain Camp, 6.0 millimeters.

Determining the values of the exponents by table 153 we have—

$$(W)^{\cdot 613.7} \div (W)^{\cdot 6.0} = .916 \qquad \log (W)^{0.71} = -.038 \qquad W = .6037$$

We now proceed to the comparison of homogeneous rays in the spectrum as measured by the bolometer at Lone Pine and Mountain Camp, in order to determine how these results are affected by the agent in question. Here, on account of the limited number of measurements, we are obliged to compare observations in which both the air-masses and the transmission of the unit of mass have changed, since the meteorological conditions have altered in the interval, as described on page 183. We must therefore consider not only the air between the stations but also that above the level of the higher. We may express the change in that portion of the transmission which is affected the aqueous components of the atmosphere by

$$(W_A)^{(\delta)\,[w'\,(h''' - h')\,\sec\zeta']} - {(\delta)\,[w'\,(h' - h'')\,\sec\zeta']}$$

where, however, the transmission is that pertaining to the difference of water masses above the stations and not necessarily to the water included between the stations. Besides this, as we have just said, there is a certain part of the absorption produced by atmospheric constituents, other than water, whose mass has also changed. At present, we are unable to separate these effects, and shall here provisionally consider the second part negligible.

To compare mountain observations on certain days with *hypothetical* valley results obtained by reducing those of other days by an arbitrary rule, which does not take into account the variation of the aqueous component of the atmosphere in the interval, can only lead to imperfect results. The *form* as well as the *area* of the energy curve has been changed by the new conditions of absorption, and both must be preserved for the present purpose.

The determinations of the energy of homogeneous radiations made with the spectro bolometer having only a relative value, the energy curves, drawn with these figures, must have their area adjusted to correspond with the more reliable indications of actinometric instruments, as is described elsewhere. For the present purpose, the ratio of measurements, made with the pyrheliometer on or near the dates of bolometric work, has been adopted as the criterion for bolometric reductions.

Below is given the solar radiation as determined by a water pyrheliometer, uncorrected for non-conductivity or loss by convection or imperfect absorption. The application of these corrections is here unnecessary, since we wish merely the ratio of measurements made with one and the same instrument.

At Lone Pine the radiation registered at noon, August 11, 1.253 calories; August 12, 1.141 calories; August 14, 1.225 calories; mean, 1.207 calories.

No pyrheliometer observations are recorded on September 2 or 3; but it is fair to assume that those made on September 1 and 5, under almost identical circumstances, furnish a close approximation to the results that might have been obtained on the former dates.

At the Mountain Camp the pyrheliometer gave at noon September 1, 1.447 calories; September 5, 1.462 calories; mean, 1.455 calories.

The areas of the energy curves have therefore been made conformable to the ratio $\frac{1.455}{1.207} = 1.206$, and the resulting ordinates are given in the second and third columns of the accompanying Table 155.

The values of sec ζ were, for Lone Pine, 1.08; for Mount Whitney, 1.15. And the exponent of W_A becomes:

$$(\varphi)\,[30.6 \times 1.08] - (\varphi)\,[6.0 \times 1.15]$$

which is equal to

$$2.05 - 1.51 = 0.54$$

by Table 153. The fourth column of Table 155 contains the coefficients of transmission, according to the assumption that the entire absorption is aqueous, computed for an amount of water vapor which, if condensed, would produce a layer 1^{mm} deep. In this result is included the effect produced by the layer of dry air between the stations, which is here provisionally assumed to be negligible.

This table is inserted in order that the reader may compare the results with those furnished by observations at different seasons and judge for himself how far atmospheric moisture may be considered to have affected them.

TABLE 155.

Transmission (possibly in some way connected with atmospheric moisture) from observations made at different altitudes.

	Mount Whitney	Lone Pine	W
.350	43.4	25.4	.37
.375	47.3	28.4	.39
.400	77.2	50.1	.45
.450	187.8	119.6	.37
.500	236.9	155.9	.42
.600	262.1	217.0	.58
.700	271.6	191.1	.70
.800	172.0	155.5	.84
1.000	108.2	100.2	.87
1.200	77.8	76.4	.97

We may succinctly repeat here with special reference to our third and present argument what has been already given in another connection.

The observations just cited are made by the pyrheliometer on the heat rays as a whole.

The observations with the spectro-bolometer discriminate between different spectral rays, and if they were numerous enough would show, by the comparison of those taken at Lone Pine and Mount Whitney camp, what particular rays the action of this water vapor has most affected. Unfortunately, the observations with the spectro-bolometer, under the difficulties of the expedition, are too few to settle so delicate a point as the one immediately in question. They do, however, bring independent testimony to the fact that in proportion to the presence of water vapor the heat radiation as a whole is diminished; and they give some indication, though not a conclusive one, as to the particular spectral rays which it has most affected.

We, draw, then the general deduction from all our preceding arguments that the previous comparisons, whether made between observations taken at different seasons of the year, at different hours of the day, or at different altitudes above sea-level, all point to the same conclusion, namely, that there is a large absorption of solar radiation which depends upon and increases with the prevalence of atmospheric moisture (as we have defined the word), and that this effect is most marked for the rays of *short* wave-length.

We shall not attempt to deduce any absolute values of aqueous transmission from the above measures.

CHAPTER XIX.

BAROMETRIC HYPSOMETRY.

INTRODUCTION.

The instruments used by the expedition in the barometric work were the three Signal Service barometers Nos. 1890, 2018, and 1935, the errors of which have been found to be +0.002, +0.002 and —0.008, respectively.

It was the intention, on leaving one of these instruments at Lone Pine, to establish a series of simultaneous readings at Mountain Camp and the Peak of Whitney with the other two. One of the barometers (1935) being injured, however, in its transit to Mountain Camp, synchronous observations between that point and the Peak were necessarily rendered impossible. Simultaneous observations were therefore obtained only between Lone Pine and Mountain Camp, and between Lone Pine and the Peak.

The persons engaged in barometric observations, with their initials as used for abbreviations in the tables, were the following: Capt. O. E. Michaelis, U. S. A. (O. E. M.); Sergeant J. J. Nanry, U. S. S. S. (J. J. N.); Sergeant A. C. Dobbins, U. S. S. S. (A. C. D.); Corporal H. Lamonette, U. S. A. (H. L.); and Mr. J. E. Keeler, Allegheny Observatory (J. E. K.).

The very trying observations at the Peak are due to Captain Michaelis, and Mr. J. E. Keeler, who volunteered this service, as well as to Sergeant Nanry.

A series of special tri-hourly observations had been organized for comparison between Lone Pine and the Peak, which was carried out efficiently at the former station by Sergeant Dobbins and Corporal Lamonette, who volunteered their services for this extra duty. It was found impossible to continue the same tri-hourly observations at the Peak without fire or shelter. From the eighteen observations obtained there, it will be seen, however, that if all are not absolutely synchronous with those at Lone Pine, all are so nearly so, that they may be treated as such, without sensible error, when we have interpolated values between the closely contiguous actual observations.

In the tables following, the original readings of the barometer are given for the three stations, with a synopsis of each set, as well as of the temperature and relative humidity, since the latter enters into the hypsometric formula of Bessel, employed in the reduction.

The altitude of Lone Pine above the sea level was first obtained by comparison with the stations of San Diego, and San Francisco on the sea-coast. Subsequently, through the courtesy of Mr. George Davidson, the value for Lone Pine was separately given from the levelings for the proposed Carson and Colorado Railroad through Inyo Valley. The heights of Mountain Camp and Whitney Peak are then severally referred to Lone Pine.

The general arrangement of the tables may be stated as follows: (1) Summaries of barometric, thermometric, and relative humidity readings at San Diego and San Francisco; (2) the same for Lone Pine; (3) the same for Mountain Camp; (4 the same for Peak of Whitney.

In the reductions the formula adopted is that of Bessel, with Plantamour's modifications in the values of the barometric constants, but others have been used for comparison.

This formula, as adopted by Guyot (Smithsonian Tables, D, p. 75), is as follows:

$$H - H = \frac{[\log B - \log B'] \times 398.25}{[1 - 0.0026257 \cos \varphi]} \times \frac{1.(1 + KT]}{397.25 - KT]} \times$$

$$1 - \frac{(a + a') \times 0.34807}{(397.25 - KT) \sqrt{BB'}} \times 10^{0.00028\, T - 0.00003\, T'}$$

where the various quantities have the following signification :

h = the elevation of the lower station, and
h' = the elevation of the upper station.
r = the radius of the earth.

$$H = \frac{rh}{r + h}$$

$$H' = \frac{rh'}{r + h'}$$

B = atmospheric pressure at lower station.
B' = atmospheric pressure at upper station.
L = constant barometrical coefficient depending on the relative density of the mercury and the air.
K = the coefficient of the expansion of the air.
T = the mean temperature of the layer of air between the two stations.
a = relative humidity at lower station.
a' = relative humidity at upper station.
φ = mean latitude of the two stations.

The formula was applied by means of Plantamour's tables, as given by Guyot.

I.—LONE PINE.

TABLE 156.

Comparison of means between sea-level and Lone Pine.

[Means of the 12ʰ 35ᵐ and 8ʰ 15ᵐ p. m. observations from August 17 to September 8.]

	Atmospheric pressure		Temperature		Relative humidity.
	In.	*mm.*	*F.*	*C.*	
SEA LEVEL.					
12ʰ 35ᵐ p. m..	29.534	760.21	66.85	19.33	0.645
8ʰ 15 p. m..	29.909	759.79	60.17	15.67	0.810
LONE PINE.					
12ʰ 35ᵐ p. m..		663.29	78.58		0.171
8ʰ 15 p. m		662.59	18.60		0.291

Applying Bessel's formula to these data we obtain from mean 12.35 p. m. observations, 3,921.50 feet; from mean 8.15 p. m. observations, 3,843.65 feet; general mean, 3,882.57 feet.

TABLE 157.

Results obtained by Hazen's tables* for comparison.

[Computer, F. W. V.]

Date	Results		Date	Results.	
	12ʰ 35ᵐ p. m.	8ʰ 15ᵐ p. m.		12ʰ 35ᵐ p. m.	8ʰ 15ᵐ p. m.
1881.	*Feet.*	*Feet.*	1881.	*Feet.*	*Feet.*
Aug. 17	3920	3965	Aug. 30	3880	3790
18	3880	3860	31	3890	3790
19	3820	3710	Sept. 1	3895	3860
20	3890	3920	2	3920	3885
21	3880	3760	3	4020	3940
22	3815	3750	4	3880	3835
23	3875	3815	5	3880	3770
24	3940	3915	6		3820
25	3910	3880	7	3720	3625
26	4140	4030	8	3795	
27	3820	3760			
28	3905	3855	Sums	8725	8030
29	3960	3905	Means	3896.6 +15.3	3820.4 +20.5

General mean . 3862.8 + 36.0.

*Professional Papers of the Signal Service, No. XI.

II.—MOUNTAIN CAMP.

TABLE 158.

Results by Bessel's formula.—Altitude above Lone Pine.

[Computer A. B. S.]

Date.	Results from observations at—		
	8ʰ 1ᵐ a. m.	12ʰ 35ᵐ p. m.	8ʰ 15ᵐ p. m.
1884.			
Aug. 22 ...	7935.3	8057.7	7729.2
23 ...	7901.0	8072.9	7734.2
24 ...	7876.5	7996.4	7753.0
25 ...	7830.7	7963.0	7713.9
26 ...	7848.9	8088.5	7712.7
27 ...	7967.8	7998.5	7670.7
28 ...	7910.7	7991.4	7679.5
29 ...	7960.3	8052.8	7732.5
30 ...	7925.2	8028.6	7681.7
Sept. 1 ...	7891.4	7996.2	7643.8
2 ...	7866.8	....	....
Sums	86454.6	88177.9	84852.9
Means	7850.6 ± 17.1	8016.1 ± 6.4	7713.9 ± 6.3

General mean, 7864.6 ± 102.6.

As the reader will observe, the great probable error in the general mean arises from the combination of observations at different hours, and which separately considered have, relatively, small errors.

III.—THE PEAK OF WHITNEY.

TABLE 159.

Results by Bessel's formula.—Altitude above Lone Pine.

[Computer, A. B. S.]

Date.	Local mean time.	Results.	Date.	Local mean time.	Results.
1884.		Feet.	1884.		Feet.
Sept. 2	6ʰ 00ᵐ p. m.	10759.37	Sept. 5	10ʰ 22ᵐ p. m.	10854.00
	9 00 p. m.	10683.85		Midnight	10648.40
	Midnight.	10923.90	6	1ʰ 00ᵐ a. m.	10544.90
3	3ʰ 00ᵐ a. m.	10685.85		3 00 a. m.	10549.30
	6 00 a. m.	10675.84		5 00 a. m.	10557.70
4	8 15 p. m.	10750.60		8 17 a. m.	10934.10
5	8 30 a. m.	10970.80		9 00 a. m.	11025.90
	12 40 p. m.	11233.80			
	5 07 p. m.	10903.00	Sum	....	193974.71
	6 30 p. m.	10769.90			
	8 20 p. m.	10823.40	Mean	....	10776.30 ± 27.4

The special mean of the first five results (which meet all requirements, inasmuch as those observations are exactly synchronous with the corresponding tri hourly ones at Lone Pine), with its probable error, is 10724.7 ± 29.8, and giving this last value double weight, we have: *Final mean*, 10762.

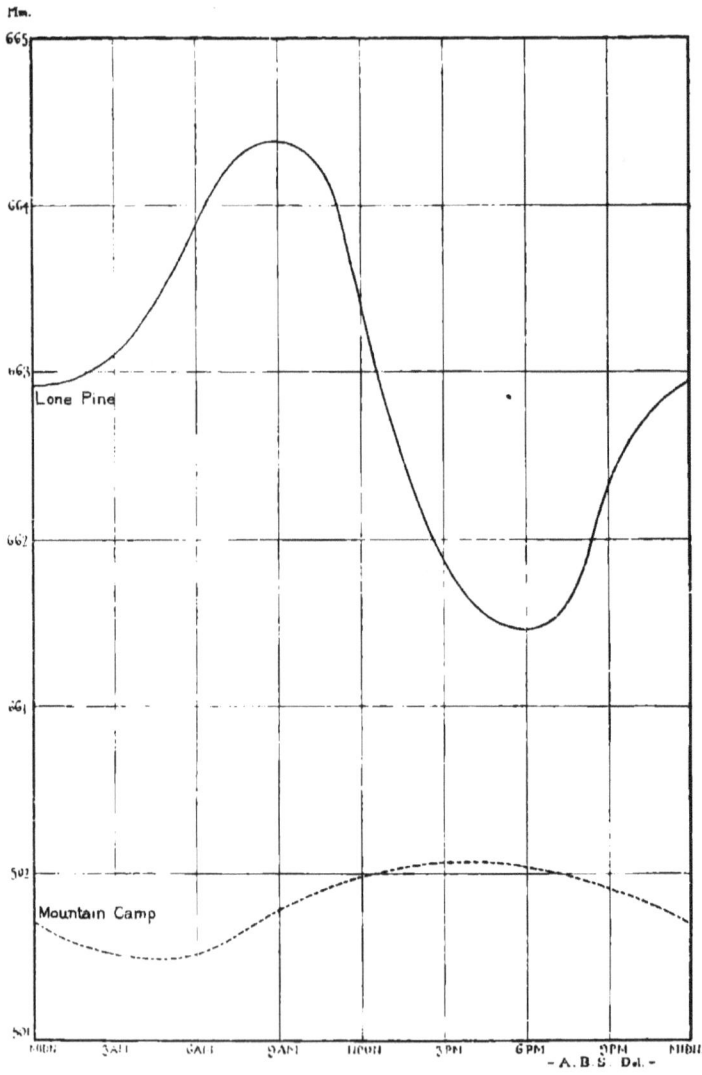

PLATE XVIII.
DIURNAL VARIATION OF THE BAROMETER.

TABLE 160.

Special results obtained by Loomis's tables for comparison.—Altitude of Whitney Peak above Lone Pine.

Computer, J. E. K.

Date	Local mean time	Results in feet
1881.		
Sept. 2	9 — p. m.	10,729.0
	6 — p. m.	10,685.5
	8 — a. m.	10,700.5
	12 — p. m.	10,98— .0
	8 29 p. m.	10,528.7
9	8 17 a. m.	10,910.5
Sum		63,625.0
Mean		10,564 . 71.3

In order to compare the different hypsometrical formulæ in use, the same data, obtained by observations on the same day and at the same hour, were reduced by the various methods.

The following table of results obtained will show the variation due to the different manners of treating the material:

TABLE 161.

Reductions of single observations September 2, 1881 9 p. m.). Altitude of Whitney Peak above Lone Pine.

Results.	Methods.	Computers.
10,611.8	Bessel's formula	A. B. Schanz.
10,671.0	Engineer's Prof. Pap. No. 12	Do.
10,40 .3	Belcos's tables	Do.
10,40 .4	Guyot's tables	Do.
10,729.0	Loomis's tables (Pr. Ast.)	J. E. Keeler.
10,50 .5	Loomis's tables (Smiths. Coll.)	A. B. Schanz.
10,613.7	Pappe's tables	Do.
10,80 .8	Bailey's tables	Do.

A point worthy of notice is the fact that the results from observations made at different times of the day by no means agree. If, for example, we take the reduction of Mountain Camp observations by Bessel's formula, we have for the 8.15 a. m. observation the mean altitude above sea-level (11 results) 11,509±17 feet; for the 12.35 p. m. observation, the mean (11 results) 11,836±6 feet; for the 8.15 p. m. observation, the mean (11 results) 11,565±6 feet (with the assumption Lone Pine altitude = 3,850 feet).

From our tables of relative humidity we see that both at Lone Pine and Mountain Camp the percentage of saturation is at a minimum at noon, while the morning reading is between the noon and evening readings. It might, therefore, be suggested that the relative humidity, besides its primary effect on the results, had a secondary influence due to its diurnal variation. In order to test this possibility, the means of the morning, noon, and evening observations at Lone Pine and Mountain Camp were again reduced by another method, viz, Hazen's table extended : Professional Papers of the Signal Service, No. VI.* The results came out : For 8.15 a. m., assuming the height of Lone Pine to be 3,850 feet, altitude of Mountain Camp above sea-level = 11,625 feet ; for 12.35 p. m., 11,775 feet ; for 8.15 p. m., 11,450 feet.

It becomes immediately apparent that these quantities, though separately less than the corresponding numbers above, bear almost exactly the same ratio to one another as the results by Bessel's formula, although Hazen's formula has no term depending on the humidity. It is therefore very possible that the temperature and relative humidity, though they must be taken into account, are not taken into account in the best way in our formulæ. The investigation of this problem is not, however, part of our purpose.

Another observation made in this case, as often before, is the small scope of the diurnal variation of the barometer on the top of a mountain in comparison with that at its base. This is readily shown by examination of Plate XVIII.

_____

* This table depends on a formula developed by Professor Angot, but modified by Hazen, from the results of observations at high and low stations.

Still another fact is brought out very prominently when the parallel series of observations at Lone Pine and Mountain Camp are plotted together, namely, that on the mountain the principal maxima and minima of atmospheric pressure fall from 6 to 10 hours behind the corresponding periods at Lone Pine. In general, the maxima appear to be less retarded than the minima, and the mean of 15 comparisons indicates that the retardation for a maximum amounts to one hour for an elevation of about 1,500 feet, while the minima fall an hour behind when the elevation is but 800 feet. Loomis (American Journal of Science, CXVII, p. 11) concludes from his comparisons, "that over the United States both maxima and minima of atmospheric pressure generally occur first near the surface of the earth, and they occur later as we rise above the surface, the retardation amounting to one hour for an elevation of from 900 to 1,500 feet." Here, however, Professor Loomis is referring to those larger maxima and minima which accompany storms. It is interesting to note the similarity of the phenomenon in both the larger and the smaller atmospheric fluctuations.

The following letter was subsequently received, further comments being superfluous:

DAVIDSON OBSERVATORY,
San Francisco, October 25, 1883.

MY DEAR SIR: Following up one trail and another, I am able to answer your question about the height of Lone Pine.

The top of the rail at Lone Pine station of the Carson and Colorado Railroad Company is 3,760 feet above the sea.

Yours, very truly,

GEORGE DAVIDSON.

If, therefore, we take the altitude of Lone Pine from barometric determinations, we have the altitude of Mount Whitney

10,762 feet + 3,883 feet = 14,645 feet

while, with the above value from recent railroad levelings furnished by Mr. Davidson, of the Coast Survey, we have the altitude of Whitney Peak

10,762 feet + 3,760 feet = 14,522 feet

The discrepancy between these two values is not at all surprising when we consider the distance from the ocean and the intervening elevations, but the latter value is probably the more trustworthy.

The adopted results, then, arrived at by this department of the reductions are

	Feet.
Lone Pine, above sea-level	3,760
Mountain Camp, above sea-level	11,625
Whitney Peak, above sea-level	14,522

Former results obtained for the height of the Peak are those of

Whitney	14,898
Captain Wheeler	14,448

TABLE 162.

Summary of barometric readings at San Diego and San Francisco.

(Reduced to sea-level. Observers, United States Signal Service. Computer, A. B. S.)

Date	12ʰ 35ᵐ p. m. Lone Pine M. T.			8ʰ 15ᵐ p. m. Lone Pine M. T.			Date	12ʰ 35ᵐ p. m. Lone Pine M. T.			8ʰ 15ᵐ p. m. Lone Pine M. T.		
	San Francisco.	San Diego.	Mean.	San Francisco.	San Diego.	Mean.		San Francisco.	San Diego.	Mean.	San Francisco.	San Diego.	Mean.
1884							1884						
Aug. 17	30.046	29.947	29.996	30.013	29.921	29.967	Aug. 31	30.068	29.958	29.998	29.981	29.898	29.939
18	30.001	29.908	29.956	29.966	29.901	29.965	Sept. 1	29.981	29.858	29.919	29.961	29.817	29.874
19	29.896	29.810	29.853	29.863	29.901	29.916	2	29.954	29.846	29.895	29.934	29.836	29.884
20	30.015	29.897	29.956	29.966	29.980	29.973	3	29.975	29.844	29.883	29.765	29.765	29.824
21	29.968	29.847	29.912	29.913	29.885	29.904	4	29.755	29.754	29.754	29.667	29.746	29.746
22	29.964	29.915	29.938	29.939	29.904	29.921	5	29.739	29.743	29.736	29.764	29.764	29.762
23	29.901	29.925	29.957	29.953	29.897	29.925	6	29.886	29.848	29.870	29.930	29.881	29.896
24	30.046	29.958	29.997	29.995	29.944	29.971	7	29.979	29.832	29.905	29.928	29.833	29.870
25	30.111	29.980	30.050	30.045	29.980	30.017	8	29.923	29.877	29.899	29.889	29.829	29.859
26	30.101	29.980	30.129	30.129	29.924	30.024							
27	30.160	29.885	29.992	30.003	29.840	29.921	Sums	268.493			687.905		
28	29.988	29.841	29.925	29.943	29.875	29.904							
29	29.941	29.909	29.925	29.928	29.885	29.906	Means	29.944			29.909		
30	30.040	29.971	29.999	30.015	29.960	29.987							

TABLE 163.

Summary of temperature and relative humidity at San Diego and San Francisco, Cal.

(Observers, United States Signal Service. Computer, A. B. S.

12ʰ 35ᵐ p. m.   8ʰ 15ᵐ p. m.

Date.	Temperature.			Relative humidity.			Temperature.			Relative humidity.		
	San Francisco.	San Diego.	Mean.	San Francisco.	San Diego.	Mean.	San Francisco.	San Diego.	Mean.	San Francisco.	San Diego.	Mean.
1881.	°F.	°F.	°F.	Per C.	Per C.	Per C.	°F.	°F.	°F.	Per C.	Per C.	Per C.
Aug. 17	62	72.0	67.0	72	57	64.5	56	65.4	60.7	90	78	84.0
18	61	72.5	66.7	75	62	68.5	55	65.3	60.1	91	78	85.5
19	62	73.0	67.5	72	64	68.0	55	66.0	60.5	91	84	87.0
20	61	72.5	66.7	75	66	70.5	55	67.2	61.1	90	78	84.0
21	59	71.5	65.2	70	66	72.0	55	65.0	60.0	86	84	85.0
22	62	71.3	66.6	70	55	63.5	54	65.1	59.0	80	84	82.0
23	59	71.5	65.2	75	66	70.5	55	66.0	60.5	86	73	79.5
24	61	72.0	66.5	72	62	67.0	56	65.0	60.3	80	75	75.0
25	60	72.0	66.0	67	57	62.0	50	65.2	62.1	88	84	85.0
26	60	71.0	65.5	72	57	64.5	53	64.5	58.7	68	84	72.0
27	65	70.3	67.6	54	57	55.5	54	67.0	62.5	80	60	70.0
28	65	73.5	69.2	58	59	58.5	55	63.3	61.7	77	74	77.5
29	62	71.3	66.6	67	57	62.0	57	64.9	60.9	70	74	71.5
30	67	70.5	68.7	55	57	56.0	55	64.3	59.6	83	78	80.5
31	62	69.8	65.9	67	57	62.0	57	64.1	60.5	83	78	80.5
Sept. 1	59	70.6	64.8	73	62	68.5	56	63.9	60.9	88	74	82.0
2	62	72.3	67.1	72	53	62.5	56	65.1	60.5	90	84	87.0
3	64	72.0	68.5	75	62	68.5	57	67.0	62.0	80	78	79.0
4	68	70.3	69.2	51	57	54.0	62	65.2	63.7	75	73	74.0
5	63	70.5	66.2	70	57	63.5	53	65.3	59.2	76	73	74.5
6	65	70.5	67.7	60	66	63.0	53	65.5	59.2	88	84	87.0
7	61	73.0	67.0	75	66	70.5	53	60.5	61.2	83	76	84.5
8	59	73.5	66.2	70	66	68.0	51	67.2	59.1	96	84	90.0
Sums		1553.0			1453.5			1396.7			1863.0
Means		66.85			64.50			60.17			81.0

TABLE 164.

Barometric observations at Lone Pine.

(Barometer No. 1690, Signal Service. Correction +0ˢ.002. Observer, A. C. D. Computer, F. W. V.)

Date.	Local mean time.	Original.		Barometer.				Date	Local mean time	Original.		Barometer.			
		Barometer reading.	Attached thermometer.	Corrected for instrumental error.	Correction for temperature.	Reduced to freezing point.	Reduced to millimeters.			Barometer reading.	Attached thermometer.	Corrected for instrumental error.	Correction for temperature.	Reduced to freezing point.	Reduced to millimeters.
		Inch.	Fah.	Inch.	Inch.	Inch.	mm.			Inch.	Fah.	Inch.	Inch.	Inch.	mm.
1881.								1881							
Aug. 5	5ʰ 17ᵐ a. m.	26.380	88.5	26.382	−.141	26.241	666.51	Aug. 19	12ʰ 45ᵐ p. m.	26.364	94.5	26.366	−.153	26.213	665.80
5	17 p. m.	26.271	82.4	26.273	−.127	26.146	664.09	19	15 p. m.	26.396	90.0	26.270	−.074	26.196	665.33
6	17 a. m.	26.360	82.0	26.362	−.140	26.223	666.03	20	15 a. m.	26.372	82.0	26.374	−.126	26.248	666.68
6	17 p. m.	26.327	82.0	26.329	−.128	26.201	665.50	20	15 p. m.	26.333	89.5	26.335	−.151	26.174	664.91
7	17 a. m.	26.386	85.0	26.388	−.133	26.255	666.67	20	15 p. m.	26.220	88.0	26.222	−.095	26.127	663.82
7	17 p. m.	26.321	77.5	26.323	−.115	26.208	665.67	21	15 a. m.	26.344	86.5	26.342	−.136	26.206	665.62
8	17 a. m.	26.372	94.0	26.374	−.154	26.220	665.90	21	12 p. m.	26.204	86.0	26.206	−.137	26.149	664.17
8	17 p. m.	26.226	73.6	26.270	−.106	26.124	663.54	21	8 p. m.	26.298	78.2	26.291	−.112	26.123	664.70
9	17 a. m.	26.300	97.0	26.302	−.161	26.141	663.97	22	15 a. m.	26.343	87.0	26.345	−.134	26.211	665.73
9	17 p. m.	26.162	83.5	26.164	−.129	26.035	661.28	22	12 p. m.	26.312	92.5	26.320	−.150	26.140	664.06
10	17 a. m.	26.263	85.5	26.265	−.157	26.108	663.13	22	15 p. m.	26.230	69.0	26.232	−.081	26.151	664.32
10	17 p. m.	26.129	74.0	26.131	−.107	26.024	661.00	23	15 a. m.	26.350	83.0	26.352	−.153	26.219	665.93
11	15 a. m.	26.276	94.5	26.277	−.155	26.123	663.51	23	12 p. m.	26.315	92.5	26.320	−.149	26.171	664.74
11	15 p. m.	26.151	75.3	26.153	−.110	26.045	661.48	23	15 p. m.	26.222	72.3	26.224	−.103	26.121	663.47
12	15 a. m.	26.310	84.5	26.312	−.155	26.157	664.38	24	15 a. m.	26.317	88.0	26.310	−.112	26.175	664.84
12	15 p. m.	26.150	76.0	26.152	−.111	26.041	661.48	24	12ʰ 35ᵐ p. m.	26.204	90.5	26.264	−.146	26.138	663.80
13	15 a. m.	26.212	94.1	26.214	−.154	26.060	661.91	24	15 p. m.	26.200	75.5	26.202	−.110	26.092	662.72
13	15 p. m.	26.073	79.5	26.073	−.119	25.956	659.27	25	15 a. m.	26.345	82.0	26.347	−.126	26.221	666.01
14	15 a. m.	26.240	91.2	26.242	−.148	26.093	663.23	25	12 p. m.	26.290	87.0	26.299	−.129	26.170	664.94
14	15 p. m.	26.119	77.5	26.121	−.114	26.007	660.57	25	15 p. m.	26.222	67.0	26.224	−.090	26.134	663.79
15	15 a. m.	26.250	89.6	26.252	−.143	26.109	663.16	26	15 a. m.	26.253	84.0	26.255	−.132	26.123	663.52
15	12 p. m.	26.210	82.0	26.212	−.148	26.061	662.01	26	12 p. m.	26.185	89.3	26.187	−.142	26.045	661.48
16	15 p. m.	26.127	78.0	26.129	−.110	26.019	660.87	26	15 p. m.	26.005	84.0	26.007	−.082	26.005	660.42
16	15 a. m.	26.276	86.0	26.279	−.131	26.144	664.04	27	15 a. m.	26.254	70.6	26.254	−.099	26.155	663.82
16	12ʰ 35ᵐ p. m.	26.256	89.5	26.258	−.143	26.115	663.31	27	12 p. m.	26.252	80.0	26.254	−.127	26.127	663.82
16	15 p. m.	26.104	72.8	26.106	−.102	26.002	660.30	27	8 p. m.	26.163	30.0	26.165	−.142	26.137	663.82
17	15 a. m.	26.272	86.0	26.274	−.136	26.192	665.26	28	12 p. m.	26.220	86.0	26.232	−.134	26.098	663.67
17	12ʰ 35ᵐ p. m.	26.208	88.5	26.300	−.141	26.159	664.45	28	15 p. m.	26.230	86.0	26.232	−.134	26.098	662.87
17	15 p. m.	26.114	74.2	26.106	−.103	26.003	660.57	29	15 a. m.	26.110	—	26.193	−.092	26.101	661.38
18	15 a. m.	26.312	87.5	26.314	−.126	26.188	665.16	29	15 p. m.	26.150	79.5	26.157	−.110	26.040	662.06
18	15 p. m.	26.290	89.3	26.392	−.142	26.150	664.20	29	12 p. m.	26.167	84.0	26.169	−.139	26.030	661.38
18	15 a. m.	26.230	64.4	26.232	−.084	26.148	664.14	30	15 a. m.	26.136	74.8	26.058	−.113	26.382	663.57
19	15 a. m.	26.385	81.9	26.387	−.123	26.202	667.04	30	15 a. m.	26.313	78.8	26.315	−.113	26.202	665.52

* Probably this reading should have been 26.188.

TABLE 164.

Barometric observations at Lone Pine—Continued.

		Original.		Barometer.						Original.		Barometer.			
Date	Local mean time.	Barometer reading.	Attached thermometer.	Corrected for instrument error.	Correction for temperature.	Reduced to freezing point.	Reduced to millimeters.	Date	Local mean time.	Barometer reading.	Attached thermometer.	Corrected for instrument error.	Correction for temperature.	Reduced to freezing point.	Reduced to millimeters.
		Inch.	*Fah.*	*Inch.*	*Inch.*	*Inch.*	*mm.*			*Inch.*	*Fah.*	*Inch.*	*Inch.*	*Inch.*	*mm.*
1881.								1881.							
Aug. 30	12 35 p. m.	26.303	83.6	26.305	—.129	26.176	664.86	Sept. 4	1 27 p. m.	26.044	79.8	26.046	—.119	25.927	658.54
30	8 15 p. m.	26.244	63.5	26.246	—.082	26.164	664.55	4	8 15 p. m.	26.034	73.0	26.035	—.103	25.932	658.15
31	8 15 a. m.	26.350	79.6	26.361	—.120	26.241	666.51	5	8 15 a. m.	26.100	82.7	26.102	—.126	25.976	659.78
31	12 35 p. m.	26.310	85.0	26.312	—.193	26.179	664.94	5	12 35 p. m.	26.079	87.6	26.081	—.138	25.943	658.94
31	8 15 p. m.	26.192	80.0	26.194	—.072	26.122	663.49	5	8 15 p. m.	26.112	74.4	26.114	—.107	26.007	660.57
Sept. 1	8 15 a. m.	26.310	80.0	26.312	—.142	26.170	664.71	6	8 15 a. m.	26.351	77.9	26.382	—.117	26.236	666.38
1	12 35 p. m.	26.223	86.5	26.225	—.136	26.093	662.65	6	12 35 p. m.						
1	8 15 p. m.	26.075	63.0	26.077	—.080	25.997	660.51	6	8 15 p. m.	26.270	85.5	26.272	—.064	26.204	665.07
2	8 15 a. m.	26.194	82.7	26.196	—.126	26.070	662.37	7	8 15 a. m.	26.430	80.5	26.432	—.123	26.309	668.24
2	12 35 p. m.	26.152	90.6	26.154	—.145	26.009	660.62	7	12 35 p. m.	26.369	86.2	26.371	—.136	26.235	666.36
2	8 15 p. m.	26.120	71.0	26.122	—.098	26.023	660.90	7	8 15 p. m.	26.292	86.0	26.254	—.065	26.189	665.19
3	8 15 a. m.	26.162	82.8	26.164	—.126	26.058	661.35	8	8 15 a. m.	26.381	86.0	26.383	—.135	26.248	666.64
3	12 35 p. m.	26.089	88.0	26.091	—.141	25.989	659.37	8	12 35 p. m.	26.411	87.7	26.313	—.139	26.174	664.81
3	8 15 p. m.	25.905	66.5	25.907	—.088	25.908	658.08	8	18 15 p. m.						
4	8 15 a. m.	26.102	78.6	26.104	—.117	25.987	660.06								

Omitted on account of illness of observer.   (Station closed by direction of Professor Langley.)

TABLE 165.

Summary of Lone Pine barometric readings.

[Corrected, reduced to freezing point and to millimeters. Observer, A. C. D. Computer, F. W. V.]

Date.	8ʰ 15ᵐ a. m.	12ʰ 35ᵐ p. m.	8ʰ 15ᵐ p. m.	Date.	8ʰ 15ᵐ a. m.	12ʰ 35ᵐ p. m.	8ʰ 15ᵐ p. m.
1881.	*mm.*	*mm.*	*mm.*	1881.	*mm.*	*mm.*	*mm.*
Aug. 17	668.26	664.43	662.37	Aug. 31	666.51	664.94	663.49
18	665.16	664.20	664.14	Sept. 1	664.71	662.65	660.51
19	667.04	648.80	665.36	2	662.37	660.62	660.90
20	666.68	664.01	663.82	3	661.35	659.37	658.08
21	665.62	664.17	664.79	4	664.86	658.54	658.15
22	665.75	664.96	664.22	5	659.78	658.94	660.57
23	665.95	664.74	663.47	6	666.38		665.07
24	664.84	663.89	662.77	7	668.24	666.36	665.19
25	666.01	664.94	663.79	8	666.64	664.81	
26	664.52	661.48	660.47				
27	664.33	663.62	663.82	Sums....	15289.17	14592.48	14577.01
28	664.74	662.67	661.18				
29	662.87	661.38	660.06	Means...	664.75	663.29	662.50
30	665.52	664.86	664.55				

TABLE 166.

Summary of Lone Pine thermometrical and hygrometrical observations.

[Observer, A. C. D. Computer, F. W. V.]

Date.	Dry bulb thermometer reduced to Centigrade.			Relative humidity by Smithsonian tables			Date.	Dry bulb thermometer reduced to Centigrade.			Relative humidity by Smithsonian tables.		
	8ʰ 15ᵐ a. m.	12ʰ 35ᵐ p. m.	8ʰ 15ᵐ p. m.	8ʰ 15ᵐ a. m.	12ʰ 35ᵐ p. m.	8ʰ 15ᵐ p. m.		8ʰ 15ᵐ a. m.	12ʰ 35ᵐ p. m.	8ʰ 15ᵐ p. m.	8ʰ 15ᵐ a. m.	12ʰ 35ᵐ p. m.	8ʰ 15ᵐ p. m.
1881.	°	°	°	*Per c.*	*Per c.*	*Per c.*	1881.	°	°	°	*Per c.*	*Per c.*	*Per c.*
Aug. 17	24.33	30.22	23.17	30.7	13.7	21.7	Aug. 31	19.28	27.11	15.17	42.9	16.0	44.3
18	22.22	29.61	14.78	36.7	14.6	39.3	Sept. 1	16.72	27.00	18.84	33.5	23.8	35.2
19	23.17	31.56	14.50	76.5	14.1	65.3	2	23.47	31.00	26.94	30.8	11.5	17.8
20	22.72	33.06	20.78	38.0	12.7	44.0	3	24.61	29.61	20.94	14.4	8.4	15.3
21	25.11	32.89	26.83	38.8	16.4	31.5	4	18.83	25.25	22.44	30.4	17.0	13.5
22	24.17	31.80	17.56	41.0	15.2	47.0	5	22.33	27.22	29.72	27.0	24.4	20.2
23	25.00	31.50	22.06	30.3	12.4	35.8	6	18.72	.....	11.50	30.0	.....	67.0
24	26.28	30.33	24.89	38.7	20.3	30.0	7	19.50	26.96	11.01	38.4	18.2	77.0
25	22.61	28.80	20.39	41.1	17.0	30.0	8	20.83	29.06	.....	42.9	18.6	.....
26	24.44	30.17	18.44	39.7	11.4	32.3							
27	18.44	24.67	11.78	38.4	21.7	51.6							
28	19.17	26.39	15.89	38.8	28.9	60.1	Sums	505.15	639.39	499.28	792.2	376.4	859.6
29	20.67	26.50	15.56	34.5	17.0	36.3							
30	19.83	25.44	15.39	36.7	28.1	41.5	Means	24.06	28.79	18.60	34.4	17.1	39.1

TABLE 167.

Special tri-hourly observations of barometer No. 1-901—S, S. Station, Long Peak.

			Original.			Barometer		
Date.	Local time	Observer	Barometer reading.	Attached thermometer	Corrected barometer reading	Correction for temperature	Reduced to freezing point.	Reduced to multiplier feet.
1861.			*Inches*	*Fahr.*	*Inches*	*Inches*	*Inches*	*feet*
Aug 15	Noon	H. L.	26.230	96.5	26.232	1.9	26.052	662.2
15	3 p. m.	A C D	26.132	87.0	26.144	137	26.007	660.4
15	6 p. m.	A C D	26.112	89.0	26.111	112	25.972	650.68
15	9 p. m.	A C D	26.112	74.5	26.144	108	26.016	661.30
15	Midnight	A C D	26.131	65.5	26.136	080	26.050	661.45
16	3 a. m.	H L	26.120	56.0	26.122	069	26.051	661.74
16	6 a. m.	A C D	26.160	49.5	26.171	049	26.122	663.49
16	9 a. m.	A C D	26.208	67.8	26.260	150	26.150	663.20
16	Noon	H L	26.270	92.3	26.372	149	26.123	662.52
16	3 p. m.	A C D	26.101	85.0	26.106	174	26.062	661.06
16	6 p. m.	A C D	26.181	88.0	26.185	154	26.031	661.69
16	9 p. m.	A C D	26.201	71.5	26.203	107	26.101	662.06
16	Midnight	A C D	26.200	64.0	26.202	087	26.119	662.41
17	3 a. m.	A C D	26.178	52.4	26.180	056	26.124	662.74
17	6 a. m.	A C D	26.216	51.0	26.218	052	26.165	664.60
17	9 a. m.	A C D	26.331	80.0	26.276	113	26.193	663.23
17	Noon	H L	26.310	91.0	26.312	148	26.164	664.55
17	3 p. m.	A C D	26.252	85.5	26.254	133	26.101	662.50
17	6 p. m.	A C D	26.198	84.0	26.200	139	26.070	662.17
17	9 p. m.	A C D	26.194	68.0	26.196	097	26.106	662.58
17	Midnight	A C D	26.192	57.5	26.191	068	26.126	661.59
18	3 a. m.	A C D	26.160	51.8	26.182	055	26.127	662.02
18	6 a. m.	A C D	26.221	52.5	26.223	056	26.167	664.05
18	9 a. m.	A C D	26.322	88.0	26.271	141	26.194	665.29
18	Noon	A C D	26.312	91.5	26.314	155	26.159	664.13
18	3 p. m.	A C D	26.216	86.0	26.248	136	26.112	663.23
18	6 p. m.	A C D	26.236	87.2	26.238	138	26.100	662.30
18	9 p. m.	A C D	26.155	62.4	26.157	079	26.158	661.40
18	Midnight	A C D	26.211	54.8	26.216	062	26.181	663.00
19	3 a. m.	A C D	26.254	49.5	26.256	056	26.200	663.47
19	6 a. m.	A C D	26.204	51.2	26.206	058	26.248	666.68
19	9 a. m.	A C D	26.101	88.2	26.106	140	26.266	662.69
19	Noon	A C D	26.288	89.0	26.090	165	26.225	666.41
19	3 p. m.	A C D	26.211	87.5	26.216	199	26.157	664.69
19	6 p. m.	A C D	26.201	84.0	26.265	148	26.164	664.55
19	9 p. m.	A C D	26.272	62.0	26.271	079	26.195	665.31
19	Midnight	A C D	26.250	57.4	26.272	068	26.204	665.57
20	3 a. m.	A C D	26.251	52.5	26.253	057	26.193	665.44
20	6 a. m.	A C D	26.360	55.5	26.362	064	26.298	666.13
20	9 a. m.	A C D	26.290	89.0	26.292	147	26.245	666.64
20	Noon	A C D	26.236	89.0	26.238	165	26.157	665.14
20	3 p. m.	A C D	26.271	90.0	26.273	144	26.129	662.92
20	6 p. m.	A C D	26.256	85.0	26.258	132	26.096	662.62
20	9 p. m.	A C D	26.230	71.0	26.232	160	26.032	663.71
20	Midnight	A C D	26.220	64.2	26.222	084	26.141	664.92
21	3 a. m.	A C D	26.216	56.5	26.218	065	26.145	664.60
21	6 a. m.	A C D	26.211	56.0	26.211	065	26.178	664.91
21	9 a. m.	A C D	26.350	92.8	26.352	151	26.202	664.28
21	Noon	H. L.	26.320	98.0	26.322	147	26.157	661.28
21	3 p. m.	A C D	26.254	91.5	26.256	115	26.096	662.50
21	6 p. m.	A C D	26.206	80.5	26.208	142	26.066	662.78
21	Midnight	H. L.	26.200	78.0	26.202	111	26.091	662.70
22	3 a. m.	A C D	26.201		26.213	070	26.143	662.77
22	6 a. m.	A C D	26.216	54.3	26.218	050	26.184	663.19
22	Noon	A C D	26.365	92.5	26.360	140	26.220	664.05
22	Noon	H L	26.313	90.4	26.350	150	26.200	665.44
22	3 p. m.	A C D	26.272	88.4	26.274	140	26.134	663.58
22	6 p. m.	A C D	26.250	86.0	26.252	134	26.118	664.42
22	9 p. m.	A C D	26.212	65.0	26.214	045	26.170	664.44
22	Midnight	A C D	26.206	60.1	26.218	073	26.175	664.64
23	3 a. m.	A C D	26.265	56.0	26.167	069	26.106	664.63
23	6 a. m.	A C D	26.270	51.0	26.252	059	26.193	665.29
23	9 a. m.	H. L.	26.368	90.4	26.320	147	26.223	665.24
23	Noon	H L	26.311	90.5	26.316	147	26.190	665.60
23	3 p. m.	A C D	26.244	88.7	26.246	140	26.106	662.08
23	6 p. m.	A C D	26.210	86.0	26.212	139	26.083	662.30
23	9 p. m.	A C D	26.222	70.5	26.234	073	26.120	663.70
23	Midnight	H. L.	26.204	61.5	26.206	067	26.129	663.67
24	3 a. m.	H. L.	26.180	55.5	26.192	065	26.127	662.92
24	6 a. m.	A C D	26.216	52.0	26.218	058	26.160	664.45
24	9 a. m.	A C D	26.334	91.8	26.326	148	26.178	663.01
24	Noon	H. L.	26.302	91.5	26.350	155	26.140	664.17
24	3 p. m.	H. L.	26.220	87.5	26.272	128	26.144	663.52
24	6 p. m.	A C D	26.185	85.0	26.187	138	26.049	662.79
24	9 p. m.	A C D	26.292	73.2	26.204	102	26.102	662.90
24	Midnight	A C D	26.194	56.5	26.196	072	26.124	663.52
25	3 a. m.	A C D	26.196	55.5	26.198	070	26.112	663.60
25	6 a. m.	A C D	26.225	54.0	26.227	065	26.162	663.70
25	9 a. m.	A C D	26.304	91.0	26.306	147	26.159	665.20
25	Noon	H. L.	26.310	90.5	26.247	133	26.114	665.44
25	3 p. m.	H. L.	26.231	88.5	26.272	130	26.142	664.79
25	6 p. m.	A C D	26.212	85.4	26.214	148	26.154	664.44
25	9 p. m.	A C D	26.220	69.0	26.222	072	26.112	663.86
25	Midnight	A C D	26.211	61.0	26.213	076	26.157	663.86
26	3 a. m.	A C D	26.175	53.2	26.177	067	26.120	663.14

*Observer failed to awaken.

TABLE 167—Continued.

Special tri-hourly observations of barometer No. 1200—S. S. Station, Lone Pine—Continued.

Date.	Local time.	Observer.	Original Barometer.				Barometer.	
			Barometer reading.	Attached thermometer.	Corrected for instrument error.	Correction for temperature.	Reduced to freezing point.	Reduced to millimetres.
1881.			*Inches.*	*Fahr.*	*Inches.*	*Inches.*	*Inches.*	*mm.*
Aug. 26	6 a. m.	A. C. D.	26.166	51.6	26.168	−.054	26.114	663.28
26	9 a. m.	A. C. D.	26.260	90.7	26.262	.146	26.116	663.33
26	Noon	A. C. D.	26.208	94.0	26.210	.153	26.057	661.84
26	3 p. m.	H. L.	26.146	85.5	26.188	.132	26.055	661.79
26	6 p. m.	H. L.	26.112	86.3	26.114	.135	25.979	659.86
26	9 p. m.	A. C. D.	26.088	67.0	26.090	.089	26.001	660.42
26	Midnight	A. C. D.	26.131	62.0	26.133	.078	26.055	661.79
27	3 a. m.	A. C. D.	26.141	55.0	26.143	.067	26.076	662.32
27	6 a. m.	A. C. D.	26.196	55.5	26.198	.063	26.135	663.82
27	9 a. m.	A. C. D.	26.262	72.8	26.264	.104	26.160	664.45
27	Noon	H. L.	26.280	85.8	26.282	.135	26.147	664.12
27	3 p. m.	H. L.	26.212	76.0	26.214	.111	26.103	663.01
27	6 p. m.	A. C. D.	26.198	69.0	26.200	.094	26.106	663.08
27	9 p. m.	H. L.	26.184	55.5	26.186	.063	26.123	663.52
27	Midnight	H. L.	26.204	52.5	26.206	.056	26.150	664.20
28	3 a. m.							
28	6 a. m.	H. L.	26.210	44.5	26.212	.037	26.165	664.68
28	9 a. m.	A. C. D.	26.289	80.8	26.301	.123	26.178	664.91
28	Noon	H. L.	26.200	90.2	26.208	.143	26.120	663.52
28	3 p. m.	H. L.	26.160	80.8	26.162	.122	26.040	661.40
28	6 p. m.	A. C. D.	26.136	83.5	26.138	.128	26.010	660.64
28	9 p. m.	A. C. D.	26.110	62.0	26.112	.078	26.034	661.25
28	Midnight	A. C. D.	26.118	51.2	26.120	.052	26.068	662.11
29	3 a. m.	A. C. D.	26.124	46.0	26.126	.041	26.085	662.55
29	6 a. m.	A. C. D.	26.150	40.0	26.152	.048	26.104	663.03
29	9 a. m.	A. C. D.	26.224	84.2	26.226	.127	26.099	662.90
29	Noon	H. L.	26.192	86.8	26.194	.136	26.058	661.89
29	3 p. m.	A. C. D.	26.088	82.0	26.090	.124	25.966	659.52
29	6 p. m.	A. C. D.	26.060	72.0	26.062	.101	25.961	659.10
29	9 p. m.	H. L.	26.068	64.5	26.070	.084	25.986	660.03
29	Midnight	H. L.	26.128	63.8	26.130	.082	26.048	661.80
30	3 a. m.	H. L.	26.150	57.3	26.152	.066	26.086	662.57
30	6 a. m.	H. L.	26.212	54.5	26.214	.061	26.153	664.30
30	9 a. m.	H. L.	26.320	78.4	26.322	.119	26.203	665.55
30	Noon	H. L.	26.326	89.2	26.328	.142	26.186	665.11
30	3 p. m.	A. C. D.	26.256	79.0	26.258	.118	26.140	663.84
30	6 p. m.	A. C. D.	26.234	74.0	26.236	.107	26.129	663.67
30	9 p. m.	A. C. D.	26.235	61.0	26.237	.076	26.161	664.44
30	Midnight	H. L.	26.276	51.8	26.278	.055	26.223	666.06
31	3 a. m.	H. L.	26.222	44.0	26.224	.045	26.179	664.94
31	6 a. m.	A. C. D.	26.246	44.0	26.248	.037	26.221	666.01
31	9 a. m.	A. C. D.	26.366	82.2	26.370	.126	26.244	666.58
31	Noon	A. C. D.	26.340	90.3	26.342	.145	26.197	665.39
31	3 p. m.	A. C. D.	26.250	82.0	26.252	.125	26.127	663.62
31	6 p. m.	A. C. D.	26.223	73.2	26.225	.105	26.120	663.44
31	9 p. m.	A. C. D.	26.200	60.7	26.202	.075	26.127	663.62
31	Midnight	H. L.	26.182	50.2	26.184	.051	26.133	663.77
Sept. 1	3 a. m.	H. L.	26.190	48.5	26.192	.047	26.145	664.07
1	6 a. m.	H. L.	26.194	46.5	26.196	.043	26.153	664.28
1	9 a. m.	A. C. D.	26.299	86.3	26.301	.133	26.168	664.65
1	Noon	A. C. D.	26.252	91.8	26.254	.144	26.108	663.68
1	3 p. m.	A. C. D.	26.150	84.0	26.152	.118	25.993	660.98
1	6 p. m.	A. C. D.	26.094	73.0	26.096	.103	25.993	660.21
1	9 p. m.	A. C. D.	26.084	63.0	26.083	.085	25.998	660.33
1	Midnight	A. C. D.	26.073	58.5	26.075	.070	26.005	660.32
2	3 a. m.	A. C. D.	26.070	50.0	26.072	.050	26.022	660.95
2	6 a. m.	A. C. D.	26.094	47.0	26.096	.043	26.053	661.74
2	9 a. m.	A. C. D.	26.200	86.5	26.202	.136	26.066	662.06
2	Noon	A. C. D.	26.162	93.6	26.164	.152	26.012	660.68
2	3 p. m.	A. C. D.	26.114	85.0	26.116	.113	26.003	660.47
2	6 p. m.	A. C. D.	26.100	77.0	26.102	.112	25.990	660.13
2	9 p. m.	A. C. D.	26.128	72.0	26.130	.101	26.029	661.13
2	Midnight	A. C. D.	26.110	60.0	26.112	.094	26.018	660.64
3	3 a. m.	A. C. D.	26.095	56.0	26.097	.087	26.010	660.44
3	6 a. m.	A. C. D.	26.120	65.0	26.122	.085	26.037	661.33
3	9 a. m.	A. C. D.	26.162	84.0	26.164	.130	26.034	661.25
3	Noon	A. C. D.	26.115	91.5	26.117	.147	25.970	659.63
3	3 p. m.	H. L.	26.022	82.5	26.024	.125	25.895	657.72
3	6 p. m.	A. C. D.	26.000	74.6	26.002	.107	25.895	657.72
3	9 p. m.	A. C. D.	26.000	70.5	26.002	.097	25.903	657.93
3	Midnight	A. C. D.	26.029	65.0	26.031	.085	25.945	658.90
4	3 a. m.	A. C. D.	26.031	62.8	26.033	.080	25.953	659.20
4	6 a. m.	A. C. D.	26.064	62.0	26.066	.078	25.988	660.88
4	9 a. m.	A. C. D.	26.112	81.0	26.114	.122	25.992	660.18
4	Noon	A. C. D.	26.101	88.0	26.103	.140	25.963	659.15
4	3 p. m.	A. C. D.	26.003	88.0	26.005	.139	25.881	657.37
4	6 p. m.	A. C. D.	25.951	77.5	25.953	.112	25.841	656.35
4	9 p. m.	A. C. D.	26.033	76.5	26.035	.108	25.927	656.54
4	Midnight	A. C. D.						
5	3 a. m.	A. C. D.	26.022	54.5	26.024	.063	25.961	659.45
5	6 a. m.	A. C. D.	26.029	60.0	26.031	.073	25.958	659.32
5	9 a. m.	A. C. D.	26.115	87.0	26.117	.136	25.981	658.91
5	Noon	A. C. D.	26.100	83.7	26.102	.129	25.973	659.81
5	3 p. m.	A. C. D.	26.032	82.5	26.034	.125	25.909	658.08
5	6 p. m.	A. C. D.						

L. Observer failed to awaken.   † Omitted, observer sick.   ‡ Discontinued by directions of Professor Langley. Dated September 3, 1881.

TABLE 168.

Summary of special tri-hourly observations of the barometer at Lone Pine.

[Corrected. Reduced to freezing point and to millimeters. Barometer 1890, S. S. Observers, A. C. D. and H. L. Computer, E. H. W.]

Date.	3 a. m.	6 a. m.	9 a. m.	Noon.	3 p. m.	6 p. m.	9 p. m	Midnight.
1881.	mm.	mm.	mm.	mm.	mm.	mm.	mm.	mm.
Aug. 15	......	......	......	662.25	660.31	659.68	661.50	661.66
16	661.74	661.19	664.20	663.52	661.96	661.69	662.88	663.44
17	663.54	664.60	665.29	664.55	662.96	662.17	662.80	663.70
18	663.62	664.63	665.29	664.43	663.23	661.93	664.40	665.06
19	663.47	664.68	664.74	665.11	664.89	664.55	665.34	665.57
20	665.44	666.41	666.61	665.34	663.67	662.82	663.74	662.97
21	664.02	664.91	665.50	664.38	662.70	662.86	662.70	......
22	663.77	665.19	665.95	665.44	661.79	663.58	664.45	664.81
23	664.05	665.29	664.06	665.21	663.08	662.50	663.30	663.67
24	663.62	664.45	664.91	664.17	662.52	661.79	661.98	661.72
25	663.80	664.96	665.70	665.11	663.77	663.79	663.86	663.86
26	663.44	664.28	665.33	664.84	661.79	659.86	660.42	661.70
27	662.32	661.82	664.15	664.12	663.01	663.08	663.32	664.20
28	......	664.68	664.91	663.52	664.40	660.64	661.25	662.11
29	662.55	663.03	662.90	663.86	639.52	659.40	660.05	661.60
30	662.37	664.50	665.60	665.11	663.94	663.67	664.48	666.06
31	664.91	666.01	666.58	665.79	663.62	664.14	663.62	663.77
Sept. 1	664.07	664.28	664.65	663.08	660.95	660.21	660.23	661.52
2	660.95	661.74	662.06	660.69	660.47	660.13	661.13	661.84
3	660.64	661.33	664.25	659.61	658.08	657.72	657.93	658.99
4	659.20	660.08	660.15	658.47	657.57	656.35	658.54	......
5	659.45	659.32	658.91	659.81	658.08	......	......	......
Sums..	13250.09	13942.53	13952.47	11790.84	14561.11	13901.86	13903.48	12580.14
Means	663.00	664.40	664.40	663.40	661.87	661.52	662.33	663.17

If in the above table the missing readings be supplied by interpolation, the mean value for 3 a. m. will be 663.10 and that for midnight 662.95.

For a summary of the special tri-hourly observations of the relative humidity at Lone Pine the reader may consult the Table 149*a*.

We give here a summary of the thermometric measurements.

TABLE 169.

Summary of special tri-hourly observations of the thermometer at Lone Pine.

[Dry bulb. Reduced to Centigrade. Observers, A. C. D and H. L. Computer, F. W. V.]

Date	3 a. m.	6 a. m.	9 a. m.	Noon.	3 p. m.	6 p. m.	9 p. m.	Midnight.
1881.								
Aug. 15	......	......	......	31.83	32.11	28.22	23.72	18.85
16	14.78	10.11	26.28	29.50	31.11	26.28	20.83	16.72
17	10.11	9.83	25.67	29.50	31.28	26.56	16.72	11.33
18	12.33	11.50	25.56	30.84	32.11	26.56	12.61	13.39
19	10.94	11.22	26.56	32.94	30.50	28.22	14.22	15.11
20	10.11	10.84	26.67	33.61	34.28	27.83	22.22	19.89
21	14.78	14.78	27.67	30.22	32.94	29.61	22.50	......
22	14.78	12.16	27.11	32.39	33.50	27.39	13.72	14.50
23	15.72	12.89	26.67	32.00	33.44	27.11	19.28	15.06
24	13.17	11.44	27.50	31.44	32.56	27.67	19.56	11.78
25	12.33	11.11	26.28	28.67	31.22	23.44	20.11	17.56
26	10.39	10.94	25.83	31.00	28.72	26.72	18.72	18.44
27	14.50	13.44	26.09	33.44	27.39	18.11	10.56	9.44
28	......	9.09	21.22	27.83	29.06	24.61	11.61	12.44
29	7.44	9.81	22.61	27.00	28.94	19.28	18.33	10.00
30	14.94	13.67	21.00	28.67	28.89	21.50	13.11	9.56
31	8.89	5.00	23.17	29.09	28.89	19.44	14.22	8.44
Sept. 1	9.28	9.00	23.22	28.50	30.17	16.39	13.72	17.00
2	10.33	8.67	24.67	31.00	32.00	21.00	22.06	23.17
3	20.94	18.72	25.72	29.67	29.17	22.72	20.61	16.72
4	17.39	16.67	21.06	23.44	26.89	24.89	......	......
5	12.61	15.33	23.72	28.22	29.17	......	......	......
Sums..	253.76	248.15	518.25	653.97	674.51	516.61	375.87	290.44
Means	12.69	11.82	24.68	29.73	30.84	24.60	17.90	15.29

TABLE 170.

Barometric observations at Mountain Camp.

[Barometer No. 2018 Signal Service. Correction +0ⁱⁿ.002. Observer, J. J. N. Computer, R. H. W.]

Date	Local mean time	Original. Barometer read. inc.	Attached ther- mometer.	Corrected for in- strumental error.	Correction for tem- perature.	Reduced to freez- ing point.	Reduced to milli- meters.	Date	Local mean time	Original. Barometer read. inc.	Attached ther- mometer.	Corrected for in- strumental error.	Correction for tem- perature.	Reduced to freez- ing point.	Reduced to milli- meters.
		Inches.	Fah.	Inches.	Inches.	Inches.	mm.			Inches.	°Fah.	Inches.	Inches.	Inches.	mm.
Aug. 17	8ʰ 15ᵐ p.m.	19.716	40.0	19.718	—.023	19.697	500.30	Aug. 26	12 35 p.m.	19.732	60.0	19.734	—.005	19.670	499.84
18	8 15 a.m.	19.763	41.0	19.765	—.059	19.706	500.52	26	8 15 p.m.	19.655	40.0	19.657	—.020	19.637	498.77
18	12 35 p.m.	19.780	63.0	19.782	—.061	19.721	500.91	27	8 15 a.m.	19.591	50.5	19.593	—.058	19.535	496.89
18	8 15 p.m.	19.777	44.0	19.779	—.028	19.751	501.67	27	12 35 p.m.	19.650	56.5	19.652	—.049	19.603	497.71
19	8 15 a.m.	19.817	66.5	19.838	—.067	19.772	502.29	27	8 15 p.m.	19.661	34.5	19.603	—.011	19.652	499.15
19	12 35 p.m.	19.847	67.5	19.853	—.063	19.784	502.50	28	8 15 a.m.	19.680	57.0	19.695	—.050	19.645	498.98
19	8 15 p.m.	19.856	45.5	19.858	—.060	19.828	503.62	28	12 35 p.m.	19.711	59.5	19.713	—.055	19.658	499.30
20	8 15 a.m.	19.670	63.5	19.672	—.062	19.810	503.18	28	8 15 p.m.	19.663	38.0	19.665	—.016	19.649	499.08
20	12 35 p.m.	19.882	68.0	19.884	—.071	19.814	503.26	29	8 20 a.m.	19.663	53.5	19.665	—.044	19.621	498.37
20	8 15 p.m.	19.819	45.0	19.821	—.030	19.791	502.40	29	12 35 p.m.	19.680	62.0	19.682	—.050	19.632	498.42
21	8 15 a.m.	19.849	47.0	19.851	—.068	19.782	502.45	29	8 15 a.m.	19.623	30.0	19.625	—.013	19.612	498.13
21	12 35 p.m.	19.854	68.5	19.856	—.071	19.785	502.53	30	8 15 a.m.	19.702	56.0	19.704	—.048	19.656	499.25
21	8 15 p.m.	19.782	45.0	19.784	—.030	19.754	501.74	30	12 35 p.m.	19.755	61.0	19.747	—.057	19.680	500.12
22	8 15 a.m.	19.838	63.5	19.840	—.065	19.775	502.28	30	8 15 p.m.	19.743	40.5	19.735	—.021	19.714	500.72
22	12 35 p.m.	19.851	67.0	19.853	—.068	19.785	502.53	31	8 15 a.m.	19.770	55.5	19.762	—.047	19.715	500.75
22	8 15 p.m.	19.801	46.0	19.803	—.021	19.772	502.20	31	12 35 p.m.	19.804	65.5	19.796	—.065	19.731	501.16
23	8 15 a.m.	19.853	65.5	19.855	—.063	19.770	502.15	31	8 15 p.m.	19.749	41.0	19.741	—.021	19.720	500.88
23	12 35 p.m.	19.842	68.0	19.844	—.070	19.774	502.25	Sept. 1	8 15 a.m.	19.762	59.0	19.754	—.055	19.699	500.35
23	8 15 p.m.	19.787	44.0	19.789	—.026	19.763	501.89	1	12 35 p.m.	19.749	62.5	19.741	—.061	19.680	499.86
24	8 15 a.m.	19.818	62.5	19.820	—.060	19.760	501.89	1	8 15 p.m.	19.682	40.0	19.674	—.020	19.654	499.20
24	12 35 p.m.	19.821	65.5	19.826	—.065	19.761	501.92	2	8 15 a.m.	19.708	54.0	19.700	—.045	19.655	499.23
24	8 15 p.m.	19.793	44.0	19.795	—.028	19.767	502.07	6	8 15 p.m.	19.563	39.5	19.595	—.020	19.745	501.52
25	8 15 a.m.	19.794	65.5	19.796	—.068	19.754	501.23	6	8 20 a.m.	19.802	55.5	19.804	—.047	19.757	501.82
25	12 35 p.m.	19.820	54.5	19.822	—.064	19.759	501.87	7	12 35 p.m.	19.839	65.5	19.831	—.065	19.766	502.04
25	8 15 p.m.	19.771	41.0	19.773	—.022	19.751	501.67	7	8 15 p.m.	19.791	41.0	19.793	—.022	19.771	502.18
26	8 15 a.m.	19.747	55.0	19.749	—.046	19.703	500.45								

* Barometer No. 1955, Signal Service, correction — 0ⁱⁿ.008 after this.
Instruments on summit, September 2 to 6, 1884. Barometer No. 2018, Signal Service, hereafter, correction = +0ⁱⁿ.002.

TABLE 171.

Summary Mountain Camp barometric readings.

[Corrected, reduced to freezing point and to millimeters. Barometers No. 2018 and 1955, S.S. Observer, J. J. N. Computer, R. H. W.]

Date.	8ʰ 15ᵐ a.m.	12ʰ 35ᵐ p.m.	8ʰ 15ᵐ p.m.	Date.	8ʰ 15ᵐ a.m.	12ʰ 35ᵐ p.m.	8ʰ 15ᵐ p.m.
1884.	mm.	mm.	mm.	1884.	mm.	mm.	mm.
Aug. 17			500.30	Aug. 29	498.37	498.42	498.13
18	500.52	500.91	501.67	30	499.25	500.12	500.72
19	502.29	502.50	503.62	31	500.75	501.16	500.88
20	503.18	503.26	502.40	Sept. 1	500.35	499.86	499.20
21	502.45	502.53	501.74	2	499.23	....	....
22	502.28	502.53	502.20	6			501.52
23	502.15	502.25	501.89	7	501.82	502.04	502.18
24	501.89	501.92	502.07				
25	501.23	501.87	501.67	Sums ...	8541.77	8016.22	8017.56
26	500.45	499.84	498.77				
27	496.89	497.71	499.15	Means ...	500.69	501.01	500.98
28	498.98	499.30	499.08				

TABLE 172.

Summary of Mountain Camp thermometer and hygrometrical readings.

[Observer, J. J. N. Computer F. W. V.]

Date.	Dry-bulb thermometer reduced to Centigrade.			Relative humidity by Smithsonian tables.			Date.	Dry-bulb thermometer reduced to Centigrade.			Relative humidity by Smithsonian tables.		
	^{h} 15^m a. m.	12^h 35^m p. m.	^{h} 15^m p. m.	^{h} 15^m a. m.	12^h 35^m p. m.	^{h} 15^m a. m.		^{h} 15^m a. m.	12^h 35^m p. m.	^{h} 15^m p. m.	^{h} 15^m a. m.	12^h 35^m p. m.	^{h} 15^m a. m.
1881.				Per c.	Per c.	Per c.	1881.				Per c.	Per c.	Per c.
Aug. 22	....	16.4	7.8	....	14.7	46.1	Aug. 30	7.2	13.3	3.9	23.6	20.7	34.3
23	10.6	15.0	8.8	4.4	12.6	39.9	31	6.7	11.4	3.9	35.8	17.7	40.9
24	9.4	15.0	6.7	15.4	20.3	26.9	Sept. 1	8.6	13.6	3.3	14.8	24.7	37.3
25	8.1	14.4	5.0	17.8	18.0	32.3	2	6.7	....	....	26.9	....	....
26	6.1	12.5	4.4	33.2	21.8	23.1							
27	2.8	11.1	2.2	51.4	22.8	57.8	Sums	78.3	148.5	47.1	304.9	226.5	449.4
28	5.0	11.4	2.2	52.5	24.5	44.3							
29	6.9	11.7	2.2	28.3	26.7	67.5	Means	7.1	14.5	4.3	27.6	20.6	40.9

TABLE 173.

Barometric observations at Peak of Whitney.

[Barometer No. 2018, S. S. Correction 0.002 inch. Computer, R. H. W.]

Date.	Local mean time.	Observer.	Original.		Barometer.			
			Barometer reading.	Attached thermometer.	Corrected for instrument error.	Correction for temperature.	Reduced to freezing point.	Reduced to millimeters.
1881.	$h.$ $m.$		Inches.	° Fah.	Inches.	Inches.	Inches.	mm.
Sept. 2	6 00 p. m.	J. J. N.	17.600	30.0	17.602	-.001	17.599	447.00
2	9 00 p. m.	J. J. N.	17.567	26.5	17.569	-.004	17.601	447.10
3	3 00 a. m.	J. J. N.	17.509	23.5	17.521	-.005	17.506	446.40
3	6 00 a. m.	J. J. N.	17.522	22.3	17.531	-.009	17.540	445.50
3	9 15 p. m.	J. E. K.	17.518	23.5	17.520	-.009	17.529	445.22
4	8 30 p. m.	J. E. K.	17.514	28.2	17.516	+.000	17.516	444.80
5	12 40 p. m.	J. E. K.	17.627	52.8	17.629	-.008	17.591	446.79
5	5 07 p. m.	J. E. K.	17.600	62.5	17.602	-.056	17.514	445.65
5	6 30 p. m.	O. E. M.	17.680	61.8	17.682	-.054	17.628	447.73
5	8 20 p. m.	O. E. M.	17.690	42.0	17.642	-.020	17.622	447.54
6	1 00 a. m.	O. E. M.	17.560	36.0	17.601	-.013	17.588	446.72
6	10 22 a. m.	O. E. M.	17.558	32.0	17.560	-.005	17.555	445.88
6	12 00 midnight	O. E. M.	17.558	31.5	17.560	-.005	17.555	445.88
6	1 00 a. m.	O. E. M.	17.610	30.0	17.612	+.002	17.610	447.27
6	3 00 a. m.	O. E. M.	17.616	30.0	17.612	-.002	17.610	447.27
6	6 00 a. m.	O. E. M.	17.610	28.0	17.612	+.003	17.615	447.33
6	8 17 a. m.	O. E. M.	17.692	32.0	17.694	-.037	17.657	448.47
8	9 00 a. m.	O. E. M.	17.680	51.4	17.642	.042	17.640	448.09

TABLE 174.

Simultaneous observations of the barometer, thermometer, and of relative humidity between Lone Pine and Whitney Peak.

[Observers J. J. N. and A. C. D. Computer, A. B. S.]

Date.	Local mean time.	Barometer reading.		Temperature °C.		Relative humidity.	
		Lone Pine.	Whitney Peak.	Lone Pine.	Whitney Peak.	Lone Pine.	Whitney Peak.
1881.	$h.$ $m.$	mm.	mm.				
Sept. 2	6 00 p. m.	660.13	447.00	21.06	-1.10	0.503	0.307
2	9 00 p. m.	661.13	447.10	22.06	-2.10	0.170	0.452
2	Midnight	660.81	446.40	21.17	-3.60	0.175	0.316
3	3 00 a. m.	660.63	445.50	20.94	-3.70	0.121	0.509
3	6 00 a. m.	661.53	445.22	18.72	-5.40	0.276	0.714
4	8 35 p. m.	658.15	444.80	23.10	-2.11	0.145	0.409
5	12 70 p. m.	659.78	446.79	19.50	11.60	0.352	0.307
5	5 00 p. m.	658.94	445.65	28.20	16.90	0.230	0.154
5	6 57 p. m.	657.08	447.73	28.03	16.40	0.231	0.637
5	8 20 p. m.	660.00	447.54	29.10	5.69	0.235	0.545
5	8 20 p. m.	660.57	446.72	22.70	3.10	0.200	0.573
5	10 22 p. m.	660.81	445.88	16.00	0.00	0.167	0.452
5	Midnight	660.90	445.88	14.30	-0.30	0.180	0.316
6	1 00 a. m.	660.81	447.27	14.40	-1.10	0.200	0.428
6	3 00 a. m.	660.63	447.27	17.27	17.10	0.200	0.671
6	6 00 a. m.	661.40	447.33	15.20	-2.70	0.310	0.502
6	8 17 a. m.	660.58	448.47	14.00	11.10	0.292	0.304
6	9 00 a. m.	663.81	448.09	16.10	12.40	0.273	0.714
Sums		11893.33	8046.79	357.05	-4.52.40	4.100	8.742
Means		660.85	446.68	19.87	+2.80	0.228	0.486

CHAPTER XX.

REPORT OF W. C. DAY ON CARBONIC ACID IN LOCALITY VISITED BY EXPEDITION.

[When the services of Mr. Day were secured for the expedition, it was intended that they should be chiefly given in his capacity as a professional chemist, to the determination of the amount of carbonic acid in the air at the various stations. The exigencies of the service essentially modified this plan, and a large part of Mr. Day's time was necessarily diverted by me to the physical experiments, for which we were short-handed. Accordingly, for the fact that a larger number of chemical determinations was not made I am responsible rather than Mr. Day. I present those which were secured, together with an interesting *résumé* of our previous knowledge on the subject by him.—[S. P. LANGLEY.]

REPORT ON WORK DONE IN DETERMINING THE AMOUNT OF CARBONIC ACID CONTAINED IN THE ATMOSPHERE OF THE LOCALITY VISITED BY THE EXPEDITION.

By Mr. W. C. DAY, of Johns Hopkins University.

Before submitting an account of the work performed, it will perhaps be best to present a brief statement of our knowledge in reference to the following questions:

First, what is the proportion of carbonic acid in the air, and is this proportion constant?

Second, what is the action of atmospheric carbonic acid upon solar radiation?

The chief causes tending to increase the atmospheric carbonic acid are as follows:

(1.) The respiration of animals.

(2.) Combustion of carbonized material.

(3.) Exhalations of carbonic acid caused by volcanoes and other infra-terrestrial agencies.

The causes of decrease in the amount of this gas are chiefly—

(1.) The decomposition of carbonic acid by living vegetables under the influence of sunlight; oxygen being thereby liberated, while carbon is assimilated by the plant.

(2.) The formation of carbonate of lime by the absorption of atmospheric carbonic acid through the agency of certain animals, giving rise to coral reefs and animals and the whole of the vast limestone deposits.

(3.) The absorption or fixation of carbonic acid by inorganic chemical processes.

Owing to insufficient data, it cannot be said whether atmospheric carbonic acid is increasing or decreasing; but certainly if any essential change is going on, such change must be very slow, and years of the most accurate and systematic analyses would be necessary to reveal it. From such knowledge as we have, however, the total amount of this gas in the atmosphere seems to be constant.

With regard to local variations and their causes there has been much discussion within the past few years. As it is with the question of local variation that we are chiefly concerned, let us consider, in a general way, the causes capable of producing alterations in the amount of carbonic acid contained in the atmosphere of any given locality.

The large amounts of the gas emitted by volcanoes would naturally tend to raise the proportion of carbonic acid contained in the surrounding atmosphere. Air in the vicinity of densely populated cities would also be expected to contain an excess of this gas. The atmosphere of one place non-productive of carbonic acid, and separated by miles from another characterized by a large consumption of fuel, might, nevertheless, contain a large proportion of this product of combustion, the excess being due to prevailing winds sweeping from the latter place over the former and carrying with them the contaminated air.

Such are some of the causes tending towards local accumulation.

What, now, are the causes tending to oppose those just considered?

The diffusion of gases is, of course, the cause of primary importance. The stirring up of the air by winds also tends to equalize the proportions of the atmospheric constituents. These two causes may be considered the chief ones, which act in direct and immediate opposition to those of local accumulation. Another cause, which not only tends to prevent accumulation, but which may even be the cause of a local deficiency, is the ultimate consumption of carbonic acid by vegetable life. We would naturally expect to find less of this gas in the atmosphere of a region characterized by abundant vegetation than in one more barren or one inhabited to a greater extent by animals.

Other causes bearing upon the question of the uniform composition of the air have been supposed by individual investigators to exist. These causes, more or less generally accepted as such, will be presented later, together with the facts which either support or oppose them.

The figures which in all probability most nearly represent the actual average proportion of carbonic acid in our atmosphere at sea level are: Three parts by volume of carbonic acid to 10,000 parts of air; the extreme limits of local variations may be considered to be 2 and 4.5 parts carbonic acid to 10,000 of air, although some analyses on record would extend these limits in either direction.

An interesting theory, bearing upon the constancy of the proportion of atmospheric carbonic acid, has lately been advanced by M. H. Schloesing. (Comptes Rendus, tome 90, page 1410.) This investigator maintains that the variations of the atmospheric carbonic acid is contained between very narrow limits, and in this connection expresses his firm confidence in the opinions of M. Reiset (to be referred to further on). Schloesing has found that pure water, placed in contact with an earthy carbonate and surrounded by an atmosphere containing carbonic acid, becomes charged with a quantity of bicarbonate, which increases, according to a mathematical law, with the tension of the carbonic acid in this atmosphere. The bicarbonate is found as a result of the absorption of carbonic acid by the earthy carbonate employed.

When a neutral salt of soda, lime, or magnesia in solution is introduced into the water the quantity of bicarbonate found differs from that found when pure water is used; but it increases with the amount of carbonic acid, and a state of equilibrium is produced between the amount formed and the tension of the gas. As a result of a series of analyses of sea-water, Schloesing has also found that the greater part of the carbonic acid contained in a given amount of water is present in bicarbonates. These experiments have led the author to the belief that the sea acts as a reservoir and regulator of atmospheric carbonic acid.

Owing to the continual motions going on in the air and in the sea, the state of equilibrium above referred to is not perfectly realized in nature. However, there is a tendency toward that condition.

If the variation of the carbonic acid is positive, the sea-water absorbs the excess with the formation of bi- or acid carbonates; if negative, carbonic acid is released to the air and neutral carbonates are precipitated. The sea is regarded by the same investigator as a regulator also of atmospheric ammonia.

M. Marié-Davy (Comptes Rendus, tome 90, p. 32) claims to have established a relation between the carbonic acid of the atmosphere and the great general movements of the latter.

The above author has in his possession the results of four years of daily determinations of atmospheric carbonic acid. These determinations were made at Montsouris by M. Lévy and his assistant, M. Allaire. They show variations between the limits 22 and 36 parts of carbonic acid to 100,000 of air.

In attempting to account for these variations M. Marié-Davy first thought that they were due to the influence, on the one hand, of Paris as a fruitful source of carbonic acid, on the other, of the cultivated fields as agents of consumption. But, contrary to this belief, it was noticed that the northern winds, which blow from Paris over Montsouris contain less carbonic acid than those from the south, which come directly from the country.

The local influence is thus subordinated to another of a higher order. Generally the winds from the south or southwest graze the surface of the soil, while those from the north or northwest plunge down from the elevated atmospheric regions. If, then, it can be supposed that the latter contain less carbonic acid than the former, the matter is explained.

If his views on this question are true, the great value of including among the ordinary meteorological observations systematic determinations of carbonic acid in the air, is at once apparent.

The following views of M. Reiset (Comptes Rendus, tome 90, page 1144), upon the proportion of carbonic acid in the air, introduce a controversy between the latter and M. Marié-Davy, the point of difference being the more or less wide limits of atmospheric carbonic acid variation.

In the month of June, 1879, M. Reiset commenced in the country a series of investigations which was continued up to the first frost in November. The general mean deduced from ninety-one determinations made during the day or night during this period is 29.78 volumes of carbonic acid to 100,000 of air, dry at 0° and 760 ᵐᵐ. In a preceding communication it had been announced that from the 9th of September, 1872, to the 20th of August, 1873, the mean was 29.42. These latter determinations are therefore confirmed by those made six years later thus showing that the limits of variation are much smaller than as revealed by the investigations of M. Marié-Davy.

The conclusions arrived at by M. Reiset are as follows:

After an interval of six years the same proportion of carbonic acid is found, i. e., 29.78 volumes carbonic acid to 100,000 of air.

The air analyzed during the night contains more carbonic acid than that during the day; 28.91 to 100,000 is the proportion found for the day between 9ʰ a. m. and 4ʰ p. m.; 30.81 is the proportion for the night; several foggy nights are included among the number for which this mean was found. M. Reiset then refers to the results obtained by M. Marié-Davy; he makes no comment upon the value of the hypothesis advanced by the latter, but says that the variations between the limits 22 to 36 per 100,000 of air shown to exist by M. Marié-Davy's determinations are in total disagreement with the results of his own investigation. M. Reiset further expresses doubt as to the degree of precision attainable by the method employed by M. Marié-Davy, and justifies this doubt by a criticism of the latter's methods and apparatus.

The author concludes this paper by quoting some remarks of Gay-Lussac advocating the idea that the proportion of carbonic acid in the air is constant, i. e., varies between very narrow limits, this constancy being due to the state of incessant motion which, it is reasonable to suppose, exists in the atmosphere, thus tending to distribute uniformly its accidental constituents.

M. Marié-Davy replies to the remarks of M. Reiset, and reasserts his confidence in the results arrived at by his collaborator, M. Lévy, and he emphasizes the importance of continued investigations, in various places, with the aim of settling the question of the connection between the proportion of atmospheric carbonic acid and the general movements of the atmosphere (Comptes Rendus, tome 90, page 1287).

The principal point on which these eminent observers differ, is whether the extent of variation in the amount of atmospheric carbonic acid is contained between the limits believed to exist by Marié-Davy or between less widely separated limits as those advocated by Reiset.

The recent investigations of MM. Müntz and Aubin (Comptes Rendus, tome 92, page 247), on the proportion of atmospheric carbonic acid, had for their object the solution of the following questions:

For a given place are the variations in the proportion of carbonic acid considerable or merely insignificant?

Is the carbonic acid uniformly distributed in the various layers of the atmosphere, or is there an excess in the lower portions?

Müntz and Aubin first possessed themselves of a method which would give results upon which entire dependence could be placed. Their method may briefly be described as follows: The principle involved consists in fixing the carbonic acid by an absorbing body, from which it is subsequently disengaged and measured directly. The absorbing body is pumice stone, impregnated with a solution of potassium hydrate. This is contained in a glass tube drawn out at both ends; the tube thus prepared is sealed at both ends and opened on the spot when the determination is to be made. Air to the amount of at least 200 liters is passed through the tube by means of an aspirating gasometer. The tube, again sealed, may be preserved indefinitely. The remainder of the determination is made in the laboratory. Thus one end of the tube is connected with a mercury pump, through the other end dilute sulphuric acid is allowed to enter, the carbonic acid thus disengaged is determined by direct measurement in a graduated receiver.

The evidences of accuracy given in the memoir are certainly satisfactory, at the same time the method is simple throughout. The operation of passing the air through is one which may be performed by a person unaccustomed to delicate manipulation. The method is specially valuable in its application to determinations in places difficult of access. For accuracy and simplicity this method seems to be equaled by no other at present in use. All work connected with its application involving delicacy and careful manipulation is performed in the laboratory; the remaining operations may be performed by an ordinarily intelligent person at the place chosen for the determinations.

In the present state of meteorological science the indications are that the functions of atmospheric carbonic acid are more important and more complex than has been supposed until within the last few years. Assuming the correctness of the views of Marié-Davy, a knowledge of the variations in the amount of this atmospheric constituent would be of the greatest practical value. But in order to prove the correctness or the incorrectness of these views, it is necessary that simultaneous analyses should be systematically made at a number of different places widely separated and so situated as to secure various conditions. The Signal Service system of this country seems to embrace conditions most favorable for work of this nature. The men stationed at the various observatories over our vast area of territory are amply well fitted for the portion of the work which would fall to their lot. From a single laboratory tubes prepared as above described might be sent to the different stations, where the necessary volume of air could be simply passed through them, the atmospheric conditions at the time noted, and the tubes returned to the laboratory, where the operation would be completed.

Let us now consider some of the results which this method in the hands of its originators, Müntz and Aubin, has already developed (Comptes Rendus, tome 92, page 1229). Two stations were established, one at Paris, 6ᵐ above the ground, the other at an open place in the country. The atmosphere of a large city like Paris is incessantly contaminated by the products of combustion and those of the respiration of its numerous inhabitants. The atmosphere of the country is without these abundant sources of carbonic acid. At the Paris station a large number of determinations were made during the interval between December, 1880, and May, 1881.

The differences in the proportion of acid are notable. They varied between the limits 2.88 to 4.22 volumes carbonic acid to 10,000 of air.

The maxima correspond always with weather cloudy and calm, the air being disturbed by no energetic agitation; also with a predominance of the local influence. The minima, on the other hand, are evident with an atmosphere pure and agitated.

The quantities of carbonic acid found during weather cloudy and calm vary between the limits 3.22 and 4.22 volumes per 10,000 air. Those found for clear weather are comprised between the limits 2.89 and 3.4. These figures do not differ sensibly from those by M. Boussingault.

The largest quantities were observed during abundant falls of snow or during thick fogs, conditions which fetter the movements of the atmosphere. The results obtained at the station in the country confirm those of M. Reiset. During the day the quantities are comprised between the limits 2.70 and 2.99 volumes per 10,000 air. During the night there is an increase and the mean approaches 3.

The variations observed during a single day, the weather meanwhile undergoing change, are given as follows:

Volumes of 10,000.

```
April 1, 9ʰ a. m. Sky clear, air agitated .......................................  2.73
          1ʰ 30 p. m. Sky cloudy.................................................  2.90
          4ʰ p. m. Sky very cloudy, commencement of rain ....................  2.99
```

These variations, although contained between narrow limits, are nevertheless significant.

Müntz and Aubin (Comptes Rendus, tome 93, page 797) have also applied their valuable method to the analyses of air in elevated regions. In these determinations the precaution of taking the air through metallic tubes, about 30 feet from the observer, rendered any error due to respiration of the operator null. The point chosen for observation was the summit of the Pic du Midi in the Pyrenees Mountains, altitude, 2,877ᵐ. This peak is separated from other elevated mountains. The air which circulates there is that of the upper currents; the rapidity of the winds removes any suspicion of local influence.

The following table shows the results:

Date.	Hours.	Direction of wind.	Volume of air employed.	Carbonic acid per 10,000 of air.
			Liters	
Aug 8	12ʰ 50ᵐ to 1ʰ 00ᵐ	SW., strong ...	200.2	2.91
9	4 00	7 35	do ...	197.6	2.95
10	8 51	2 07	NW. weak	247.3	2.91
10	2 07	6 2	do	206.7	2.75
11	9 34	11 18	SE., variable	207.2	2.95
11	11 19	3 01	do	215.4	2.80
11	3 02	5 43	do	206.8	2.76
12	7 19	10 50	SE., very feeble	206.2	2.82
12	10 51	2 24	do	207.5	2.85
12	2 25	6 07	do	205.1	2.79
13	11 10	2 10	SW., strong	204.2	2.87
14	2 43	6 2	do	221.5	2.69
14	9 31	1 17	do	242.1	2.91
14	1 17	4 55	do	204.6	2.89

For the sake of comparison analyses were made in two valleys at the base of the mountains, i. e., Pierrefitte, altitude 507ᵐ, and Luz, altitude 730ᵃ, as follows:

Pierrefitte, August 5, 2ʰ to 5ʰ p. m., carbonic acid per 10,000 of air 2.79
Pierrefitte, August 6, 8ʰ to 11ʰ a. m., carbonic acid per 10,000 of air (foggy) 3.00
Luz, August 7, 8ʰ to 11ʰ a. m., carbonic acid per 10,000 air 2.69

The last determination was made in the midst of abundant vegetation.

MM. Müntz and Aubin arrive at the following conclusions:

All these figures are very nearly identical with those found in the lower portions of the atmosphere by ourselves, by M. Reiset, and M. Schultze.

We believe, then, that we are authorized in making the statement that the carbonic acid is uniformly distributed throughout the atmosphere, and we confirm thus the ideas of M. Reiset upon this subject and those of M. Schloesing on the circulation of carbonic acid at the surface of the globe.

Our present knowledge as to the proportion of carbonic acid in the air and the limits of variation of this proportion may, so far as it is based on experimental evidence, perhaps be stated as follows:

(1) Carbonic acid produced and consumed as shown in the beginning of this report is present in our atmosphere in the *nearly* constant proportion of three volumes of carbonic acid to 10,000 of air.

(2) The variations of this proportion are contained between narrow limits, but are significant.

Maxima of these variations are caused by predominance of causes of production; they are also coincident with weather cloudy and calm, with falling rain or snow.

Minima are noticed simultaneously with an agitated atmosphere and clear sky. Minima have also been attributed to the immediate proximity of abundant vegetation.

(3) The nature and direction of winds have been found to influence the proportion.

(4) The sea appears to exert a controlling influence upon the proportion of carbonic acid, acting as a reservoir for the gas, absorbing or liberating it according as it varies, increasing or diminishing, within the narrow limits which have been determined.

(5) The influence of altitude is as yet not established beyond the possibility of doubt.

By some experimenters the proportion has been found to increase with the altitude; by a greater number to diminish with the altitude; by still a third class of investigators the proportion has been found to be constant or independent of altitude.

In order to secure the best possible result of analyses which have already been made by various investigators, and by use of various methods, the latter should be compared by simultaneous application to the analyses of the same air, and under the same conditions, the standard test for each method being its ability to determine the amount of carbonic acid in a certain volume of air artificially prepared by introducing into air first deprived of carbonic acid a known volume of the latter.[*]

We will now present the results of observations and determinations made in connection with the Mount Whitney Expedition.

CONDITIONS LIABLE TO AFFECT THE QUANTITY OF CARBONIC ACID IN THE ATMOSPHERE SURROUNDING MOUNT WHITNEY.

The territory bordering upon this portion of the Sierra Nevada Range is, owing to the scarcity of water, of a desolate character. The vegetation of the Owens River Valley is exceedingly scanty, consisting of plants capable of extracting the little water they need from a considerable depth.

The country is naturally sparsely inhabited, occasional oases having been selected for the location of mining camps. One of the latter, known by the name of "Lone Pine," is situated on an elliptically-shaped oasis whose dimensions may be stated as 3½ miles by about 2.

The soil throughout the valley is fertile enough for abundant vegetation, and when supplied with water, as is the case in the oasis of Lone Pine, amply repays cultivation.

The mountains are, up to an elevation of ten thousand feet, covered by pine forests. Some of the individual trees are of large size.

The soil, naturally more sterile than in the valleys, is apparently the condition which imposes a limit upon the development of trees or other vegetation capable of withstanding extremes of heat and cold.

Water among the mountains is abundant, accumulating in numerous little lakes and rushing in streams down the steep sides to the thirsty soil below, from which it is speedily removed by evaporation to the still thirstier atmosphere.

The sky at the time the determinations were made was, with one exception, beautifully clear.

Occasional forest fires in the mountains are a source of carbonic acid capable of producing a temporary local excess in the amount of this gas in the atmosphere.

We see, then, from the above considerations, that the causes both of production and consumption of carbonic acid are apparently somewhat less active than in territory less barren.

METHOD EMPLOYED.

The method used was a modification of Pettenkofer's process, which consists in passing a known volume of air through baryta water, the strength of which solution is determined before and after the operation by means of a standard solution of oxalic acid. In the following determinations standard hydrochloric acid was used instead of oxalic acid and litmus solution was used for the color reaction.

[*] An extremely important function of atmospheric carbonic acid in the view of many investigators is its action on radiant heat, but the laboratory experiments upon the subject have not yet brought certainty, and we here can do hardly more than to refer the reader to the best known original memoirs.

Tyndall ("Philosophical Transactions," 1859, and "Investigations in the domain of radiant heat") finds that the absorptive power of carbonic-acid gas is about 90 times that of oxygen, and reaches well-known conclusions as to the absorptive power of aqueous vapor.

Magnus (translated, "Philosophical Magazine," vol. 26) finds absorptive effect of air the same whether dry or saturated with aqueous vapor.

Lecher and Pernter ("Wiedemann's Annalen," 1881, 12), after extreme precautions, find results not greatly different from Tyndall as regards gases, but reached wholly opposite ones as regards aqueous vapor, believing that the absorptive power of the atmosphere for radiant heat is due mainly to carbonic acid, which is present with the water, and to whose association with the water the effects attributed to aqueous vapor are really due.

Reiset and Müntz and Aubin (already cited) also find a connection between the amount of aqueous vapor and that of carbonic acid in the atmosphere.

Determination at Lone Pine, August 9, 1881.

1 c. c. HCl solution contained   0.003230 gramme Cl.
1 c. c. Ba (OH)₂ solution before operation = 0.663 c. c. HCl.
1 c. c. Ba (OH)₂ solution after operation = 0.600 c. c. HCl.
0.063 c. c. HCl = .0002035 gramme Cl., equivalent to carbonic acid precipitated from 1 c. c. Ba (OH)₂

$$71 \; Cl_2 : 44 \; CO_2 : : .0002035 \; Cl. : .0001261 \; CO_2$$

60 c. c. Ba (OH)₂ solution were employed in the operation.
60 c. c. Ba (OH)₂ = .007566 gramme CO₂ found:
Volume of air passed through solution, 20.18 liters.
Barometer 638ᵐᵐ. Mean temperature of air, 27° Centigrade.
Volume of air reduced to 760ᵐᵐ and 0° Centigrade, 16.16 liters.
Volume of CO₂ found .00383⁸ liter.
0.003838 ÷ 16.16 = .0002377, i. e. there were 23.77 parts by volume of CO₂ to 100000 of air at 760ᵐᵐ
and 0° Centigrade.
The sky during this determination was clouded.

Determinations at Mountain Camp, Mount Whitney.

I.—SEPTEMBER 7, 1881.

1 c. c. HCl = .003230 gramme Cl.
Before operation added 4.69 c. c. HCl to neutralize 10 c. c. Ba (OH)₂.
After operation added 4.99 c. c. HCl to neutralize 10 c. c. Ba (OH)₂.
50 c. c. Ba (OH)₂ solution were employed in the operation.
Weight of CO₂ found = .000505 gramme.
Volumes of CO₂ found = .000256 liter.
Volume air passed through Ba (OH)₂ = 30.24 liters.
Volume air corrected for temperature and pressure = 18.52 liters.
Volumes CO₂ per 100000 of air found = 16.44 liters.
Sky clear. Barometer = 502ᵐᵐ. Mean temperature = 21° Centigrade.

II.

Weight of CO₂ found .00729 gramme = .003699 liters CO₂ at 0° and 760ᵐᵐ.
Barometer = 502ᵐᵐ. Mean temperature = 21.6° Centigrade.
Volume air passed through = 30.24 liters = 18.49 liters at 0° and 760ᵐᵐ.
CO₂ per 100000 volumes of air = 20.00.

III.—SEPTEMBER 8, 1881.

Weight of CO₂ found in 60.48 liters air = .015838 gramme.
Volume CO₂ found at 0° Centigrade and 760ᵐᵐ = .008058 liters.
Barometer = 502ᵐᵐ. Mean temperature = 20.4° Centigrade.
Volume air used = 60.48 liters; reduced to 0° Centigrade and 760ᵐᵐ = 37.16 liters.
Volumes carbonic acid per 100000 of air = 21.68.

SUMMARY.

Place.	Date.	Volumes carbonic acid per 100000 air.
Lone Pine	Aug. 9, 1881	23.77
Mountain Camp	Sept. 7, 1881	16.44
Do.	Sept. 7, 1881	20.00
Do.	Sept. 8, 1881	21.68
Mean for Mountain Camp, Mount Whitney		19.37

According, then, to these analyses there is a greater amount of carbonic acid at the lower than at the higher altitude.

It is to be regretted that the method proposed by Müntz and Aubin had not appeared in the journals in time to have been applied in the above work.

CHAPTER XXI.

GENERAL SUMMARY OF RESULTS.

PRESENT CONDITION OF KNOWLEDGE ON THE POINTS INVESTIGATED.

It is the present belief that we know with something like accuracy the amount of heat which the sun sends the earth, and also that we know in general how the atmosphere acts in keeping the earth warm by letting the solar rays pass into it and keeping back others from the soil. It has been usual to state that the extreme violet rays are not readily transmitted by our atmosphere; that of the light rays about ⅓ are absorbed and ⅔ transmitted, while that as we go on through the extreme red to the dark heat rays we find the absorption growing greater and greater, the dark heat rays being found in very small quantity because they are absorbed almost wholly. It is commonly added that it is owing to this cause that the heat, which freely enters as light, escapes with difficulty when returned as dark heat in the longer rays corresponding to the lowest portion of the solar heat spectrum, and that thus the atmosphere acts to the earth a part somewhat like that of the glass cover of a hot-bed, materially aiding the solar radiation in maintaining the temperature of the planet. The ordinary conception of this heat-storing action then involves the conclusion that the dark heat of the known solar spectrum is less transmissible than the light heat, and this presumed necessity may possibly have in some degree biased physicists in their investigations on this region, where experiments are difficult. However this may be, they have been on this point provided with little evidence, so that it was rather taken for granted as a supposed necessity, than indubitably demonstrated by sufficient experiment, that the dark heat region of the solar spectrum was comparatively non-transmissible by the terrestrial atmosphere. A confirmatory circumstance to this belief was the fact that Tyndall, and after him others, had proved, by actual experiment, that, to such kinds of heat as come from terrestrial sources of very low temperatures, vapors and gases known to be important constituents of our atmosphere were almost impermeable. Such investigations, it will be remembered, had been made almost solely by the prism, and there was no way known of learning what the wave-lengths of this dark heat really were, physicists depending for their knowledge of these wave-lengths on certain formulæ which had never been verified, as they did also in a much more important matter, the determination of the amount of heat which the earth's atmosphere diverted from the direct radiation of the sun; a determination which was often made in a way which seemed to assume that nature had spared us all the trouble possible, by here conducting the whole train of her ordinarily mysterious operations, in a way so simple that the formula expressing them was itself as elementary as we could possibly wish.

To know what kind of heat was radiated from the soil we should need to know the wave-lengths of heat of this quality, of which even now we remain in ignorance. Draper, in 1881, gives the limit of the solar heat spectrum at a wave length of about 0.001 of a millimeter. M. Becquerel, in a memoir in the Annales de Chimie et de Physique, so late as August, 1883, places the ultimate limit of the known spectrum at less than .0015 millimeters, and most explicitly approves the statement that these heat rays are less transmissible by the atmosphere than others. Our own measures, here given, add the very remarkable absorption band Ω with others, and extend the directly observed spectrum to a wave length of nearly .003 millimeters, while making it probable that the transmissibility of the atmosphere *increases* up to nearly this point, where it suddenly ceases, as if all beyond were an unlimited cold body.

Having been led by the study of selective absorption to think that the portion of the sun's radiation reflected by particles of dust or mist, or otherwise dispersed in our atmosphere, is far larger than is commonly supposed, and that the little-known processes by which it is thus withheld are of importance in their bearing on problems of the widest interest, we commenced in 1880, at the Allegheny Observatory, the study of the solar heat by an instrument (the bolometer) specially invented with the object of doing for this heat what the eye in the visible prismatic spectrum does for light, that is, of discriminating between one heat ray and another, and we have been able to use it so as to determine, together with the hitherto unknown wave-length of a great number of dark-heat rays, the hitherto unknown amount of heat actually observed in each of these near the sea-level, and to tell approximately the hitherto equally unknown amount of heat in each of these dark rays before it was absorbed by our atmosphere. The results of these investigations went to show that the heat in most of the known dark-heat wave-lengths, instead of being absorbed by this atmosphere, was most freely transmitted by it, a conclusion directly opposed to the common belief, and, if true, of importance, for all the few known observations of physicists seemed to prove the contrary, and meteorologists had generally accepted these supposed observations. Continuing the heat measurements in the "light" region, I found that the heat existed there indeed in greater quantity than in the "dark-heat" region, and yet that it had been already greatly more absorbed, so that the original quantity of heat here must have been enormous as compared with that in the dark-heat region. All this was studied by narrow pencils of heat of different wave-lengths, each one of which was found to be acted on in a different degree by the atmosphere, so that the law of its absorption was not simple, but extremely complex. Taking such a partial account of this complexity as was possible, I found that the amount absorbed was much greater than had been supposed, and that the *solar constant* or heat outside the atmosphere was much greater than had been commonly stated, and the primary distribution of the rays so totally different from what we see that it seemed that they had originally been heaped together towards the blue end of the spectrum, or that the color of the sun, could we see it outside the atmosphere, would be bluish, so that media in our atmosphere, which we commonly think of as transparent, had been "playing a part analogous to that of a yellowish or reddish glass whose impure color is not a monochromatic yellow or red, but a compound of many or even all the spectral tints in unaccustomed proportions. Had we in all our lives had no light but the electric light, seen only through such a reddish glass shade, we should doubtless believe this reddishness the 'natural' color of the glowing, naked carbons, and the sum of all radiations. It would apparently answer (to a race brought up in ignorance of any other light) to our notion of *whiteness*. Its color would then seem to be no 'color' at all, and the medium would, in this case questionless, be deemed transparent (as we believe our air transparent); and if this medium were removed, and the electric light seen in its true whiteness, it could not but seem that *it* was strongly colored."[†]

*Let me be permitted, for the use of any reader unfamiliar with the subject, a very elementary illustration : Our observations at Allegheny had appeared to show that the atmosphere had acted with *selective* absorption to an unanticipated degree, keeping back an immense proportion of the blue and green, so that what was originally the strongest had when it got down to us become the weakest of all, and what was originally weak had become relatively strong, the action of the atmosphere having been just the converse of that of an ordinary sieve, or like that of one which should keep back small particles analogous to the short wave-lengths (the blue and green) and allow freely to pass the large ones (the dark-heat rays). It seemed from these observations that the atmosphere had not merely kept back a part of the solar radiation, but had totally changed its composition in doing so, not by anything it had put in, but by the selective way in which it had taken out, as if by a capricious intelligence. The residue that had actually come down to us thus changed in proportion was what we know familiarly as "white" light, so that white is *not* "the sum of all radiations," as used to be taught, but resembles the pure original sunlight less than the electric beam which has come to us through reddish-colored glasses resembles the original brightness. With this visible heat was included the large amount of invisible heat, and if there was any law observable in this "capricious" action of the atmosphere, it was found to be this, that, throughout the whole range of the known heat spectrum, what I have compared to the atmospheric sieve acted in the opposite way to the common one, or that large wave-lengths passed it with greater facility than the smaller ones.

†See Young's "The Sun" (first edition) for the views which had been reached on this point at Allegheny before the expedition started.

The simple law which we seem to have established is, then, that (with the exception of the cold bands and interruptions analogous to the black lines of the spectrum) the larger the wave-length the greater is the transmissibility, down to the utmost limit of the solar heat spectrum which has been observed, and that here the transmissibility *suddenly* ceased.

I was slow to admit this myself at first, and much rather disposed to believe that I had made some miskake than that there could have been so much previous error in the matter. My conclusions were then tested in every possible way without changing the result, and the object of this expedition was to test still otherwise and to put if possible beyond a doubt conclusions which if true must considerably affect our whole present view as to the region of the action to which we owe the preservation of the organic life of the globe.

The following summary of the results of the expedition, so far as they refer to the determination of the *solar constant* alone, will direct the reader's attention to certain conclusions to be gathered from what has already been stated at large in previous pages. First, as regards methods of observation. It had seemed to me most probable, after nearly two years of observation on the transmissibility of different kinds of heat through the comparison of high and low sun observations, that this method was affected with some constant error of a kind which had never been determined, or, in other words, that there was a daily systematic change in the transmissibility of the atmosphere between the high and the low sun observations, calculated to affect their results, and this quite independently of any error which might have been due to inability to study the heat in separate rays. At the goal of our expedition we found one of the most perfect climates of the globe for our purpose, where the observations of high and low sun at either station could sometimes be made in seemingly almost ideal conditions of tranquillity and constancy throughout the day. Apparently we have found that there is such a systematic error in high and low sun observations at one station. This seems to be demonstrated in a convincing manner by calculating at the lower station, from our high and low sun observations there, the heat which should be found at a certain height in the atmosphere; then actually ascending to this height, and finding the observed heat there conspicuously and systematically greater than the calculated one. (See page 114 and elsewhere.)

But we find also, by direct ascension in the atmosphere, that there is another important point which apparently can never be settled by observations at a single station. It is not the question whether the upper air is thinner than the lower, but whether if we take equal weights or equal masses as samples of the upper and the lower air, we shall find these equal masses possessing equal transmissibilities. The experience of previous observers has rather tended to give us a contrary opinion, and our own leaves this important point in no doubt. We find, as we ascend in the atmosphere, its transmissibility (*weight for weight*) markedly changing, so that on the whole, quite independent of its lessening density, the air becomes more and more transmissible as we directly ascend in it. (See page 117, etc.) Our present observations further bring us information about the way in which the atmosphere here behaves, not only as a whole, but also as regards each spectral ray. We find the transmissibilities, as determined only by actual observation, and without the use of any formula, confirm in the most absolute manner the conclusion that, with the exception of the cold bands already noticed, *the transmissibility increases throughout the spectrum when the wave-length increases*, the violet being more transmissible than the ultra violet, the blue than the violet, the green than the blue, the yellow than the green, the red than the yellow, and the border of the dark heat than the red, until when we have gone still farther down the spectrum the transmissibility will become almost complete, except for sudden great interruptions of it—the cold bands, the largest of which was discovered on Mount Whitney, and will be found delineated here in our spectral charts. (See page 133 and chapters on spectro-bolometer generally.) We find also, however carefully determined the transmissibilities are at any single station, that the ratio of these transmissibilities to each other may be very different when determined by direct comparison of the results above and below. We mean that, for instance, the transmissibility of the red ray may be nearly the same whether determined at one station or by the comparison between two, but that the transmissibility of the blue ray as determined at any single station, even an elevated one, is altogether greater than that determined by combination of observations made at an elevated station with those at a lower one. In other words, when we have actually ascended to the elevated station we find there a greater proportion of the blue rays even than we had been prepared to expect from our observations at Allegheny, and it seems probable that all of our previous conclusions as to the predominance of these and the composition of the light outside the atmosphere have been rather within than without the truth.

We find, then, both that the ordinary method of high and low sun comparisons, under exceptionally good conditions, gives us too large coefficients of transmission, and that the error is greater as the ray is more refrangible, the error of the most assiduous observation at one station on the more refrangible rays being very marked indeed, and always of a nature to give too small a value to the solar constant. These remarks apply not only to heterogeneous radiations such as are noted by the thermometer, but even to pencils of rays almost physically linear. It is observed at the same time as we ascend that the transmission of each ray on the whole grows greater for like air masses, but that the proportion in which it grows greater appears also to vary very much between the extremities of the spectrum. The fact that like air masses grow more transmissible as we ascend, and the fact that non-homogeneous rays as a whole are less transmissible than we calculate from our present formulæ, are not isolated from each other, but have a bond of union in a third fact, that the selective absorption of our atmosphere is largely due to distinct particles present in the lower air in greater quantity than in the upper.

From all that has been observed, whether at Allegheny, Lone Pine, or Mountain Camp, we conclude that the action of our atmosphere is incomparably more complex than the ordinary theory assumes it to be, and, even were we provided with a much better theory, we repeat that causes not yet fully understood, introduce systematic errors into high and low sun observations, tending to impair the results of the best observer at one station, and on the whole to lead him to underrate the amount of the absorption. It will not be surprising, then, to find that, while our results for the amount of this absorption are much larger than those of most previous observers, we cannot assert their accuracy within limits as narrow as might be wished. Pouillet gives his result in terms of thousandths of a calorie, and so do some of his successors, though probably the use of a third place of decimals is not intended by them to represent the accuracy of their results, even in their own opinion. However this may be, it seems certain that the earlier results have been changed by amounts significant even in the unit's place; or if we suppose that only the second decimal is significant in the opinion of these writers, we shall find that their error is still fully one hundred times as great as what they themselves supposed it likely to be. Warned by this, I shall not ask for confidence in the new value of the solar constant beyond at most the first decimal place.

We have from the observations at Allegheny 2.84 calories. There seems to be very little doubt, in view of subsequent experiences, that this value is larger than would have been obtained under more perfect observing conditions. We believe, in other words, that during the low sun observations at Allegheny the atmosphere is systematically less transmissible as compared with the noon observations than it proved at Mount Whitney. The actually observed heat at Mountain Camp we have shown was about 2 calories. We have already remarked that an extremely useful check upon our observations is the value to be found by adding to the heat received directly from the sun that received at the same time from an exceptionally pure sky. This (in view of the sun's altitude at Mount Whitney) will give us a value of about 2.6 calories. The result which we have deduced as most probable from our comparisons of the heat at the upper and lower station is 3.07. (See page 118.) These are the principal means for our final determination, unless we include one other of a wholly different nature.

The earth's actual mean surface temperature being about 15° or 16° (Centigrade), and it being admitted that the heat from the interior, from the stars, from the dynamic effect of meteorites, and in general from all other sources, is negligible, it will follow that if we know the laws under which this heat enters and escapes from our atmosphere, we can determine what amount must be supplied to the earth from without to maintain this known annual temperature of 15 or 16 degrees. The time has not yet probably come for doing this with certainty, but this method is so wholly independent of the others that it may be interesting to us to know its results. Pouillet's data in this respect have been modified somewhat by recent observation. Accepting them, however, as approximately true, we must admit, if we follow his ingenious course of reasoning (but reject his hypothesis of an enormons heat radiation from the stars), that the solar radiation is represented by 3.13 calories. We have fixed already, from the nature of the observations on Mount Whitney, as an upper limit to the solar constant the value 3.5 calories, and as a lower limit the value 2.6 calories. Between these limits we have three independent determinations, without including the method of Pouillet. I have already given the reasons which make me deem it unadvisable to attempt to assign weights to these determinations and to combine them by any conventional rule.

The reader has had before him in the preceding pages a detailed statement of the observations and methods which have led to these results, and my own inference from them is that in the present condition of our knowledge it is impossible to fix any value of the solar constant with the precision which used to be assigned to it ere the difficult conditions of the actual problem were known, though I think that it has been clearly shown in the preceding pages that this solar constant is greater than has ordinarily been believed.

My conclusion is that in view of the large limits of error we can adopt THREE CALORIES *as the most probable value of the* SOLAR CONSTANT, by which I mean that at the earth's mean distance, in the absence of its absorbing atmosphere, the solar rays would raise one gramme of water three degrees Centigrade per minute for each normally exposed square centimeter of its surface.

This is approximately 126,550,000 ergs per square centimeter per minute. Expressed in terms of melting ice, it implies a solar radiation capable of melting an ice shell 51.45 meters deep annually over the whole surface of the earth. Somewhat less than two-thirds of this amount reaches us at the sea level ordinarily from a zenith sun, but unless very great precautions are exercised we are apt to undervalue this directly received amount. It follows, then, that the selective absorption of our atmosphere is not only more diverse in kind, but that the total atmospheric absorption is far greater in amount than has been commonly supposed.*

On other important points our conclusions are as follows: (1) That although the actual solar radiation is thus largely increased, yet the temperature of the earth's surface is not due principally to this direct radiation, but to the quality of selective absorption in our atmosphere, *without which the temperature of the soil in the tropics under a vertical sun would probably not rise above* −200 C. Nearly all the 215 or more degrees of difference between this and the actual mean temperature of the planet's surface is due to this selective absorption, which accumulates the heat, though in a manner which has not been hitherto correctly understood. It should be understood that these researches have here a practical bearing of great consequence. The temperature of this planet, and with it the existence, not only of the human race, but of all organized life on the globe, appears, in the light of the conclusions reached by the Mount Whitney expedition, to depend far less on the direct solar heat than on the hitherto too little regarded quality of *selective* absorption in our atmosphere, which we are now studying. (2) Generally speaking, the radiation which we see enter we see escape within the utmost limits of the known solar spectrum. The heat-storing action, from checked re-radiation, to which the surface temperature of this planet is due, apparently goes on *beyond* these limits where no spectral measurements have yet been made. *No such wave-lengths as those belonging to the heat radiated from the soil, we believe, have ever entered our atmosphere from the sun,* though we admit their existence in the solar spectrum before absorption. These statements must not be understood as at all implying a denial of the action of water vapor, which we find probably plays an important part in the phenomena with which we have been dealing.

The preceding considerations I have limited to the bearing of these observations on meteorology, but I need hardly point out to the student of solar physics how greatly the knowledge of the relative increment of the blue, violet, and ultra-violet region must raise our estimate of the *temperature* at which such radiations were first emitted, or enlarge on the relation of the preceding work to celestial physics in other ways.

The observations which have been detailed could hardly have been made at a better site than they were, and I know of none in the country as good. All the comparisons of the work herein cited as done at the sea level (at Allegheny), with that at a great altitude, enhance our estimates of the importance to meteorology of systematic observations at a very great elevation. It would

---

* September 15, 1884.—The reader who may be indisposed to accept so great an absorption without further discussion is referred for it to an article by the writer in the American Journal of Science for September, 1884.

be of the greatest service to solar physics and to meteorology if an observatory for these objects could be established on Mount Whitney, and I strongly recommend the site, for the purposes named, as among the most desirable on the North American continent. Until such a permanent observatory is established observations made under the extreme difficulties attendant on such an expedition as the present cannot possess the accuracy otherwise attainable, and everything which has been described as to the hard conditions under which these were carried out will dispense me from claiming for the results, where novel, a higher character than that of useful first approximations in a field of research where so much of the highest interest yet remains.

I have already acknowledged my indebtedness to the military members of the expedition, but to all, both military and civil, I cannot but remember how much it owes for earnest and helpful service beyond what the line of strict duty demanded. To Mr. J. E. Keeler, in particular, whose ability and fertility of resource were tested in many varied capacities, and without whom, on more than one occasion, we should have failed, particular acknowledgment is due. After the return of the expedition, Mr. F. W. Very was joined with him and others in the very long reductions. To Mr. Very's conscientious care, and to his acquaintance with the subject of meteorology, I desire to express my great indebtedness, as well as to others who have assisted me in the preparation of the volume.

I cannot close without remembering that the expedition itself, and this account of it, are first owing to the generosity of a citizen of Pittsburgh, a friend of the observatory, who does not desire the mention of his name as a donor, but to whom anything of use to knowledge here is primarily due.

S. P. LANGLEY.

ALLEGHENY OBSERVATORY, ALLEGHENY, PA., *December* 21, 1883.

APPENDICES.

APPENDIX I.

DISCUSSION OF THE METHOD EMPLOYED IN THE REDUCTION OF PSYCHROMETER OBSERVATIONS.

It is well known that the indications of the psychrometer are only comparable with each other when certain conditions of environment are complied with. Regnault himself has experimentally investigated the magnitude of the influence produced by changed surroundings, and the consequent deviation from his theory, and has found it to be great in many instances. As the conditions which have been found most conducive to accuracy were seldom present in the psychrometer observations made during the expedition, it seemed desirable to examine the probable extent of their influence, with a view to the further improvement of the results, if this should be found possible. Regnault's formula, adopted in the "Smithsonian Tables" (B, page 12), is

$$x = f' - \frac{0.480}{610 - t'}(t - t')h = f' - A(t - t')h,$$

in which, for slight variations of temperature, the coefficient A is nearly constant.

In the Lone Pine observations t' averages about 15° Centigrade, whence $\frac{0.480}{610 - 15} = .000807$ is the value of A used in the Lone Pine reductions. Regnault's experiments show that in the open air a smaller value than this may be used, but in inclosed positions a greater one is necessary, on account of the radiation from the walls of the inclosure; and in a small inclosed chamber A may become as great as .001280. This is the same thing as saying that the depression of the wet bulb under these circumstances may need to be increased, for the purpose of our calculation, in the proportion of $\frac{1280}{807} = 1.59$.

The surroundings of the psychrometer at Mount Whitney appear to have been very similar to the "inclosed chamber" of Regnault, and it is worth noting that a comparison of its indications with those of a Regnault's hygrometer showed the necessity of a nearly identical correction.

At Lone Pine the case was more open, allowing a freer circulation of air; but, owing to the constant invasion of dust, it is probable that the wick was seldom in a condition to allow of free evaporation. A number of comparisons of this instrument with a Regnault's hygrometer seemed to show that a correction, similar to that deduced for the mountain apparatus, was also desirable for the psychrometer used at Lone Pine.

TABLE A.

Comparison* of psychrometer and Regnault's hygrometer.

[Wind, gentle or calm.]

Instrument.	Mount Whitney.*	Lone Pine.†
	C.	C.
Dry-bulb thermometer	1° 62	27° 62
Dew-point (psychrometer)	— 0 45	15 45
Dew-point (Regnault's hygrometer)	—12 5	8 72
Depression of dew-point (psychrometer)	19 07	12 18
Depression of dew-point (Regnault's hygrometer)	31 12	18 91
Ratio	$\frac{31.12}{19.07} = 1.63$	$\frac{18.91}{12.18} = 1.55$

*Observer, O. E. M.   †Observer, A. C. D.

* It is to be noted that Regnault's hygrometer itself indicates too low a dew-point when it is exposed to a dry wind. (See an article by Crova, "Sur l'hygrométrie," Journal de Physique, tome II, 2ᵉ série, p. 450.)

The question of the influence of the wind upon the psychrometer observations is one which requires more careful consideration.

A few comparisons of the psychrometer and Regnault's hygrometer, taken during the prevalence of a wind described by the observer as "fresh," indicate a diminished factor for reduction of depression of dew-point by the psychrometer to correspond with the same quantity as given by the Regnault hygrometer.[*]

TABLE B.

Comparison of psychrometer with Regnault's hygrometer.

(Wind, fresh. Station, Lone Pine. Observer, A. C. D.)

Dew-point (psychrometer).	Dry-bulb.	Dew point (Reg. hygr.).
C.	C.	C.
12 .62	19 .80	6 .50
11 .44	24 .50	9 .50
4 .93	27 .39	3 .39
14 .25	22 .72	8 .72
1 .28	25 .94	2 .56
Mean.. 8 .90	23 .79	6 .67

Depression of dew-point (psychrometer) .... 14°.89
Depression of dew-point (Regnault's hygrometer)... 17 .12

Ratio .... 1 .15

The summary of Lone Pine observations, in Table 150, shows an apparently systematic variation in the computed force of vapor with varying velocities of the wind as follows:

Average values from a smooth curve:

	m m.
Calm	8.8
Gentle	7.5
Fresh	6.1
Brisk	4.5
Gale	3.8

The extreme values of 8.8 mm. and 3.8 mm. for "calm" and "gale," respectively, correspond to dew-points of 9°.4 C. and —2°.4 C., and if the reduction were due to air-currents about the psychrometer, assuming an air-temperature of 30° C., we should have 20°.6 C. and 32°.4 C. for the depressions of the dew-point in the two cases, the ratio of which is 1.57, a smaller change than that observed by Regnault, who has demonstrated the effect of the wind upon the psychrometer by passing dry air with various velocities through a tube containing a wet and a dry bulb thermometer. (See Annals de Chimie et de Physique, 3° sér., tome XV, p. 201.).

The following are some of his figures:

	t.	t'.	t—t'.	Flow of air in one second.
	C.	C.	C.	Ce ctm.
(1)	14°.96	7° 58	7 .38	18.5
(2)	14 .96	4 .33	10 .63	84.5

As reduced by the ordinary tables, no account being taken of the wind, these values would be obtained:

	Temperature of air.	Temperature of dew-point.	Depression of dew-point.
	C.	C.	C.
From (1)	15 .0	—3 .3	18 .3
From (2)	15 .0	—37 .7	52 .7

* For complete record of above observations see pages 171, 172.

The above numbers illustrate the wide divergence of results computed by the tables from such unusual experiments, as well as the marked action of the wind when very dry.

A comparison of (1) and (2) shows that with a velocity of wind 6.35 times as great the depression of the wet bulb was 1.41 times as great, and the calculated depression of the dew point was 2.88 times as great as when a moderate velocity of the air current was maintained.

It must be remembered that the absolute dryness of the air used in these experiments of Regnault increases the errors in question to their greatest possible extent.

If the diminution of aqueous vapor with increasing wind which was observed at Lone Pine had been altogether an instrumental effect, produced, for example, by imperfect evaporation from a wick incessantly clogged with dust, the difference between the indications of the psychrometer and Regnault's hygrometer should have progressively diminished with the increment of the wind, and should even have been obliterated during a gale; but while the comparisons in the table give some support to this explanation, it is probable that some part of the observed diminution in the indications of the psychrometer with increasing wind is not an instrumental peculiarity, but an actual fact, since the prevailing winds at Lone Pine have blown across a wide expanse of desert, whence it may well happen that an increasing wind should systematically diminish the absolute amount of moisture by bringing into this arid region increasing quantities of the still drier air of the desert.

The observations on top of the mountain are not sufficiently numerous to afford much help in regard to this subject. So far as they go, they afford no indication of diminished moisture with increasing wind. Whether this is owing to the instrument case being "as well shielded from air currents as could be obtained" is an open question, but it raises a suspicion in regard to the magnitude of the supposed influence.

It is to be noted that when the southeast wind, so prevalent at Lone Pine, blew on the mountain (which happened but seldom) it produced a diminution in the absolute moisture. (See, however, the description of the environment of the Mountain Camp and its probable influence on the wind, page 180.)

It must, therefore, be concluded that a part of the diminution in the psychrometer indications observed to occur at Lone Pine with increasing wind must be attributed to instrumental peculiarities, and a part to the desert origin of the wind, but the proportional part to be ascribed to either influence remains uncertain.

While the dew-points observed with Regnault's hygrometer were probably hardly numerous or accurate enough to be used for correcting Regnault's psychrometer formula, it seemed possible that they might be employed to determine an empirical correction which, applied to the indications of the psychrometer, would counteract the effect of the prejudicial influence from the instrumental environment. The opinion of the Signal Service was therefore sought.

Letter of Prof. S. P. Langley to the Chief Signal Officer, United States Army.

ALLEGHENY OBSERVATORY,
November 21, 1884.

SIR: It was the opinion of Professor Abbe that results obtained on Mount Whitney by the wet and dry bulb might be of doubtful precision, owing to the exceptional atmospheric conditions. I therefore made observations simultaneously with the Regnault hygrometer, the results so far justifying Professor Abbe's prediction that they are quite discrepant with each other. As the treatment of the case involves questions of some importance, I send a copy of our preliminary reductions of these few cases where we have simultaneous readings of both instruments.

An opinion from the Signal Office as to the relative value of these instrumentally discrepant determinations I should receive with interest.

S. P. LANGLEY.

General W. B. HAZEN, *Chief Signal Officer, United States Army.*

In regard to the above the following remarks by Professor Abbe were received from the Chief Signal Office:

Memorandum dated November 26, 1884. Remarks on 1623 Mic., 1884.

The observations at Mount Whitney and Lone Pine, as communicated by Professor Langley, show larger discrepancies between the dew-points observed by Regnault's apparatus and those computed by the psychrometer than are usually found. For instance, on page 2, at the lower camp, altitude 12,000 feet (barometric pressure not stated, but assumed to be about 20 inches), for dry-bulb 63° F. and for wet-bulb 45° F., the Regnault formula gives a computed dew-point of about 26° F., while the observed dew-point is 9.5° F., or lower by 16.5° F. The similar comparative observations made at Lone Pine give observed dew-points almost always lower than computed according to Professor Langley by quantities varying between 0 and 18° F. The former of these observations relates to an atmosphere whose relative humidity is 10 or 25 per cent., but for observations at Lone Pine the relative humidity is not given.

These discrepancies are much larger than those found by Regnault, Blanford, Belli, and others who have made similar observations in dry atmospheres and at low pressures, and I think that the explanation must consist partly in the errors incident to the use of the dew-point apparatus. Any more definite statement than this is impossible, owing to the fact that nothing is said by Professor Langley as to whether he has corrected for barometric pressure and the errors of his thermometers. The apparent effect of the wind is such as to indicate that he used thick wicking in place of thin muslin. His empirical reductions for wicking and for wind are plausible, but should be founded on a large number of observed dew-points before being adopted. If he still has the apparatus he should repeat these observations in the dry winter air.

Unless I know by personal inspection how carefully the dew point apparatus was handled I should advise the adoption of the psychrometer work. If the observed dew-points are considered reliable *per se*, they can hereafter be utilized in connection with those made throughout the world to improve on Regnault's psychrometer formula, which latter work has been undertaken by several, notably Belli. (See Zeitschrift für Meteorologie, 1880 and 1881.)

In accordance with the advice of Professor Abbe, the hygrometic indications of the psychrometer have been adopted without further correction.

Note by Professor Abbe, January 31, 1884.—Not only did the meagerness of the data sent by Professor Langley forbid deducing an empirical correction applicable to his psychrometric results, but from a theoretical point of view it is probable that the use of thick wicking on the wet-bulb so protected it from the direct influence of radiation that even when t-t was large the correction for this item was reduced to a minimum; in other respects, of course, the use of thick wicking is objectionable, but as the Regnault dew-point apparatus in very dry air and low dew-points gives too low results, I think it preferable to retain Langley's psychrometric results and reserve the observed dew-points for further discussion.

APPENDIX 2.

EXPERIMENTAL DETERMINATION OF WAVE-LENGTHS IN THE INVISIBLE PRISMATIC SPECTRUM.

(NOTE.—The following investigation was made at the expense of the Bache fund, and is published here by the permission of its trustees.)

In September, 1881, while engaged upon Mount Whitney, in measuring with a linear bolometer the heat in the invisible spectrum of a flint prism, I came upon a hitherto unknown cold-band * whose deviation indicated a (probably) very great wave length. We have had up to the present time no way of measuring such wave-lengths directly, but are accustomed to determine them by more or less trustworthy extra-polation formulæ, the best known of which is Cauchy's. Accordingly, I attempted to calculate the wave-length by Cauchy's formula, but was at once conducted to an impossible result. The formula, in other words, declared that no such index of refraction as I had measured was possible in the prism in question. But the measurement was a fact beyond dispute, and this drew my attention to the grossness of the errors to which the customary formulæ may lead.

Every prism gives a different map of the spectrum, nor when we find a band or line by the prism have we any means of fixing the absolute place, except by a reference to the normal or wave-length scale, or to one derived from it.

It is desirable to define, at the outset, the sense in which the term "normal" is here used as a synonym for "wave-length" spectrum.

The amount of energy in any region of the spectrum, such as that in any color, or between any two specified limits, is a definite quantity, fixed by facts which are independent of our choice, such as the nature of the radiant body, or the absorption which the ray has undergone. Beyond this, nature has no law which *must* govern us in representing the distribution of the energy, and all maps and charts of it are conventions.

If the length of the spectra formed by any two different agents, such as a prism and a grating, be made equal, it does not then follow that the lengths of similar portions must be equal. In the

* Since designated as "Ω."

case supposed we observe, in fact, that the red portion (for instance) of the prismatic spectrum will be narrower than the red portion of the second. But since the amount of energy in the red must be really the same in both, we must, in a graphic representation of this energy, increase the height of the ordinates in the red of the prismatic spectrum, so that the areas shall remain the same.

The position of the maximum ordinate is, then (in one sense), a matter of choice, and fixed only by the scale we elect to employ. We find, for instance, in the prismatic spectrum, that this ordinate is in some part of the infra-red, depending on the particular prism used, while in the grating spectrum it is, under the same circumstances, always in one part of the yellow; and we might conceive of an apparatus which should always exhibit it in the ultra-violet, or which should even show the same energy at one wave-length as at any other, or embody any other arbitrary mental picture of it. It is certainly a practical consideration of the first importance that no such apparatus actually exists; but still, whether it exists or not, in so representing the distribution of energy we should break no law except that imposed by considerations of simplicity and convenience.

Did the word "normal" then signify "absolute" there would be no spectrum exclusively entitled to such a name; but in this connection the word is always to be understood in its radical meaning of an accepted rule or type of construction. Such a type exists in the wave-length spectrum, and it has obtained general acceptance, not only on account of its simplicity and convenience, but of its, at present, unique claim to be a "natural" one. It is properly distinguished as the "natural" scale from its not merely representing a mental picture of the distribution of the energy under a very simple law, but of actually *bring* that which we do produce by our most efficient optical apparatus and make visible and measurable at will.

While we remain at liberty, then, to represent the energy spectrum in terms of the wave-frequency or of the reciprocal of the square of the wave-length, or of any other function of it, and while we may often find occasion to use these scales for some special purposes, we are (and all the more especially that we habitually speak in terms of the wave-length) led by considerations of a very practical kind to take as our normal or standard scale that of the wave-length itself.

Since we have this normal spectrum actually before us, through the concave gratings constructed by Professor Rowland, it may seem as though we might dispense with the prism, but this is not as yet possible for the lower part of the spectrum, where overlapping spectra and feeble heat here make the use of the grating too difficult. We could use the solar energy here, not in the form of heat, but of chemical action, as in photography, a great advance might be made; and there is reason, I believe, to hope that the labors of Professor Rowland and Captain Abney will ere long do this for us with precision. At present, however, we have only heat, and the thermopile or the bolometer, which latter, though less sensitive than the camera, can be made, as I shall show, to determine experimentally within known limits of error, the actual wave-lengths corresponding to given indices of refraction, and hence to afford here valid experimental data for passing from the prismatic spectrum to the normal one. The reason why this so desirable information has never been obtained before is twofold: (1st) While the measurement in question can best be made by means of a prism and grating conjointly, the heat, which in the lower prismatic spectrum is very faint, becomes almost a vanishing quantity when it has passed the grating also, where the heat is on the average less than one tenth that from the prism. We must use, too, if possible, a narrow aperture to register this heat, for a broad one might on account of the compression of the infra-red by the prism cover the whole field, in which its work should be to discriminate; (2d) We must have not only an instrument more sensitive than the common thermopile, but we must devise some way of fixing, with an approximate precision, the point at which we are measuring when that point is actually invisible.

The apparatus I have devised for this double purpose has done its work with a degree of accuracy, which, if it may be called considerable, as compared with what we have been used to in heat measurements, is yet necessarily inferior to that obtained by the eye, and less than we may hope for at some future time from photography. Nevertheless, it has, I believe, given experimental data, very far outside the visible spectrum, by which we may either construct an empirical formula and supply its proper constants so that it will be trustworthy within extended limits, or test the exactness of such formulae as Cauchy's, Redtenbacher's, &c., which, while professing a theoretical basis, only agree in their results within the limits of the visible spectrum (from which they have been in fact derived, and where they are comparatively unneeded). They contradict each other, as will be seen, as soon as they are called on for information, in the region outside of it, where they would be chiefly useful.

The present work has been preceded by a new map of the invisible prismatic spectrum, where the abscissæ were proportional to the deviations in a certain prism (see Plate XI); and the immediate object of this research is to pass from the arbitrarily spaced prismatic scale, belonging to the particular prism in question, to a map on the normal and absolutely general one.

I should perhaps make the cautionary remark, that the general conclusions here offered, as to the relation of wave lengths and indices of refraction, have been drawn from observations on a single prism and have not been experimentally verified on others. This is on account of the extremely slow and laborious character of the process used (which has involved some months of labor for this special prism). Though there seems no reason to doubt the generality of our conclusions, it may be hoped that these experiments will be repeated with prisms of other material, and by other observers, now that the preliminary obstacles have been removed.

In order to map the spectrum on the normal scale, where the wave-lengths are equally spaced, from such a map as that shown in Plate XI ("Prismatic Spectrum"), in which the consideration of wave-lengths does not enter, it is necessary to establish some relation between the wave lengths of rays and their deviations, or between their wave-lengths and refractive indices, which are connected with the deviations by the well-known formula

$$ n = \frac{\sin \frac{(a+d)}{2}}{\sin \frac{a}{2}} $$

where a = the refracting angle of the prism, d = the deviation, and n = the corresponding index of refraction. In the visible spectrum the deviation, in any prism, of the Fraunhofer lines (whose wave-lengths have been very accurately determined) can be measured by means of an eye-piece with cross-wires; and, from a sufficient number of such measurements, by making ordinates proportional to indices of refraction (or to deviations) and abscissæ proportional to wave-lengths, a curve may be found whose equation is $n = (\varphi)\lambda$ or $d = (\varphi)\lambda$, representing the required relation to any degree of exactness.

In the invisible spectrum the difficulties are immensely greater, and demand special means, not only on account of this invisibility, but owing to the absorption of the prism and to its compressing the rays.

The prism here used was made by Adam Hilger, of London, and its optical properties are in every way satisfactory. It is of a white flint, which has proved singularly transparent to the longest solar waves. Its principal constants have already been given (p. 130).

APPARATUS FOR MEASURING OBSCURE WAVE-LENGTHS.

In 1882 an apparatus was employed in which invisible rays, after passing through the Hilger prism, at a known deviation, fell on a Rutherford reflecting grating (either of 681 lines to the millimeter or half that number), from which the diffracted invisible ray fell on the bolometer at a measured angle with the grating. By the use of the known formula ($n s \lambda = \sin i + \sin r$) connecting the angle of diffraction with the wave-length, the wave-length was then found.

Several determinations were thus made of wave-lengths in the upper part of the infra-red, where the heat is relatively great; but, though the definition of the Rutherford grating was admirable, it was not large enough to supply sufficient heat to enable measures in the lower infra-red to be made with confidence.

In May, 1882, I had the good fortune to secure one of the very large concave gratings, then newly constructed by Professor Rowland, and which he was kind enough to make for me of a very short focus, so as to give a specially hot spectrum. After many essays, during which a great number of mechanical and optical arrangements for getting rid of the superposed spectra were tried with unsatisfactory results, it became clear that, for this large and concave grating, it was necessary to let the ray fall first on it and then on the prism, thus making the wave-length the known, and the deviation the unknown, quantity.

In the use of this form of grating the slit is placed in the circumference of a circle, whose diameter is equal to the radius of curvature of the grating, and which touches its surface. The spectra are then formed, without the need of collimator, observing telescope or any further apparatus, all lying upon the circumference of the circle which contains the slit. The grating which was employed contains 18,050 lines, 142 to the millimeter (3,610 per inch), ruled on the sur-

Fig. 17.

Course of Rays through Apparatus
in the determination of lengths of waves of Obscure Heat

face of a concave mirror of speculum metal of 1m.63 (64 inches) radius of curvature, and exposes a ruled surface of 129cm (20 square inches). By this large surface a spectrum is produced sufficiently hot, even in its lower wave-lengths, to effect the bolometer strips after the various reflections and absorptions to which the heat is necessarily subjected in passing through the apparatus.

Figure 17 illustrates the means finally adopted, and the course of the rays through the apparatus; although, for the sake of distinctness, the mechanical devices used to maintain the proper arrangements of the parts are omitted. The rays of light, coming from the 12-inch flat mirror of the large siderostat, pass across the apparatus, and fall upon a 7 inch concave speculum at M, by which at a distance of about 5 feet they are converged to a focus at S_1. At this point is a vertical slit, adjustable to any desired width by a double screw, which moves both jaws at once, so as to keep the center always in the same place. This slit is protected from the great heat by a plate of iron pierced with an aperture only a little larger than the slit when open to the usual width. Beyond S_1 the rays diverge and fall upon the concave grating G. Directly opposite the grating is a second slit, S_2, also double acting, and the apparatus is so arranged that the two slits S_1, S_2, and the grating G, always lie upon the circumference of a circle, whose diameter is 61 inches; and therefore in whatever manner the slits may be placed, the light coming through S_1 forms a sharp spectrum upon S_2. A very massive arm carrying the grating, the slit S_2, and the heavy spectro-bolometer, is pivoted at the center of the circle, so that the relative positions of these parts are unchanged. The slit S_2 is automatically kept diametrically opposite the grating, and on the normal to its center.

The slit S_2 is the slit of the spectro-bolometer, provided with the same attachments as when used for mapping the visible spectrum (except that it is now fitted with simple collimating and objective lenses of the same special kind of diathermanous glass as the prism, instead of its own concave mirror). Its eye-piece and the bolometer are interchangeable.

By means of the eye-piece and graduated circle, the deviation and consequently the refractive index of the rays passing through the slit can be determined, if they are visible. If they are invisible, their exact wave-length is known by a simple ocular observation of the visible ones, on which they are superposed by the action of the grating, while their subsequent deviation is determinable by the bolometer placed at B, provided they retain sufficient energy to affect the instrument. It will be seen that according to this method all those invisible rays which are n times the definitely known length of some visible ray are caused to pass together through a slit, and then through a prism, which sorts out the ray of the first spectrum from that of the second, that of the second from that of the third, and so on, so that the corresponding index of refraction may be determined by observation with the eye in the case of the visible, with the bolometer in that of the invisible ray.

To illustrate the use of the above described apparatus under somewhat unfavorable circumstances, let us consider as an example the observations of June 13, 1882, which were taken far down in the spectrum where the heat is feeble and the galvanometer deflection small, requiring a widely open slit. The apparatus having been previously adjusted, and the sunlight properly directed by the siderostat, the visible Fraunhofer line D, of the third spectrum of the grating, was caused to fall upon the slit S_2 of the spectro-bolometer. Then, according to the theory of the grating, there passed through this slit, rays having the wave-lengths—

0μ.589 (3d spectrum, visible).
1. 178 (2d spectrum, invisible).
1. 767 (1st spectrum, invisible).

The prism having been removed, and the telescope brought into line, an image of S_2, of the same size as the slit itself, was formed in the focus of the object lens, and on testing with the bolometer, whose face was covered with a card-board screen pierced centrally with a 2ᵐᵐ slit, the heat of this image produced a deflection of the galvanometer needle of about 30 divisions. The prism was then replaced on the automatic holder and set to minimum deviation, and the image of the slit, containing superposed rays whose combined effect had produced the deflection just mentioned, was separated into three similar images (as in Fig. 18),[*] each composed of nearly homogeneous rays, and of same dimensions as the slit S_2. Of these three bands only the first or most refrangible, containing the D, line, was visible, and its deviation was found to be 172°41', agreeing with the value given by the table. It was the object of the experiment to find the place of the lower invisible band, by groping for it, i. e., to determine its deviation by trials with the bolometer at intervals sufficiently close to avoid the possibility of missing it altogether. According to Briot's formula, the deviation should be 152°21', and in the preliminary search the circle was accordingly

---

[*] These three images, being composed of rays of different wave-lengths, could not all be in focus at the same time, since the collimator and objective of the spectrometer were simple lenses. The lenses were adjusted by means of a table of focal distances previously prepared, so as to throw a sharp (invisible) image of the band to be detected.

set to this reading. Beginning at this point, and exposing the bolometer at every five minutes of deviation, it was found that the maximum effect was obtained nearer 45° 15'. The approximate position having thus been found, the slit S, was narrowed to 2ᵐᵐ, and the following measurements

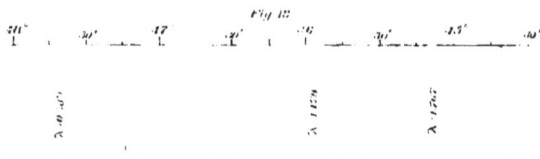

Fig. 12

Spectrum formed by Prism in determination of Wave-lengths.

taken, the horizontal line giving the mean results of a series of thirty exposures of the bolometer, as it was moved through the spectrum.

TABLE C.

Method of finding refrangibility of feeble heat rays.

Prismatic deviation	45 02	45 07	45 10	45 15	45 20
Means of galvanometer readings	4 6	5 6	6 0	5 8	2 7

The maximum reading at 45° 10' corresponds to a coincidence of the 2mm. bolometer aperture with the 2mm. invisible image of the slit, whose position is sought. From a subsidiary curve drawn through the points whose co-ordinates are respectively,

$$(x=45° 02', \quad y=4.6),$$
$$(x=45° 07', \quad y=5.6),$$
$$(x=45° 10', \quad y=6.0), \&c.,$$

it was concluded that the deviation of rays whose wave-length is 1.767 is 45° 10', and each point in this determination being obtained from the mean of five observations, the result is partly free from irregularities caused by changes in the state of the sky, and minute instrumental variations from extraneous causes, which here become of great relative importance, owing to the feeble heat measured.

Subsequent determinations, like the preceding, gave for the deviation of the same ray 45° 06' and 45° 07', and from a consideration of all the deviation adopted was (instead of 45° 21', as given by Briot's formula) 45° 08', corresponding to a refractive index of 1.5549.

By means of measurements like the one described above, the deviations of various obscure rays of known wave-lengths were determined. The indices of refraction were then computed by the usual formula

$$n = \frac{\sin \frac{1}{2}(a+d)}{\sin \frac{1}{2}a}$$

where $a=62° 34' 43''$. The results are contained in the following table, where, however, only the results of successful days are given, most of the observations having been lost through changes of the sky during the course of one determination: *

TABLE D.

Experimental determination of d or n as a function of λ (Hilger prism).

Date of observation.	λ	d	n
1882			
April 12	1 940	0 1054	46 12	1 5654
April 9	1 290	0 1069	45 54	1 5625
June 27	1 653	0 1091	45 46	1 5582
June 15 27	1 767	0 1094	45 08	1 549
July 11	2 090	0 1104	44 45	1 5 41
June 7	2 356	0 0110	44 25	1 5478

*All these observations for discovering the relation between n and λ can be conducted with at least as much advantage by a powerful and constant electric light as by sunlight. The latter only, however, was at the observer's actual command.

We observe that, where measures are taken in the prismatic spectrum alone, we can generally use with advantage a bolometer of as small an aperture as one fifth of a millimeter; but that here it is advisable to open it to 2mm., owing to the relative expansion of the spectrum, and to the very feeble heat. Where such measures are taken in the shorter wave-lengths of the infrared, i. e., in the upper (invisible) grating spectrum, they are comparatively easy from the greater heat, and can be made with a narrow aperture; but where in the lower invisible spectrum, as here, they grow more and more difficult as the wave-length increases, so that if we could repeat each observation often enough, we should determine a separate probable error for each point of the curve.

Owing to difficulties arising from the almost infinitesimal amount of heat in question, numerous subsidiary observations are requisite for a single determination, which it therefore takes long to make, the probable errors in the table being found in each case from between 20 and 100 readings. If it should possibly appear to the reader that in the three months of consecutive labor which were given to this part of the work, more than six points might have been determined in the curve, he is asked to remember that what is here difficult has till now been impossible. If we treat, in such a case as that given above, the discrepancy of the cited determinations as being fairly typical (as they appear to be), we shall obtain a probable error of about one minute of arc, and a comparison of the different points with each other on the large curve exhibited indicates a similar result. It will not appear improbable that this accuracy of setting should be attained by a bolometer whose face covers many minutes of arc, when it is noted that in the given instance nearly 100 readings are taken to fix the single determination. The error in the determination of a wave-length, again, for one and the same error in deviation, increases rapidly as we go down the spectrum. If, then, we regard the deviations as being correct, and ask in turn what the probable error of the corresponding wave-length is, as given by our curve, we find that this probable error of λ varies at each point, but that it but slightly exceeds a unit of the second decimal place, in any case, for an error of deviation of one minute of arc. The most satisfactory evidence, however, as to the degree of accuracy attained, is derived by an inspection of the curve of observation on the original charts. For we are now prepared to draw a curve expressing graphically the relation between deviations or refractive indices and wave-lengths, extending throughout both the visible and invisible parts of the spectrum. Plotting the points given by the data in Table D, and drawing a smooth curve through them, we obtain the "curve of observation" showing u as a function of λ in the lower curve of Plate XIX, and d as a function of λ in the curve of Plate XX, where the points obtained by observation are distinguished by small dots.

There would be no gain in accuracy, at this stage, in attempting to work from a formula representing the equation of the curve obtained, as the graphical construction is fully as trustworthy as the data. This I say with special reference to the large original charts * which have been drawn by Mr. J. E. Keeler, of this observatory, and which seem to me favorable specimens of the accuracy attainable by this method.

We are now prepared to test the accuracy of the various formulæ connecting refraction with wave-length, though it will be convenient to first prepare a table showing what this relation is in the visible part of the spectrum of the prism employed.

In the following table the deviations in the visible spectrum were measured by the spectrometer, reading to 10″ of arc, which has been already described, in which for this special purpose the bolometer was replaced by an achromatic observing telescope with a micrometer eye-piece, and the indices of refraction were computed by the usual formula. "O" in the ultra violet was measured by aid of a Soret fluorescent eye piece, and its wave-length is from Cornu. The other wavelengths are taken from Angstrom. But the unit is here the *micron* = $\frac{1}{1000}$ millimeter = (10,000 times the unit of Angstrom's scale). "λ" is here the symbol for the wave-length.

---

* These original charts were exhibited to the members of the National Academy of Sciences, at Washington, in April, 1883. The engraving here given in illustration, being on a much reduced scale, will merely indicate the exactness of interpolation possible by the originals.

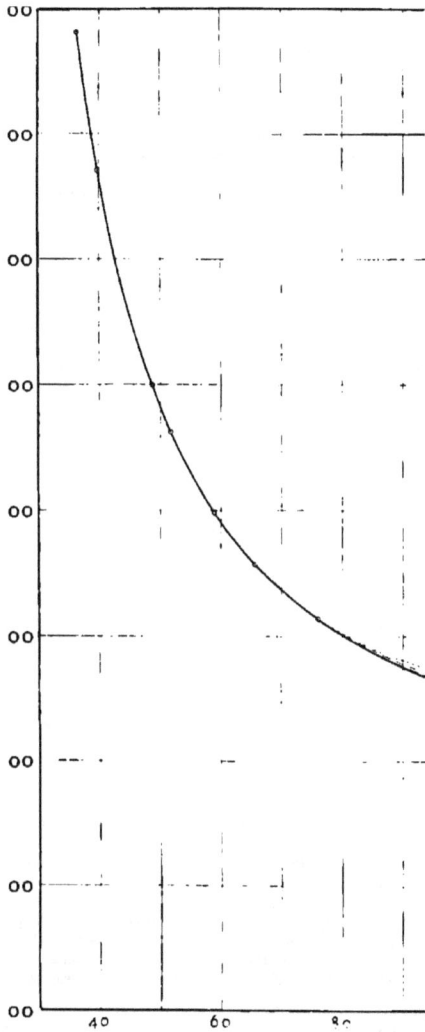

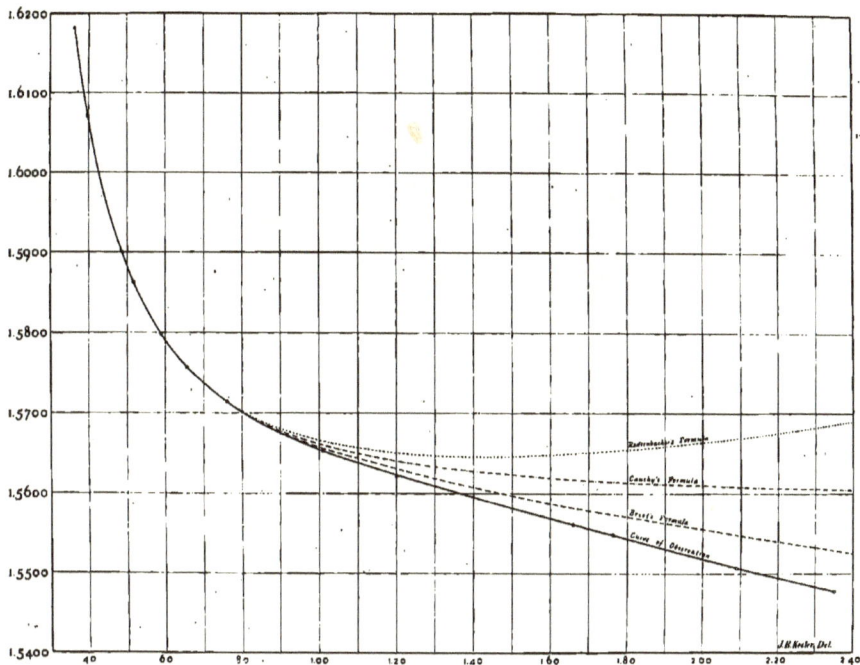

PLATE XIX.
CURVE $n = f(\lambda)$ FOR THE HILGER PRISM.

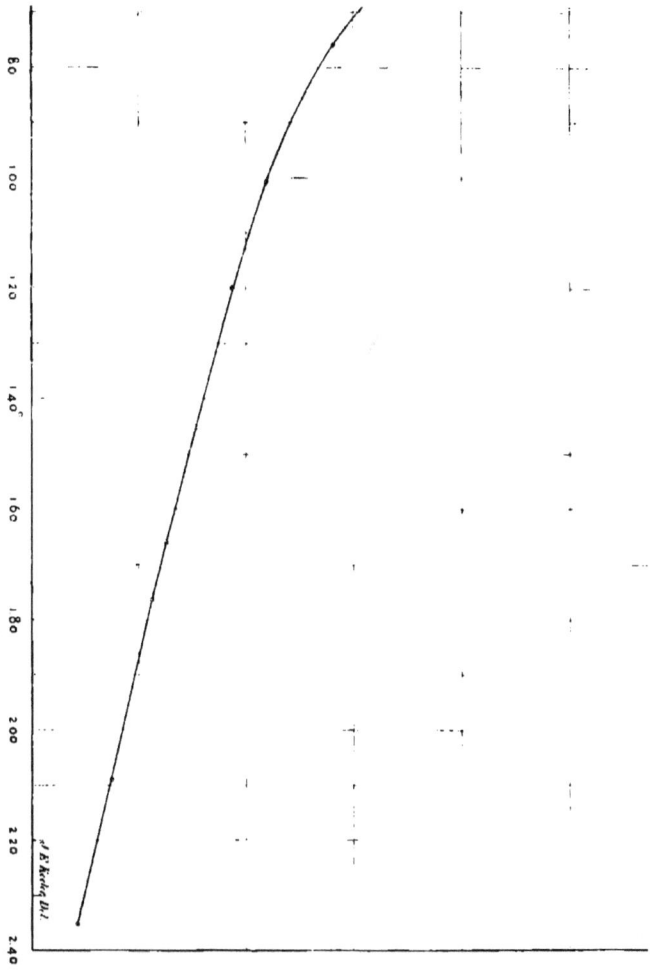

The following indices in the visible spectrum on which the computations for testing the formulæ are founded are trustworthy to the fourth decimal place here given:

TABLE E.

Observed indices in visible spectrum of Hilger prism.

Line.	λ	d	n
A	0.76009	46 49 05	1.5714
C	0.65618	47 15 45	1.5757
D	0.58890	47 41 15	1.5798
B₁	0.51607	48 21 05	1.5862
F	0.48606	48 44 15	1.5899
H₁	0.39679	50 34 05	1.6070
O	0.34400	52 43 00	1.6295

A smooth curve drawn through the points, whose positions are given by the above table, represents with accuracy the relation between n and λ in the visible part of the spectrum. This method is, however, obviously inapplicable to the very extended invisible portion below the A line; and accordingly attempts were first made to effect the determination of corresponding indices and wave-lengths by extending the curve derived from the above observations by means of formulæ. Several formulæ have, it will be remembered, been proposed by physicists, expressing n as a function of λ, and containing constants which are to be determined by observation. But it has never hitherto been possible to test these formulæ far from the visible spectrum, whence their constants have been in fact derived.

This desirable test we are now prepared to apply. The simplest as well as the most widely used formula is that of Cauchy, which, as it is commonly written,

$$\left(n = a + \frac{b}{\lambda^2} + \frac{c}{\lambda^4} \right)$$

contains three unknown quantities requiring for their determination three simultaneous equations. Selecting the lines A, D, and H for this purpose, we have from the table just given the three equations—

$$1.5714 = a + \frac{b}{(0.76009)^2} + \frac{c}{(0.76009)^4} \qquad 1.5798 = a + \frac{b}{(0.58890)^2} + \frac{c}{(0.58890)^4}$$

$$1.6070 = a + \frac{b}{(0.39679)^2} + \frac{c}{(0.39679)^4}$$

from which by elimination

$$a = 1.5593, \qquad b = 0.006775, \qquad c = 0.0001137,$$

so that for this prism, the formula becomes,

$$n = 1.5593 + \frac{0.006775}{\lambda^2} + \frac{0.0001137}{\lambda^4},$$

which, we find on trial, satisfies the observations in the visible part of the spectrum within very narrow limits. When, however, we attempt to extend the application of the formula to the infrared region, its results are not so satisfactory. Since b and c are both positive, the least value which n can have, in our prism, according to the formula, is a, or 1.5593, corresponding to a deviation of 45° 35', whereas the bolometric measurements show that in this prism the solar spectrum, after absorption, extends as low as 14°, with every sign that if it do not extend yet farther, it is not on account of the prism, but because below this point the heat is absorbed by some ingredient of our atmosphere.

We conclude, then, that Cauchy's formula gives grossly erroneous results, when extended far beyond the limits within which the observations on which it is founded are made. Its implicit assertion that the lower limit of the prismatic spectrum (however great the wave length of the ray transmitted) is not so far below A as A is below D, is absolutely contradicted by these experiments; and all extra-polations made by it, far from the visible spectrum in which its constants have been determined, are wholly untrustworthy, as will appear more fully later.

Redtenbacher proposes the formula

$$\frac{1}{n^2} = a + b\lambda^2 + \frac{c}{\lambda^2}$$

for expressing the same relation. Using the same lines as before for determining the unknown constants, we have for the Hilger prism

$$\frac{1}{n^2} = 0.412297 - 0.00003711\lambda^2 - \frac{0.0039220}{\lambda^2}$$

a formula which also satisfies the observations in the visible spectrum, but fails when extended to the invisible. The curve representing it has a minimum point, corresponding to $n = 1.5647$, for a value of λ found from the equation $\lambda^4 = \frac{c}{b}$, or, in the special case of the formula above, where $\frac{c}{b}$ is positive, $\lambda = 1.430$; so that for every value of n greater than 1.5647 there are two real values of λ. This formula, therefore, is even less satisfactory than that of Cauchy.

Briot gives a formula which has been asserted by other investigators[*] to represent satisfactorily the results of observation throughout the whole spectrum, namely :

$$\frac{1}{n^2} = a + b\left(\frac{n^2}{\lambda^2}\right) + c\left(\frac{n^4}{\lambda^4}\right) + k\left(\frac{\lambda^2}{n^2}\right)$$

From four equations like this, using values of n and λ, corresponding to the Fraunhofer lines A, C, F, and H, the values of the constants were determined[†] as follows:

$a = 0.41028,$ $b = -0.0013495,$ $c = -0.0000003379,$ $k = +0.0022329.$

With the aid of these constants, the wave-lengths corresponding to given refractive indices were computed, and a curve representing the formula was plotted. This curve, as well as those representing Cauchy's and Redtenbacher's formulæ, is shown in Plate XIX, where we may obtain by simple inspection the actual errors of all the formulæ in question, or we may take them from the following table, whose results, I hope, will supply useful data for those who are interested in theories of dispersion.

TABLE F.

Approximate errors in wave-lengths by Briot's, Cauchy's, and Redtenbacher's formulæ, for cold bands in infra-red.
(Comparison of theories with observation).

n	Observed.		From Briot's formula.		From Cauchy's formula.		From Redtenbacher's formula.			
By observation.	λ	Value.	Error.	Value.	Error.	I. Value.	Error.	II. Value.	Error.
1.5711	0.760	0.760	0.000	0.760	0.000	0.760	0.000	....	....
1.5697	0.815	0.815	0.000	0.818	0.003	0.820	0.005	....	....
1.5687	0.850	0.850	0.000	0.853	0.003	0.862	0.012	....	....
1.5678	0.880	0.881	0.001	0.890	0.010	0.915	0.035	2.230	1.340
1.5674	0.910	0.911	0.001	0.920	0.010	0.941	0.031	2.170	1.260
1.5665	0.940	0.942	0.002	0.960	0.020	0.990	0.050	2.060	1.120
1.5656	1.130	1.170	0.040	1.270	0.110	Imaginary.		Imaginary.	
1.5646	1.270	1.336	0.066	1.730	0.160	....		....	
1.5604	1.360	1.450	0.090	2.100	1.100	....		....	
1.5576	1.540	1.750	0.210	....		....		....	
1.5572	1.580	1.809	0.229	....		....		....	
1.5541	1.810	2.105	0.295	....		....		....	
to	to	to		....		....		....	
1.5515	1.870	2.200	0.330	....		....		....	
1.5529	1.980	2.460	0.480	....		....		....	
1.5515	2.050	2.324	0.494	....		....		....	

NOTE.—A part of the above values of n, where determined from observation by the bolometer are liable to error in the fourth decimal place. For probable errors of λ, as observed, see Table D. "λ Observed" is either from a direct observation or from an interpolation between two closely contiguous observations.

[*] Mouton, Comptes Rendus, tome 89, page 297, tome 88, page 1190.

[†] This formula has the practical inconvenience of leading to cubic equations either in n or λ^2, the solution of which is so tedious as to forbid its use where many places are to be independently found. I have been aided in the present lengthy numerical computations by Prof. M. B. Goff.

It is evident that Briot's formula, though not exact, yet gives results much more trustworthy than the others considered, and it was employed in constructing provisional maps of the normal spectrum from the prismatic until an apparatus was completed for determining the wave lengths of the invisible rays by direct measurement.

We must evidently conclude, from the numbers in Table F and from the curve in Plate XIX which embodies them, that we in reality can scarcely assign any limit to the extent of the infra red prismatic spectrum, and that, far from the curve having an asymptote parallel to the axis of X, as Cauchy's theory requires, our curve, so far as we can follow it, rather tends to ultimately coincide with a straight line cutting the axis at a finite angle, and (if this axis pass through the point $n = 1$) at a great distance from the origin.

With the danger of extra-polations presented to us in such examples as have been cited, we shall not attempt to generalize the results of our observations further than to remark that, for the prism in question, we find that the deviation tends within the limits of observation to become proportional to the wave-lengths as the deviation diminishes, and that, as far as we can see at present, there is scarcely any limit to the wave-length our prism can transmit except that fixed by its absorptive effect.

The approximate limit of the solar spectrum of the Hilger prism is at $n = 1.5135$, which, according to Briot's formula, corresponds nearly to 3.4, but which, according to our bolometric observations, corresponds to an actual wave-length of 2.8. For this same point, as will be seen by Table F, the values by Cauchy's formula are impossible and those by Redtenbacher's formula are imaginary.

We may add that Briot's formula gives a point of inflexion near $\lambda = 2.5$. In other words, the curve, which up to near the limits of our chart (Fig. 2) has been convex to the axis of λ, there becomes concave, and, as we find, would cut the axis near $\lambda = 16.0$ (i. e., for $n = 1$ $\lambda = 16.0$). These values of ours for Briot's formula rest, it will be remembered, on extrapolations founded on measures in the *visible* spectrum.

WAVE-LENGTHS OF COLD LINES IN INFRA-RED PRISMATIC SPECTRUM.

The following values (in Table G) from Mouton, Abney, and Draper are the only ones I know previous to my own measures where the wave-lengths of any cold lines are given with approximate accuracy. Of these it is just to distinguish those by Abney as possessing a degree of exactness before unknown. There are some doubts about the band $\lambda = 1.36$ having really been observed before, but I have included this among those whose existence was known or suspected before my measures.

The values here given were obtained by me in 1882, and first published in the Comptes Rendus for September 11, 1882, in the form of charts which were drawn from them. These charts were so much reduced by the engraver that, though these values are still determinable from them, it may be convenient to repeat them here in their original tabular form, with the addition of the probable errors:

TABLE G.

Table of dark bands in infra-red portion of spectrum.

λ	n (Hilger prism).	d (Hilger prism).	Remarks.
0.815	1.5697	46° 39′.0	This line I have certainly *seen*, but with difficulty. It is near the utmost limit of the visible spectrum. It appears to coincide with Captain Abney's Z and with Draper's a.
0.850	1.5687	46° 33′.0	On the very limit of visibility, or beyond it, there is in fact some uncertainty as to its having been seen at all. Apparently agrees with Abney's x 530
0.890	1.5678	46° 27′.5	Quite an inconspicuous line. Abney has a heavy line near here, possibly corresponds to Draper's β.
0.904	1.5671	46° 23′.0	Inconspicuous; possibly a part of Draper's β.
0.935	1.5664	46° 20′.0	Very marked heavy line, and the first notable interruption of the energy-curve. This line, if visible, would be much heavier than Fraunhofer's A. It marks the extreme limit of Draper's investigations according to his own statement. It is identifiable also with a gap in Lamansky's curve, beyond which his thermopile detected some heat beyond the lowest point reached by Draper. The Allegheny observations make it probable that it is of telluric origin.
1.130	1.5636	46° 1′.0	Still blacker and colder band than the preceding. This line is given by Mouton as having a wave-length of 1.230(?), by Desains of 1.131(?). There is a shading apparently caused by fine lines not differentiable by the bolometer on both sides of this band. It appears to agree with a gap in Lamansky's curve and with Abney's φ. Allegheny observations make it probably of telluric origin.
1.270	1.5616	45° 49′.0	Inconspicuous line.
1.300	1.5604	45° 41′.0	Very remarkable band, or group of lines, shading off on both sides slightly, but most toward the infra-red, and almost absolutely cold and black. It appears to be identifiable with Abney's ψ and with the last gap in Lamansky's curve, and possibly with Desains' 1a.43, or Mouton's 1a.48, but it is so wide that there is some difficulty in determining the point to measure from. The coldest part of it seems to have a wave-length of 1.36. From the observations at Allegheny it is suspected to be of telluric origin.
1.540	1.5570	45° 26′.0	Inconspicuous lines
1.585	1.5571	45° 22′.5	
1.805–1.870	1.5544–1.5535	45° 5′.0–45° 1′.0	This is the great cold band originally discovered on Mount Whitney. It cuts the energy-curve almost completely in two at the sea-level, but is not absolutely cold, and slight variations in the minute amount of heat recognizable in it, by the bolometer, lead us to think it probably of telluric origin, although it was originally discovered at an altitude of 12,000 feet. It is so broad that we can only give the limits of the wave-lengths at its coldest portion. We have designated this remarkable band as Ω, not that it is the last in the spectrum, but that it is the last of the great and conspicuous interruptions of the energy.
1.981	1.5520	44° 53′.0	Small but definite line.
2.037	1.5515	44° 49′.0	Small but definite line. These two lines are the last discovered by the bolometer, beyond the observable spectrum certainly extends beyond them to a wave-length of over 2a.70 without any notable interruption of its energy, which gradually declines to the point just mentioned. Beyond wave-length 2a.70 feeble isolated indications of energy have been suspected, but the study cannot be carried much further in our lower atmosphere. Beyond this point of 2a.80 the solar spectrum before absorption probably extends to much greater wave-lengths. Beyond it, very probably the earth's own radiations into space are checked by its own envelope.

DISTRIBUTION OF ENERGY IN THE NORMAL SPECTRUM.

The curve $d = (f)\lambda$ given in Plate XX enables us to mark off a wave-length scale upon the map of the prismatic spectrum without any extrapolation between our present extreme points of observation, a deviation of 52° 43′ (corresponding to λ = 0ª.344), and a deviation of 44° 25′ (corresponding to λ = 2ª.356), and also to construct a map in which the wave-length scale is an ordinary scale of equal parts, but in which the degrees of deviation, if represented, would be unequally spaced. Such a chart of the *normal spectrum* has, as we have already remarked, the advantage of being entirely independent of any particular prism or grating, and consequently of being directly comparable with all other maps of the same kind.

If, besides making a map of the normal spectrum, we wish to construct a curve representing the corresponding distribution of energy, a further consideration of the relations existing between the two charts is necessary.   The law of dispersion of the prism causes the distribution of energy in its spectrum to be quite different from what would have been observed with a diffraction grating.* Disregarding the absorbing action of the apparatus, the amount of heat between two definite wave-lengths, as between the A and B lines, should be the same in both spectra, provided

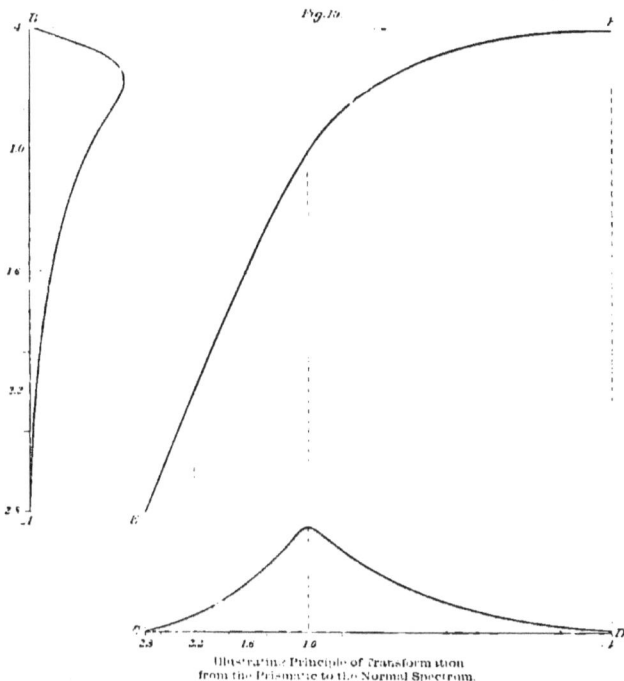

Illustrating Principle of Transformation
from the Prismatic to the Normal Spectrum.

the total quantity of heat is the same in both.   The area between any two ordinates of the curve in Plate XI (prismatic spectrum) may be considered to represent the amount of heat in the part of the spectrum included between them, and the total area of the curve represents the total amount of heat.   If, then, we suppose the area of the normal curve required to be the same as that of the prismatic, the condition to be fulfilled by the former curve is that the area, included between the ordinates at any two wave-lengths, shall be equal to that included between the same wave-lengths in the latter; and from this condition we can deduce a rule for effecting the required transformation.†

Lay off upon a line A B (Fig. 19) any convenient distance and divide it into equal spaces to represent the normal wave-length scale, and upon a line C D, at right angles to the first, lay off the same distance and divide it into the same number of parts, spaced according to the law of

* J. W. Draper, Philosophical Magazine, volume 11, page 101, 1872.
† See J. Müller, Poggendorff's Annalen, Band 105; Lundquist, Poggendorff, Annalen, Band 155, Seite 146; Mouton, Comptes Rendus, tome 79, page 256,

dispersion of the prism, as in the wave-length scale marked on the bottom of the prismatic chart (see Plate XI). Erect ordinates at the points of division, and mark them with the proper wave-lengths, beginning on both lines at the ends which lie nearest to each other, as in Fig. 19, where five ordinates are shown. Through the intersection of corresponding ordinates, draw the curve E F, and upon C D draw the curve of distribution of energy in the prismatic spectrum.

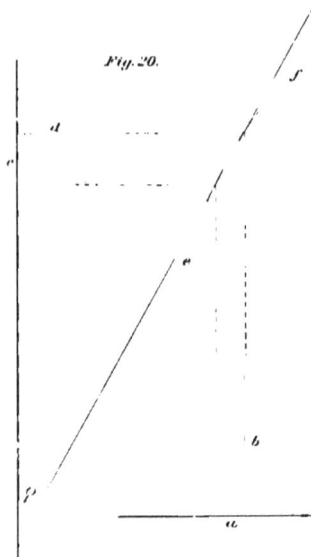

Fig. 20.

Illustrating Principle of Transformation from the Prismatic to the Normal Spectrum.

Let a, fig. 20, be a very small wave-length interval on the prismatic scale; c, the same interval on the normal scale; and b and d, the average heights of the energy curves over the two intervals respectively: the shaded part of the figure representing therefore the portion of the total area included between these limits; ef is a portion of the curve EF, fig. 19. Then according to the condition of transformation

$$cd = ab$$

whence

$$b : d :: c : a$$

From geometrical considerations,

$$c : a :: 1 : \tan \varphi$$

where φ is the angle which the chord EF joining the intersections of the two pairs of ordinates makes with AB, consequently

$$b : d :: 1 : \tan \varphi$$

from which

$$d = b \tan \varphi$$

Now when a and c are indefinitely small b and d are the ordinates of the prismatic and normal energy curves respectively, at a given wave-length, and φ is the angle formed by the tangent to EF at their point of intersection. Hence to find the height of the normal curve at a given wave-length, the corresponding ordinate of the prismatic curve must be multiplied by $\tan \varphi$.

Such a construction was applied to the prismatic energy curve of the Hilger prism, resulting in the following table (II), which exhibits for every tenth of the unit wave-length: (1), the deviations of even wave-lengths in the spectrum of the Hilger prism; (2), the ordinates of the prismatic curve of energy in this spectrum as given by observation with the bolometer—the dotted bounding curve of Plate XI), "prismatic spectrum"; (3), the values of tan φ obtained by estimation from the plotted curve EF, fig. 19; (4), the ordinates of the normal energy curve, which, on being plotted, give the outer dotted curve in Plate XII ("normal spectrum").

TABLE II.—*Table for facilitating the construction of the normal energy curve when the distribution of the energy in the prismatic curve is given.*

λ.	d.	Prismatic ordinates.	Tan φ.	Normal ordinates.
0.40	50 27 40	6 5	9 820	64
0 50	48 51 00	57	4 226	34
0 60	47 55 50	114	2 380	239
0 70	47 05 00	210	1 460	305
0 80	46 41 10	266	1 080	296
0 90	46 25 40	316	720	228
1 00	46 14 50	342	600	205
1 10	46 04 50	336	503	183
1 20	45 55 00	311	420	162
1 30	45 46 40	280	435	159
1 40	45 35 40	259	380	123
1 50	45 29 50	222	465	103
1 60	45 21 00	194	445	87
1 70	45 17 50	159	436	71
1 80	45 05 40	147	429	63
1 90	45 28 10	125	425	55
2 00	44 50 40	103	415	43
2 10	44 4 25	81	414	35
2 20	44 48 00	67	411	28
2 30	44 28 40	51	409	21
2 40	44 23 40	36	407	15
2 50	44 14 00	22	405	9
2 60	44 07 50	10	40.	4
2 70	44 0 50	3	401	1 2

The true normal energy curve with all its inflections, maxima and minima, is easily drawn after this dotted curve is plotted, for the parts of the ordinate of the latter below and above its intersection with the former irregular curve bear the same proportion to each other as in the prismatic spectrum, and we thus finally attain the object of the preceding labor.

If, now, it is desirable to map the distribution of the energy on any other scale, such as that on which the abscissæ are proportional to the times of vibration, this can be done with facility. Thus in the supposed instance we have only to find $\frac{1}{\lambda}$ corresponding to each wave-length in order to get the abscissa, and (observing that since x now $= \frac{1}{\lambda}$, $\frac{dx}{d\lambda} = -\frac{1}{\lambda^2}$) to use the multiplying factor $\frac{1}{\lambda^2}$ to obtain the length of the new abscissæ from the old in each instance. If the length of the new energy curve between the limiting perpendiculars (which now represent the reciprocals of the wave-length) is to be the same as in the old we must introduce a constant multiplier, n, writing the equation of the interpolating curve $x = \frac{n}{\lambda}$ so that the multiplying factor becomes $-\frac{n}{\lambda^2}$. Thus if the limiting ordinates of the wave-length energy curve are λ_o, λ_i, and we are to have the condition,

$$\left(\frac{1}{\lambda_o} - \frac{1}{\lambda_i}\right) n = \lambda_i - \lambda_o,$$

then,

$$n = \lambda_o \times \lambda_i.$$

If the mean ordinate of any small area of the normal energy curve between any given limits, λ_o, λ_i is denoted by y_i, and that of the corresponding area of the new curve by y, since the areas are to be the same, we have $n \left(\frac{1}{\lambda_o} - \frac{1}{\lambda_i}\right) y = (\lambda_i - \lambda_o) y_i$, whence $y = \frac{\lambda_o \lambda_i}{n} \times y_i$, which at the limit becomes $y = \frac{\lambda^2}{n} y_i$. Hence to obtain the new ordinates, the old ones must be multiplied by the reciprocals of the factors for abscissæ, or by $\frac{\lambda^2}{n}$.

The curve EF, Fig. 19, if represented by a formula would give rise to an expression of the form $d = (\varphi) \lambda$, the abscissæ measured along AB being proportional to the wave-lengths, and the ordinates parallel to CD, to the deviations. It is therefore a curve similar to that in Plate XX, except that the abscissæ and ordinates are drawn on different scales. Since $\tan \varsigma = \dfrac{dd}{d\lambda} = \dfrac{du}{d\lambda}\dfrac{dd}{du}$, the factors for multiplying the prismatic ordinates may be computed, provided the curve EF can be exactly expressed by a formula, and for the preliminary reduction this was done, the values of $\dfrac{du}{d\lambda}$ being computed from Briot's formula, and $\dfrac{dd}{du}$ from the relation $n = \dfrac{\sin \frac{1}{2}(a+d)}{\sin \frac{1}{2}a}$. When, however, it was shown by the measurements of obscure rays that Briot's formula, obtained by observations in the visible spectrum, does not exactly express the law of dispersion, the table of factors thus prepared, was of course abandoned and the graphical method described above was substituted.

I have drawn in this way (on a smaller scale than that of the normal or prismatic curves, and following the smooth curve in the former as my original) four different schemes for the distribution of the energy. (See Plate XXI.) Fig. B represents the distribution of solar energy after absorption by our atmosphere on the scale of wave-frequency (general equation of interpolating curve $x = \dfrac{1}{\lambda}$) proposed by Mr. Stoney. Fig. C represents the distribution according to a proposal $(x = \log \lambda)$ of Lord Rayleigh.

Fig. D is quite different from any of the preceding. It gives the distribution on a scale I have never seen proposed, but which I have found useful. In this the bounding curve is a *straight line* parallel to the axis of X ($y=$constant). This is not merely suggestive as illustrating what has already been remarked here as to the conventional character of the methods of showing the distribution of the energy, but it has more practical uses. In this particular construction it is evident in fact that the sums of the energies, between any two wave lengths whatever, are directly proportional to the distances between the ordinates, measured on the axis of X. If, then, we desire (for instance) to know what relation the invisible bears to the visible heat, or to inquire about what point in the spectrum the energy is equally distributed, &c., these and similar problems are solved through Fig. D by simple inspection.

I have not been able as yet to repeat the preceding determinations upon the lower part of the spectrum as often as I could wish. They are susceptible of improved accuracy by still longer experiment, but I think that within the limits of error indicated they may already be useful. I should add that throughout this investigation I have received constant and valuable aid from Mr. J. E. Keeler, not only in the graphical constructions, but in the experiments and in the computations, through all the details of which his aid has been more that of a coadjutor than an assistant.

ALLEGHENY OBSERVATORY,
Allegheny, Pa., October, 1883.

NOTE.

Since the above was in type I have seen the interesting article by Mr. H. Becquerel in the Annales de Chimie et de Physique, for September, 1883.

The wave-lengths assigned by M. Becquerel to the band at the limit of his researches, 1,440 to 1,500, appear to me too great, for this limit corresponds to the band whose wave-length is given at $1^\mu.36$ to $1^\mu.37$ on my chart, published in the Comptes Rendus of the previous year (September 11, 1882), and on a larger scale in the American Journal of Science, for March, 1883, and in the Annales de Chimie et de Physique, for August of this year. I regret that M. Becquerel has not read the article in the Comptes Rendus. Had he done so he would have seen that the wave-lengths there given were not conjectural, but directly determined by a very laborious but the only practical method from the direct use of a grating. They were the result, in fact, of the measurements I have just described, and were specially intended to give information about the unknown region extending beyond the limit of M. Becquerel's researches, such as the great newly discovered band Ω, for instance, which stretches from wave-lengths $1^\mu.80$ to $1^\mu.90$, while M. Becquerel's furthest band, as I have said, is at $1^\mu.48$. The present memoir will show what degree of reliance may be placed on these measurements.

A.

A. B. Schanz, Del.

It is understood that a photographic map of the spectrum to 1ᵘ.6 (and therefore covering the ground of M. Becquerel's paper, but not extending as far as my "ᵘ.O", will shortly be published from the joint labors of Professor Rowland and Captain Abney, and as their results will probably be accepted on all hands as more exact than the preliminary explorations in which M. Becquerel and myself have been engaged, we may await its appearance for the determination of a part of the points in question.

I would call attention to the fact that M. Becquerel has stated that the furthest band known to him in September, 1883 (except from my own researches), had a wave-length of not over 1ᵘ.50, according to his own estimate.

APPENDIX 3.

EXPERIMENTAL DETERMINATION OF THE INFLUENCE OF CONVECTION CURRENTS UPON THE LOSS OR GAIN OF TEMPERATURE BY A THERMOMETER BULB.

Several series of experiments were made to ascertain the difference in the rate of heating or cooling of a thermometer bulb in air and in *vacuo*. The bulb was inclosed in the center of a thin copper globe 5 centimeters in diameter, the stem and an exhausting tube of glass being sealed. The apparatus was connected to an ordinary Sprengel's pump, and a vacuum made to within half a millimeter of mercury. The glass tube was then sealed in the flame of a lamp and the whole was immersed in hot water kept constantly stirred, the temperatures of the water and of the inner thermometer being recorded from minute to minute. A minute portion of the heat acquired by the inner thermometer is received by conduction along the glass stem, and a very small amount by the convection of the trifling quantity of air remaining; but by far the greater part is radiated to the bulb from the copper globe, which is blackened within. The repetition of the experiment with the globe sealed but full of air enables us to discriminate between the effect of radiation and of convection. The first experiments were made with a thermometer (Green, 458) graduated from 0° to 60° C. in tenths of a degree, having a clear spherical bulb 1.227 centimeters in diameter. The succeeding ones were carried on with a blackened bulb-thermometer (Baudin, 8737) which has been used in the large Violle actinometer in the measurement of solar radiation. The observers had acquired expertness by much previous practice in reading to hundredths of a degree, and a single hundredth will represent more than the probable error of one reading.

March 27, 1883. Thermometer *in vacuo* Green 458. Observer J. E. Keeler. S. P. Langley recording. Temperature of water obtained by interpolation from a smooth curve.

	Copper globe dipped in hot water.				Copper globe dipped in cold water.			
Time	Temperature of water	Green 458	Deficiency.	First difference.	Time	Temperature of water	Green 458	Excess.	First difference.
Minute									*Minute*
0	46.50	26.50	20.00		0	4.42	31.64	27.18	
1	40.22	22.88	17.34	2.66	1	4.17	28.49	25.82	1.35
2	39.98	25.00	14.98	2.36	2	4.51	25.47	20.94	2.38
3	39.75	27.00	12.75	2.23	3	4.56	22.65	18.09	2.75
4	39.54	28.58	10.76	1.99	4	4.60	20.28	15.68	2.41
5	39.38	30.31	9.02	1.74	5	4.64	18.17	13.51	2.17
6	39.13	31.61	7.52	1.54	6	4.68	16.47	11.59	1.92
7	8.94	32.69	6.25	1.27	7	4.72	14.65	9.93	1.66
8	8.74	33.62	5.12	1.13	8	4.77	13.21	8.44	1.49
9	8.58	34.38	4.20	.92	9	4.81	12.00	7.19	1.25
10	38.42	35.00	3.42	.78	10	4.86	10.93	6.07	1.12
11	38.28	35.50	2.28	.64	11	4.89	10.01	5.10	.97
12	38.15	35.93	2.20	.52	12	4.95	9.41	4.41	.75
13	38.04	36.49	1.55	.41	13	4.98	8.67	3.69	.72
14	37.94	36.38	1.56	.29	14	5.02	8.07	3.05	.64
15	37.84	36.57	1.27	.29	15	5.07	7.60	2.53	.52
16	37.74	36.70	1.04	.23	16	5.11	7.20	2.09	.44

In the above the copper globe was dipped in the water at least 30 seconds before the first reading. The temperature of the water itself was taken every minute, and minute irregularities found corrected by a smooth curve, the ordinates of which give the values in the column headed "Temperature of water."

[Repetition of last experiment with air in copper globe. Barometer, 731 millimeters. All else as before.]

Copper globe dipped in hot water.   Copper globe dipped in cold water.

Time.	Temperature of water.	Green, 4584.	Deficiency.	First difference.	Time.	Temperature of water.	Green, 4584.	Excess.	First difference.
Minutes.					*Minutes.*				
0	41.26	18.95	22.31	....	0	4.62	31.40	26.78	....
1	41.02	23.57	17.45	4.86	1	4.65	25.87	21.22	5.56
2	40.80	26.58	14.22	3.23	2	4.69	21.52	16.83	4.39
3	40.58	29.23	11.35	2.87	3	4.72	18.11	13.39	3.44
4	40.38	31.44	8.94	2.41	4	4.76	15.40	10.64	2.75
5	40.17	32.12	7.05	1.89	5	4.80	13.28	8.48	2.16
6	39.99	34.46	5.53	1.52	6	4.83	11.65	6.82	1.66
7	39.80	35.46	4.34	1.19	7	4.86	10.30	5.44	1.38
8	39.63	36.21	3.42	.92	8	4.90	9.32	4.42	1.02
9	39.48	36.76	2.72	.70	9	4.93	8.51	3.58	.84
10	39.33	37.18	2.15	.57	10	4.97	7.78	2.81	.77
11	39.19	37.49	1.70	.45	11				

[March 27, 1883. Thermometer to cocoa. Baudin 8737. Observer, J. E. Keeler. S. P. Langley, recording. The copper globe was dipped in the water 1 minute before the first reading.]

Copper globe dipped in hot water.   Copper globe dipped in cold water.

Time.	Temperature of water.	Baudin, 8737.	Deficiency.	First difference.	Time.	Temperature of water.	B mlin. 8737.	Excess.	First difference.
Minutes.					*Minutes.*				
0	34.80	17.95	16.85	....	0	6.44	28.08	21.64	....
1	34.67	20.10	14.57	2.28	1	6.47	24.41	17.94	3.70
2	34.56	21.74	12.82	1.75	2	6.50	21.30	14.80	3.14
3	34.44	23.76	10.68	2.14	3	6.53	18.75	12.22	2.58
4	34.34	25.43	8.91	1.77	4	6.56	16.62	10.06	2.16
5	34.22	26.90	7.32	1.59	5	6.58	14.80	8.22	1.84
6	34.12	28.04	6.08	1.24	6	6.61	13.30	6.69	1.53
7	34.00	29.00	5.00	1.08	7	6.64	12.05	5.41	1.28
8	33.89	29.80	4.09	.91	8	6.67	11.00	4.33	1.08
9	33.78	30.58	3.40	.69	9	6.70	10.13	3.45	.88
10	33.68	30.84	2.84	.56	10	6.73	9.46	2.73	.72
11	33.58	31.21	2.37	.47	11	6.76	8.91	2.15	.58
12	33.47	31.47	2.00	.37	12	6.78	8.41	1.63	.52
13	33.37	31.68	1.69	.31	13	6.81	8.14	1.23	.40
14	33.27	31.82	1.43	.24	14	6.84	7.76	.92	.31
15	33.17	31.90	1.27	.18	15	6.86	7.50	.64	.28

[Repetition of last experiment with air in copper globe. All else as before.]

Copper globe dipped in hot water.   Copper globe dipped in cold water.

Time.	Temperature of water.	Baudin, 8737.	Deficiency.	First difference.	Time.	Temperature of water.	Baudin, 8737.	Excess.	First difference.
Minutes.					*Minutes.*				
0	34.80	16.42	18.28	....	0	4.73	25.70	20.97	....
1	34.70	20.96	13.74	4.64	1	4.76	20.03	15.27	5.70
2	34.60	23.53	11.07	2.67	2	4.81	16.18	11.37	3.90
3	34.50	25.80	8.70	2.37	3	4.84	13.32	8.48	2.89
4	34.41	27.76	6.65	2.05	4	4.88	11.24	6.36	2.12
5	34.41	29.18	5.13	1.52	5	4.92	9.72	4.80	1.56
6	34.22	30.20	4.02	1.11	6	4.95	8.57	3.62	1.18
7	34.12	30.95	3.17	.85	7	4.99	7.70	2.71	.91
8	34.02	31.47	2.55	.62	8	5.02	7.06	2.04	.67
9	33.92	31.90	2.02	.53	9	5.06	6.59	1.53	.51
10	33.84	32.18	1.66	.36	10	5.10	6.21	1.11	.42
11	33.74	32.39	1.35	.31	11	5.11	5.98	.84	.27
12	33.65	32.50	1.15	.20	12	5.18	5.72	.51	.30
13	34.56	32.60	.96	.19	13	5.21	5.59	.38	.16
14	33.46	32.67	.79	.17	14	5.25	5.41	.16	.22
15	33.37	32.70	.67	.12	15	5.28	5.37	.09	.07
16	33.28	32.67	.61	.06	16	5.32	5.27	−.05	.14

[March 30, 1883: Thermometer *in vacuo* Baudin, 8757. Water thermometer Green, 4586. Observer, F. W. Very, S. P. Langley recording. Temperature of room 22° to 24° Centigrade. Inclosed thermometer read when dipped.]

The copper globe having been cooled in snow is dipped in hot water. | The copper globe having been heated in water is dipped in cold water.

Time.	Temperature of water.	Baudin, 8757.	Deficiency.	First difference.	Time.	Temperature of water.	Baudin, 8757.	Excess.	First difference.
Min. Sec.					Min. Sec.				
0 00	35.26	3.00	32.26	.....	0 00	2.62	35.64	33.02	....
0 15	35.24	4.52	30.72	1.54	0 15	2.65	34.60	31.95	1.07
0 30	35.22	6.02	29.20	1.52	0 30	2.68	34.16	31.48	1.17
0 45	35.21	7.20	28.01	1.19	0 45	2.72	31.87	29.15	1.33
1 00	35.19	8.48	26.71	1.30	1 00	2.75	30.48	27.73	1.42
1 15	35.17	8.40	25.77	.94	1 15	2.78	29.25	26.47	1.26
1 30	35.16	10.39	24.77	1.00	1 30	2.82	28.00	25.18	1.29
1 45	35.14	11.41	23.73	1.04	1 45	2.85	27.00	24.15	1.03
2 00	35.13	12.50	22.63	1.10	2 00	2.88	25.78	22.90	1.25
3 00	35.06	16.02	19.04	3.59	3 00	3.02	21.65	18.63	4.27
4 00	35.00	19.19	15.81	3.23	4 00	3.15	18.38	15.22	3.40
5 00	34.94	21.91	13.03	2.78	5 00	3.28	15.69	12.41	2.82
6 00	34.87	24.18	10.69	2.34	6 00	3.41	13.15	10.04	2.37
7 00	34.81	26.04	8.77	1.92	7 00	3.54	11.66	8.12	1.92
8 00	34.75	27.57	7.18	1.59	8 00	3.67	10.15	6.48	1.64
9 00	34.70	28.90	5.80	1.78	9 00	3.80	8.98	5.18	1.70
10 00	34.64	29.58	4.86	.94	10 00	3.94	7.99	4.05	1.13
11 00	34.59	30.49	3.96	.90	11 00	4.06	7.20	3.14	.91
12 00	34.54	31.23	3.31	.63	12 00	4.18	6.60	2.42	.72
13 00	34.49	31.69	2.80	.53	13 00	4.30	6.12	1.82	.60
14 00	34.44	32.12	2.31	.48	14 00	4.42	5.70	1.28	.54
15 00	34.39	32.43	1.96	.35	15 00	4.54	5.42	.88	.40
16 00	34.34	32.64	1.70	.26	16 00	4.66	5.21	.55	.33
17 00	34.30	32.81	1.49	.21	17 00	4.77	5.02	.25	.30
18 00	34.25	32.97	1.28	.21					

[Repetition of last experiment with air in copper globe. Barometer 725mm. Temperature of room 22°. All else as before.]

The copper globe having been cooled in snow to about 3° is dipped in hot water at about 35°. | The copper globe having been heated in hot water to about 35° is dipped in cold water at about 3°.

Time.	Temperature of water.	Baudin, 8757.	Deficiency.	First difference.	Time.	Temperature of water.	Baudin, 8757.	Excess.	First difference.
Min. Sec.					Min. Sec.				
0 00	35.17	3.50	31.67	.....	0 00	3.27	35.50	32.23	....
0 15	35.15	5.35	29.80	1.87	0 15	3.29	33.22	29.93	2.30
0 30	35.13	8.02	27.11	2.69	0 30	3.32	31.10	27.78	2.15
0 45	35.11	10.05	25.06	2.05	0 45	3.34	28.28	25.84	1.81
1 00	35.10	11.98	23.12	1.94	1 00	3.36	27.26	23.70	2.04
1 15	35.08	13.63	21.45	1.67	1 15	3.38	25.08	21.70	2.20
1 30	35.06	15.18	19.88	1.37	1 30	3.41	23.46	20.05	1.65
1 45	35.04	16.35	18.69	1.19	1 45	3.43	22.48	19.05	1.00
2 00	35.02	17.60	17.42	1.27	2 00	3.45	20.99	17.54	1.51
3 00	34.95	21.95	13.00	3.92	3 00	3.55	16.52	12.97	4.57
4 00	34.87	24.48	10.39	2.21	4 00	3.64	13.38	9.74	3.23
5 00	34.80	26.92	7.84	2.51	5 00	3.71	11.02	7.29	2.45
6 00	34.72	29.70	6.02	1.86	6 00	3.82	9.31	5.49	1.80
7 00	34.65	30.01	4.64	1.38	7 00	3.91	8.13	4.22	1.27
8 00	34.57	30.97	3.60	1.04	8 00	4.00	7.20	3.20	1.02
9 00	34.50	31.62	2.88	.72	9 00	4.09	6.41	2.32	.88
10 00	34.42	32.12	2.30	.58	10 00	4.19	5.92	1.73	.59
11 00	34.34	32.48	1.86	.44	11 00	4.28	5.60	1.32	.41
12 00	34.27	32.76	1.51	.35	12 00	4.37	5.24	.87	.45
13 00	34.20	32.98	1.22	.29	13 00	4.46	5.02	.56	.31
14 00	34.12	33.07	1.05	.17	14 00	4.56	4.89	.33	.23
15 00	34.05	33.16	.89	.16	15 00	4.64	4.80	.16	.17
16 00	33.98	33.21	.77	.12	16 00	4.72	4.72	.00	.16

* Intervals irregular

In the last four series, readings every 15 seconds for the first 2 minutes were attempted; the intervals of time were, however, somewhat irregular, as the observed rates of heating or cooling show; but the observations are sufficient to prove that the copper globe acquires the temperature of the surrounding water very rapidly; indeed, it must practically do so in a single second, since the rate of heating or cooling for the first 15 seconds agrees fairly, at least within the limits of errors of observation, with the subsequent rates.

According to the law of Newton, radiation is proportional to temperature, and it is very commonly assumed that within the limited range of temperature, with which we are here dealing, *all* losses are proportional to temperature also. This cannot be really the case in theory, and it does not appear to be so in practice from the observations of Dulong and Petit. Considering the importance of the subject in our actinometric investigations, we have thought it desirable to make a series of experiments on the rates of heating and cooling in air and in vacuo with the thermometers actually used at Mount Whitney and elsewhere, and it is the results of these which we have just given. In the assumption just alluded to, the loss of heat during a time, dt, is proportional to the excess of temperature, θ, whence $\theta = Cz^{-mt}$. This is the equation of a logarithmic curve in which the temperatures of excess at equal intervals of time should bear a constant ratio to each other. We may determine whether it is a logarithmic curve by any of its characteristic properties, but conveniently here by noting whether the subtangents are constant, as in this case they should be. We can calculate these actual subtangents from the equation just given, or we can draw smooth curves through the points represented by the preceding observations, pass a logarithmic curve through three points in the smooth curve, and thus determine the logarithmic curve most nearly agreeing with the observed one. We then find the subtangents of the latter and determine how far and in what way the actual curve agrees with that which would be given if all losses were strictly in proportion to temperature. The results of this latter procedure are given in the two following tables:

Table of subtangents showing the rate of heating and cooling.

OBSERVATIONS OF MARCH 27, 1883.

Minutes	Heating in Vacuo. Subtangents from smooth curve.	Heating in air. Subtangents from smooth curve.	Minutes	Cooling in Vacuo. Subtangents from smooth curve.	Cooling in air. Subtangents from smooth curve.
0	6.10	4.10	0	5.30	2.90
1	6.30	4.15	1	5.38	3.25
2	5.84	3.90	2	5.45	3.55
3	5.75	3.75	3	5.50	3.68
4	5.64	3.75	4	5.60	3.70
5	5.10	3.75	5	5.70	3.75
6	4.90	3.75	6	5.75	3.75
7	4.80	3.75	7	5.80	3.75
8	4.70	3.75	8	5.90	3.75
9	4.60	3.75	9	5.95	3.75
10	4.60	3.75	10	6.05	3.75
11	4.65	3.75	11	6.15	3.75
12	4.80	3.75	12	6.28	3.75
13	5.00	3.75	13	6.50	3.75
14	5.10	3.75	14	6.70	3.75
15			15		3.75

OBSERVATIONS OF MARCH 30, 1883.

Minutes	Heating in Vacuo. Subtangents from smooth curve.	Heating in air. Subtangents from smooth curve.	Minutes	Cooling in Vacuo. Subtangents from smooth curve.	Cooling in air. Subtangents from smooth curve.
0	5.85	3.60	0	5.22	3.35
1	5.70	3.60	1	5.22	3.45
2	5.50	3.60	2	5.22	3.50
3	5.10	3.60	3	5.22	3.55
4	5.35	3.60	4	5.22	3.60
5	5.25	3.60	5	5.22	3.65
6	5.24	3.60	6	5.22	3.70
7	5.22	3.60	7	5.22	3.72
8	5.25	3.60	8	5.22	3.74
9	5.25	3.60	9	5.22	3.75
10	5.30	3.60	10	5.22	3.74
11	5.35	3.60	11	5.22	3.70
12	5.45	3.60	12	5.22	3.65
13	5.50	3.60	13	5.22	3.58
14	5.70	3.60	14	5.22	3.48
15	5.85	3.60	15	5.22	3.35

In these tables the progressive differences in the subtangents, though not excessive, on the whole indicate a departure from the logarithmic law.

A method, perhaps in some respects preferable, is the drawing of supplementary curves in which the rates of heating or cooling, for each excess of temperature, are made the ordinates

These ordinates should fall on a straight line in case the curve is a logarithmic one. They do so, in fact, in the case of the heating and cooling in vacuo within the limits of observation; but in the case of the heating and cooling in air, there is a slight but systematic departure, indicating that the loss by convection is *not* proportional to the excess of temperature, but that, for ordinary air pressures, the convection increases with great rapidity until the difference between the temperature of the thermometer and its enclosure is as much as 10 or 15 degrees Centigrade, after which, for still greater differences the convection increases at a smaller and nearly constant rate. It is thus shown from these observations that Newton's law, although nearly representing the loss by radiation alone for slight excesses of temperature, does not hold good for all losses, a deduction of importance in relation to the theory of the globe actinometer, of which we have made so much use.

INDEX